A FIRST COURSE IN
ALGEBRAIC GEOMETRY
AND ALGEBRAIC VARIETIES

Essential Textbooks in Mathematics

ISSN: 2059-7657

The *Essential Textbooks in Mathematics* series explores the most important topics that undergraduate students in Pure and Applied Mathematics are expected to be familiar with.

Written by senior academics as well lecturers recognised for their teaching skills, they offer, in around 200 to 400 pages, a precise, introductory approach to advanced mathematical theories and concepts in pure and applied subjects (e.g. Probability Theory, Statistics, Computational Methods, etc.).

Their lively style, focused scope, and pedagogical material make them ideal learning tools at a very affordable price.

Published:

A First Course in Algebraic Geometry and Algebraic Varieties
　　by Flaminio Flamini (University of Rome "Tor Vergata", Italy)

Analysis in Euclidean Space
　　by Joaquim Bruna (Universitat Autònoma de Barcelona, Spain &
　　Barcelona Graduate School of Mathematics, Spain)

Introduction to Number Theory
　　by Richard Michael Hill (University College London, UK)

A Friendly Approach to Functional Analysis
　　by Amol Sasane (London School of Economics, UK)

A Sequential Introduction to Real Analysis
　　by J M Speight (University of Leeds, UK)

Essential Textbooks in Mathematics

A FIRST COURSE IN
ALGEBRAIC GEOMETRY
AND ALGEBRAIC VARIETIES

Flaminio Flamini
University of Rome "Tor Vergata", Italy

World Scientific

NEW JERSEY · LONDON · SINGAPORE · BEIJING · SHANGHAI · HONG KONG · TAIPEI · CHENNAI · TOKYO

Published by

World Scientific Publishing Europe Ltd.

57 Shelton Street, Covent Garden, London WC2H 9HE

Head office: 5 Toh Tuck Link, Singapore 596224

USA office: 27 Warren Street, Suite 401-402, Hackensack, NJ 07601

Library of Congress Cataloging-in-Publication Data
Names: Flamini, Flaminio, author.
Title: A first course in algebraic geometry and algebraic varieties /
    Flaminio Flamini, University of Rome "Tor Vergata", Italy.
Description: New Jersey : World Scientific, [2023] | Series: Essential textbooks in mathematics,
    2059-7657 | Includes bibliographical references and index.
Identifiers: LCCN 2022028876 | ISBN 9781800612655 (hardcover) |
    ISBN 9781800612747 (paperback) | ISBN 9781800612662 (ebook) |
    ISBN 9781800612679 (ebook other)
Subjects: LCSH: Geometry, Algebraic--Textbooks.
Classification: LCC QA564 .F48 2023 | DDC 516.3/5--dc23/eng/20220708
LC record available at https://lccn.loc.gov/2022028876

British Library Cataloguing-in-Publication Data
A catalogue record for this book is available from the British Library.

For any available supplementary material, please visit
https://www.worldscientific.com/worldscibooks/10.1142/Q0373#t=suppl

Desk Editors: Soundararajan Raghuraman/Adam Binnie/Shi Ying Koe

Typeset by Stallion Press
Email: enquiries@stallionpress.com

Printed in Singapore

To my family

Arnaldo, Anna Maria, Ornella, Francesca, Lucia.
Their love and support are constant motivation
and always will be.

Preface

Geometry is the study of *shapes* and *spaces*. Nonetheless these two notions vary according to the different types of geometry, of which Mathematics is rich, and how their study develops over the time.

Algebraic Geometry is a branch of Mathematics whose fundamental objects to study are basically *Algebraic Varieties* which, in simple words, are geometric manifestations of solutions of systems of polynomial equations in several indeterminates. Algebraic Geometry enters into the game when equation solving leaves off, and it becomes even more important to understand the intrinsic and global properties of the solutions instead of explicitly finding solutions. This leads into one of the deepest areas in all of Mathematics, both conceptually and in terms of techniques.

Roots of Algebraic Geometry date back to Hellenistic Greeks and Medieval mathematicians, who interpreted some polynomial equations via a geometric point of view. The approach of using geometrical constructions to attack algebraic problems was also adopted by some Renaissance Italian mathematicians, such as G. Cardano and N. Fontana "Tartaglia", and then favored by most French Mathematicians in the 16th and 17th centuries, including R. Descartes and P. de Fermat, who introduced the well-known *coordinate geometry*. In the same period, B. Pascal and G. Desargues developed the synthetic notion of *projective geometry*.

It took simultaneous 19th century developments of *Non-Euclidean Geometry* and of *Abelian Integrals* in order to bring old algebraic ideas back into the geometrical realm. On the one hand, A. Cayley introduced homogeneous polynomials to study properties of projective spaces and F. Klein, in his *Erlangen Program* (1872), introduced the method of characterizing all geometries with the use of group theory and of projective geometry. Klein's approach led M. Noether and members of the 20th century Italian school of Algebraic Geometry (e.g. L. Cremona, C. Segre, G. Castelnuovo, F. Enriques, O. Chisini, P. del Pezzo, G. Fano, B. Segre, F. Severi just to mention a few) to the classification of some algebraic varieties up to *birational equivalence*. On the other hand, ideas of N. H. Abel about *Abelian Integrals* would have led B. Riemann to the development of *Riemann surfaces* and to the contribution of H. Poincaré. In the same period algebrization of Algebraic Geometry by D. Hilbert and E. Noether entered into the game and *Hilbert's Nullstellensatz* and *Noether's factorization theorem* can be viewed as first steps to connect Algebraic Geometry to Commutative Algebra.

In the 20th century, B. L. van der Waerden, O. Zariski and A. Weil developed foundations for Algebraic Geometry based on ring valuation theory, ideals and modules. In 1950–1960s, J. P. Serre and A. Grothendieck completely revised the foundations making use of *sheaf theory*. Later, from about 1960, and largely led by A. Grothendieck, the idea of *schemes* was worked out with the support of advanced techniques in homological algebra. From that moment a period of rapid developments of the theory started, which steadily continues nowadays, giving life to several new applications, such as Topology, Number Theory, Complex Analysis, Singularities, Moduli, Tropical Geometry. For a more detailed historical description of developments of Algebraic Geometry through the centuries, we refer the reader to the wonderful first chapter in Ueno (1997).

Names mentioned above, and more precisely the importance these names have had and still have, reveal why Algebraic Geometry occupies a central place in Mathematics. At the same time, constant developments of the discipline clarify its deep relationship with several branches of Natural Sciences and Engineering, for example Computer Vision, Coding Theory and Cryptography, Theoretical Physics, Statistical and Machine Learning, just to mention a few.

The aim of this book is to provide a "gentle" introduction to the study of Algebraic Varieties. The content of the book arose from courses in Algebraic Geometry that I taught at the University of Rome "Tor Vergata" beginning in 2014. Topics in the book are based on what I have learnt from Sernesi (1992), from handwritten notes from when I was a young Master's student, now published in Ciliberto (2021), as well as from Hartshorne (1977), Mumford (1988, 1995), and from several research courses taught by mathematicians, mentioned in the following Acknowledgments section, a few years later during my PhD.

The approach in this book is purely algebraic. Chapter 1 contains notation and main results in Commutative Algebra that will be used in the following chapters; the book is thereby self-contained, making it more convenient for the reader. Chapter 2 focuses on algebraic affine sets and Hilbert Nullstellensatz, analyzing in detail and with several examples the important correspondence between radical ideals and algebraic sets in an affine space and Zariski topology. The same philosophy is then used in Chapter 3 for algebraic sets in a projective space, using the notion of homogeneous elements and ideals, projective closure of an affine algebraic set, and so on.

Chapter 4 focuses on topological properties, such as irreducibility and Noetherianity, which allow us to define *Algebraic Varieties*. After a brief introduction on sheaves, Chapter 5 takes into account regular and rational functions defined on an algebraic variety and the definition of its *structural sheaf*; many consequences and examples are then discussed.

In Chapter 6, we introduce the concepts of *morphisms* and of *isomorphism classes* of algebraic varieties; we also examine important examples of morphisms such as the Veronese morphism and morphisms induced by linear systems of divisors. Chapter 7 deals with products of algebraic varieties and the Segre morphism, the graph of a morphism and fiber-products of morphisms of algebraic varieties.

This machinery is then used in Chapter 8 to introduce the definition and basic properties of *rational* and *birational maps* among algebraic varieties, which lead to the important notion of *birational classification* of algebraic varieties, a milestone in Algebraic Geometry. Several examples of rational and unirational varieties are considered, as well as fundamental birational maps, like stereographic projections and blow-ups. In particular we discuss how the blow-up process can be used for desingularization of some well-known singular plane curves.

Chapter 9 contains the fundamental result concerning *completeness of projective varieties* and its numerous important consequences. In Chapter 10, we introduce various notions of *dimension* of an algebraic variety, proving that the given notions suitably coincide. We also study the behavior of the dimension when one intersects an algebraic variety with a certain finite number of hypersurfaces, naturally arriving at the notion of *complete intersection* and *set-theoretically complete intersection*. A similar analysis occurs in Chapter 11, where *semicontinuity* for the dimension of the fibers of a dominant morphism between two algebraic varieties is proved. Chapter 12 contains the notion of tangent space, both by making use of coordinates for varieties embedded in either an affine or a projective space, and in the intrinsic definition of *Zariski tangent space* and derivations. This, together with the definition of dimension, allows to introduce the notion of *smooth*, respectively *singular*, point/locus of an algebraic variety.

At the end of each chapter the reader can find some suggested summary exercises; detailed solutions are contained at the end of the text in Chapter *Solutions to Exercises*. The book ends with a useful bibliography, for further information on the topics treated in the book, and with a detailed index of all the symbols and terminology used throughout the text.

About the Author

Flaminio Flamini is full professor in Geometry at the University of Rome "Tor Vergata". He graduated in Mathematics at the University of Rome "La Sapienza" and got his PhD in Mathematics from the Consortium of Universities of Rome "La Sapienza" and "Roma Tre". After completing his PhD, he spent a short period at the University of Illinois at Chicago as visiting scholar and then returned to Italy as an Assistant Professor at the University of L'Aquila. His research interests focus on Algebraic Geometry, and he is the author of more than 40 published papers. He is also co-author of some books concerning Commutative Algebra and its applications, and textbooks on Linear Algebra for undergraduate students. Since 2009 he has been member of the Scientific Board of the PhD School in Mathematics of the Department of Mathematics at the University of Rome "Tor Vergata". He has been Principal Investigator or a member of numerous Committee Boards which award research grants.

Acknowledgments

There are many people I should thank, perhaps first and foremost, many of my colleagues who make me daily appreciate the pleasure of collaboration, discussing Mathematics, and doing research together. On the other hand, if I take into account how my professional path began, I cannot forget those people who have turned out to be central to my academic journey from my early days as a young Master's and PhD student, those who have always believed in me and have encouraged me to do better and better. These are certainly (in alphabetical order) Ciro Ciliberto, Edoardo Sernesi ed Alessandro Verra. Besides them, I had also the privilege as a student to attend advanced courses in Algebraic Geometry taught by Enrico Arbarello, Lucia Caporaso, Fabrizio Catanese, Luca Chiantini, Igor Dolgachev, Lawrence Ein, Robert Lazarsfeld, Angelo F. Lopez, Rick Miranda and Rita Pardini. From all these people I have learnt not only concepts and techniques in Algebraic Geometry, but mainly the passion to face this amazing, elegant, and inspiring research area. I will always be grateful to them.

It is a pleasure to deeply thank my colleagues and friends Gilberto Bini, Seonja Kim, Andreas L. Knutsen, and Alessandra Sarti, not only for valuable suggestions on either the content of the text or how to produce figures with TikzPgf, but also for their words of encouragement and appreciation regarding this book.

Fianlly, I wish to warmly thank my former students Marco Carfagnini, Stefano Cipolla, Fabrizio Clementi, Daniele Di Tullio, Francesco Recupero and Emiliano Torti, who attended my Algebraic Geometry courses, for all the questions they asked me, for having carefully read the first draft of this book, and for pointing out some typos and inaccuracies.

Contents

Chapter 1

Basics on Commutative Algebra

This chapter contains fundamental definitions and results that will be frequently used in the next chapters. This choice allows the reader to have a self-contained book, avoiding to retrieve basic results in Commutative Algebra throughout the vast literature. For further information and details, the reader is referred to, e.g. Atiyah and McDonald (1969), Eisenbud (1995), Lang (2002) and Matsumura (1980).

All rings considered in this book will be commutative and with a multiplicative identity and it will be taken from granted once and for all. If R is a ring, 0_R and 1_R (or simply 0 and 1, if no confusion arises) will respectively denote its additive and multiplicative identities. Given R and S two rings, any non-zero *ring homomorphism* $\varphi : R \to S$ will be such that $\varphi(1_R) = 1_S$. An *integral domain* (or simply, *domain*) is a ring with no zero-divisors, i.e. where cancellation law holds.

A *field* $\mathbb{K}$ is a domain in which every non-zero element is invertible, i.e. a *unit*. In symbols, $U(\mathbb{K}) = \mathbb{K} \setminus \{0\}$, where $U(\mathbb{K})$ denotes the (multiplicative) group of units. For any field $\mathbb{K}$ the symbol $\overline{\mathbb{K}}$ will denote its *algebraic closure* ($\overline{\mathbb{K}}$ always exists and it is uniquely determined up to field isomorphism, cf. M. Artin's proof in, e.g. Lang (2002, cf. Chapter V, § 2. Theorem 2.5 and Corollaries 2.6 and 2.9)).

$\mathbb{Z}$ will denote the domain of integers, whereas $\mathbb{Q}$, $\mathbb{R}$ and $\mathbb{C}$ will denote the fields of rational, real and complex numbers, respectively. $\mathbb{Z}_p$ will denote the finite field with p elements, $p \in \mathbb{Z}$ a prime.

For the sake of completeness of this preliminary overview, we have to recall the classical result (first proved by Gauss in 1799), whose proof can be found in, e.g. Lang (2002), to which the interested reader is referred.

Theorem 1.0.1. *The field $\mathbb{C}$ is algebraically closed.*

Let R be a ring. An *ideal* $I \subseteq R$ is a subset of R such that, for any $a, b \in I$ and for any $r \in R$ one has $a + b \in I$ and $ra \in I$. If R is a domain, $\mathcal{Q}(R)$ will denote its *quotient field*, so that $R \subseteq \mathcal{Q}(R)$ where equality holds if and only if R is a field. For any ring R, $R[x]$ denotes the *ring of polynomials in one indeterminate* x and *with coefficients from* R. More generally, if $\underline{x} := (x_1, \ldots, x_n)$ are n indeterminates over R, $R[x_1, \ldots, x_n]$ denotes the *ring of polynomials* in the n *indeterminates and with coefficients from* R; for simplicity, sometimes we will use the symbol $R[\underline{x}]$ to denote this ring, if no confusion arises.

1.1   Ideals and Operations on Ideals

Let R be a ring. A proper ideal $\mathfrak{p} \subset R$ is said to be *prime* if $rs \in \mathfrak{p}$ implies that either $r \in \mathfrak{p}$ or $s \in \mathfrak{p}$; in particular, $\mathfrak{p}$ is prime if and only if $R/\mathfrak{p}$ is an integral domain. Analogously, a proper ideal $\mathfrak{m} \subset R$ is said to be *maximal* if no proper ideal I in R exists such that $\mathfrak{m} \subset I \subset R$, where all the inclusion are strict; in particular, $\mathfrak{m}$ is maximal if and only if $R/\mathfrak{m}$ is *simple* (i.e. with no proper ideals) which, by the assumptions on R, is equivalent to $R/\mathfrak{m}$ being a field. It is therefore clear that a maximal ideal is also prime, but the converse is not true: e.g. consider the ideal $(x) \subset \mathbb{Z}[x]$, which is prime but not maximal.

We recall some standard operations on ideals, which will be frequently used in the next chapters. Let I be an ideal of R; a set S of elements of R is called a *set of generators* for I if

$$I = \left\{ \sum_i r_i s_i \mid s_i \in S, \ r_i \in R \right\},$$

where in $\Sigma_i r_i s_i$ only finitely many r_i's are non-zero. The ideal I is said to be a *finitely generated ideal* if I admits a finite set of generators $S := \{s_1, \ldots, s_n\}$; in such a case, we will write $I := (s_1, \ldots, s_n)$. In particular, when $n = 1$, I is called a *principal ideal*.

If I_1, I_2 are ideals, their *sum* is defined as

$$I_1 + I_2 := \{x_1 + x_2 \mid x_1 \in I_1, \ x_2 \in I_2\};$$

it is the smallest ideal of R containing both I_1 and I_2. More generally, if we have a (possibly infinite) family $\{I_\alpha\}_{\alpha \in A}$ of ideals of R, their *sum* is defined as

$$\sum_{\alpha \in A} I_\alpha := \left\{ \sum_{\alpha \in A} x_\alpha \,|\, x_\alpha \in I_\alpha \right\},$$

where in the sums *almost all* (i.e. up to a finite number) of the x_α's equal 0. As above, $\Sigma_{\alpha \in A} I_\alpha$ is the smallest ideal of R containing all the I_α's. For a (possibly infinite) family $\{I_\alpha\}_{\alpha \in A}$ of ideals, their *intersection* $\cap_{\alpha \in A} I_\alpha$ is an ideal.

The *product* of two ideals I_1 and I_2 is denoted by $I_1 \cdot I_2$ and defined to be the ideal generated by all the products $x_1 x_2$, for any $x_1 \in I_1$ and $x_2 \in I_2$. The elements of this ideal are finite sums of the forms $\Sigma_i x_i y_i$, where $x_i \in I_1$ and $y_i \in I_2$, for any i. Similarly, one can define the product of a (possibly infinite) family $\{I_\alpha\}_{\alpha \in A}$ of ideals in R. As a particular case of products of ideals, for any positive integer n, one has the notion of nth-*power* of an ideal I, which is denoted by I^n (by convention, one poses $I^0 := (1) = R$). Thus, I^n is the ideal generated by all the products of the form $x_1 x_2 \cdots x_n$, where $x_i \in I$, for any $1 \leqslant i \leqslant n$.

For any ideal I of R, one defines the *radical* of I to be the ideal

$$\sqrt{I} := \{x \in R \,|\, x^n \in I, \text{ for some positive integer } n\} \qquad (1.1)$$

(some texts denote the radical with the symbol $\mathrm{rad}(I)$; see, e.g. Atiyah and McDonald (1969)). Note that, for any ideal I, one clearly has $I \subseteq \sqrt{I}$.

Definition 1.1.1. I is said to be a *radical ideal* (or simply *radical*), if $I = \sqrt{I}$, equivalently if R/I is a *reduced ring*, namely R/I has no *nilpotent elements*, i.e. there exists no $x \in R \setminus \{0\}$ such that $x^n = 0$ for some integer $n \geqslant 2$).

Lemma 1.1.2.

(i) *For any ideal $I \subseteq R$, $\sqrt{I}$ is radical.*

(ii) *Any prime (respectively, maximal) ideal is radical.*

Proof.

(i) As for any ideal, one has $\sqrt{I} \subseteq \sqrt{\sqrt{I}}$. Now, $f \in \sqrt{\sqrt{I}}$ implies there exists a positive integer n such that $f^n \in \sqrt{I}$. Thus, for some positive integer m, $(f^n)^m \in I$, i.e. $f \in \sqrt{I}$, which means that $\sqrt{\sqrt{I}} \subseteq \sqrt{I}$.

(ii) It directly follows from the definition of prime (respectively, maximal) ideal. $\qquad\square$

Let $\varphi : R \to S$ be any ring homomorphism.

Definition 1.1.3. If $J \subseteq S$ is any ideal, $\varphi^{-1}(J)$ is an ideal in R, called the *contracted ideal of J w.r.t.* φ and denoted by J^c (sometimes even by $J \cap R$, even if φ is not necessarily injective, cf. Atiyah and McDonald (1969)). Conversely, if $I \subseteq R$ is an ideal, the ideal of S generated by the subset $\varphi(I) \subseteq S$ is called the *extended ideal of I w.r.t.* φ and denoted by I^e.

Remark 1.1.4.

(i) Note that $\varphi(I) \subseteq S$ in general is not an ideal: consider, e.g. $\iota : \mathbb{Z} \hookrightarrow \mathbb{Q}$ to be the natural inclusion and I any proper ideal of $\mathbb{Z}$.

(ii) For any ideal $J \subseteq S$ one has $J \supseteq (J^c)^e$ and in general the inclusion is strict; consider, e.g. $\mathbb{Z}_2 \overset{\varphi}{\hookrightarrow} \mathbb{Z}_2[x]$, where φ the natural inclusion, and $J = (x^2 + x + 1)$, so $(J^c)^e = (0)$. Vice versa, for any ideal $I \subseteq R$ one has $I \subseteq (I^e)^c$ and in general the inclusion is strict; consider as in (i) $\mathbb{Z} \overset{\iota}{\hookrightarrow} \mathbb{Q}$ and $I = (p)$, for some prime $p \in \mathbb{Z}$, so $(I^e)^c = (1)$.

(iii) If $\mathfrak{p} \subset S$ is a prime ideal, then $\mathfrak{p}^c \subset R$ is also prime; conversely, being $\mathfrak{p} \subset R$ a prime ideal does not imply that $\mathfrak{p}^e \subseteq S$ is necessarily a prime ideal, cf., e.g. (ii) above.

(iv) Note that if $\mathfrak{m} \subset S$ is a maximal ideal then $\mathfrak{m}^c \subseteq R$ in general is only prime but not maximal. Indeed, consider, e.g. $\mathbb{Z}[x] \overset{\varphi}{\twoheadrightarrow} \mathbb{Z}$ defined by $\varphi(q(x)) = q(0)$, for any $q(x) \in \mathbb{Z}[x]$, and $J = (p)$, for any prime $p \in \mathbb{Z}$; in such a case $(p)^c$ is prime but not maximal since $\mathbb{Z}[x]/(p)^c \cong \mathbb{Z}_p[x]$ is an integral domain but not a field.

1.2   UFDs and PIDs

Recall that a ring R is said to be a *principal ring* if every ideal I of R is principal. If R is an integral domain and a principal ring, then R is said to be a *principal ideal domain* (PID). Any field $\mathbb{K}$ and $\mathbb{Z}$ are easy examples of PIDs. Recall that, when $\mathbb{K}$ is a field, the ring $\mathbb{K}[x]$ is a *Euclidean domain*, roughly speaking it is an integral domain where Euclidean division algorithm holds. This implies that $\mathbb{K}[x]$ is a PID (cf., e.g. Lang, 2002, IV § 1, Theorem 1.2).

From now on, let D denote an integral domain. Given $a, b \in D$ we say that *a divides b*, or that *b is divisible by a* (in symbols $a|b$) if there exists $c \in D$ s.t. $b = ca$; in particular, a is invertible if and only if $a|1_D$. Elements $a, b \in D$ are called *associate elements* (simply *associates*) if $a|b$ and $b|a$, i.e. $b = ea$, for some $e \in U(D)$. An element $a \in D \setminus \{0\}$ is said to be *irreducible* if it is not a unit in D and, whenever one can write $a = bc$, with $b, c \in D$, then either b or c is a unit in D; in other words a is irreducible if and only if it is divisible only by its associate elements and by the units in D. An element $a \in D \setminus \{0\}$ is said to have a *unique factorization into irreducible elements* if there exist $u \in U(D)$ and $p_i \in D$ irreducible elements, $1 \leqslant i \leqslant r$ (not necessarily all distincts), such that $a = up_1 \cdots p_r$ and, if moreover we have two factorizations

$$a = up_1 \cdots p_r = u'q_1 \cdots q_s,$$

we have $r = s$ and, after a permutation of the indexes, we have $p_i = u_i q_i$ for any $1 \leqslant i \leqslant r$, for some $u_i \in U(D)$.

D is said to be a *unique factorization domain* (UFD), if every non-zero element of D has a unique factorization into irreducible elements. In particular, any *prime element* (i.e. generating a prime ideal) of a UFD is an irreducible element (the converse being always true). At last, in a UFD one can define a *greatest common divisor* for finite set of elements in D (this is defined up to units of D). Recall that any PID is a UFD (cf., e.g. Lang, 2002, II § 5, Theorem 5.2); in particular, $\mathbb{Z}$, any field $\mathbb{K}$ and $\mathbb{K}[x]$ are UFD's.

1.3 Polynomial Rings

Here we collect some useful terminology and results on polynomials in one or more indeterminates, with coefficients from an integral domain, in particular from a UFD.

1.3.1 *Polynomials in $D[x]$, where D a UFD*

If D_0 is an integral domain, definitions in Section 1.2 apply to the integral domain $D := D_0[x_1, \ldots, x_n]$ (cf. Theorem 1.3.1 and Remark 1.3.2) giving rise to divisibility among polynomials, associate polynomials, irreducible polynomials, etc. Any polynomial of degree 1 with invertible coefficients is always irreducible. In general, the irreducibility of a polynomial depends on the domain D_0; e.g. the polynomial $x^2 + 1$ is irreducible as an element of $\mathbb{R}[x]$ but it factorizes into irreducible elements $(x - i)(x + i)$ in $\mathbb{C}[x]$, where $i^2 = -1$.

Recall the following fundamental results, whose proofs can be found in, e.g. Lang (2002, cf. Chapter IV § 2), to which we refer the interested reader.

Theorem 1.3.1 (Unique factorization theorem). *Let D be a UFD and x be an indeterminate over D. Then $D[x]$ is a UFD.*

Remark 1.3.2. When $\mathbb{K}$ is a field, $\mathbb{K}[x_1, \ldots, x_n]$ is a UFD for any $n \geqslant 1$, as one obtains by recursively applying Theorem 1.3.1: pass from $D := \mathbb{K}[x_1, \ldots, x_{n-1}]$ a UFD to $D' := \mathbb{K}[x_1, \ldots, x_n] = D[x_n]$.

Theorem 1.3.3 (Gauss' theorem). *Let D be a UFD, $\mathbb{K} := \mathcal{Q}(D)$ be its quotient field and x be an indeterminate over D. Let $f := f(x) = \Sigma_{i=0}^n d_i x^i \in D[x]$ be a non-constant polynomial. Assume that f is a primitive polynomial, namely the content $c(f) := g.c.d.(d_0, d_1, \ldots, d_n) \in D$ of f is such that $c(f) = 1$ (where the equality is meant up to a unit in D). Then f is irreducible in $D[x]$ if and only if it is irreducible in $\mathbb{K}[x]$.*

Suppose now to have $f \in D[x]$ and consider its factorization into irreducible elements in $D[x]$; merging the repeated irreducible factors of f, one gets

$$f = g_1^{e_1} \cdots g_s^{e_s}, \tag{1.2}$$

where the g_i's are all the irreducible, distinct factors of $f \in D[x]$ and e_i's are non-negative integers, $1 \leqslant i \leqslant s$. The exponent e_i, $1 \leqslant i \leqslant s$, is called the *multiplicity of the factor* g_i in f. If $e_i > 1$ then g_i is said to be a *multiple factor* of f. One obviously has $\deg(f) = \Sigma_{i=1}^s e_i \deg(g_i)$.

Definition 1.3.4. If $f \in D[x]$ is a non-constant polynomial then $f = a_0 + a_1 x + a_2 x^2 + \cdots + a_n x^n \in D[x]$, for some $a_j \in D$, $0 \leqslant j \leqslant n$ with $a_n \neq 0$. The integer $n \geqslant 1$ is the *degree of f* and $a_n \in D$ is called the *leading coefficient* of f, which is also denoted by $\mathrm{lc}(f)$. A non-constant polynomial is said to be *monic* if $\mathrm{lc}(f) = 1$.

A non-constant polynomial $f \in D[x]$, for which $\mathrm{lc}(f) \in U(D)$, is always associated to a monic polynomial $f' \in D[x]$. In particular, for $D = \mathbb{K}$ a field, any non-constant polynomial is always associated to a monic one.

Definition 1.3.5. Let $f \in D[x]$ be a non-constant polynomial. An element $\alpha \in D$ is said to be a *root* of f, if $f(\alpha) = 0$.

The following is a straightforward result.

Lemma 1.3.6. $\alpha \in D$ *is a root of $f \in D[x]$ if and only if $(x - \alpha)$ divides f in $D[x]$.*

Definition 1.3.7. $\alpha \in D$ is said to be a *multiple root* of $f \in D[x]$, if $(x - \alpha)$ is a *multiple factor* of f, i.e. if one has $f = (x - \alpha)^e g$, for some integer $e \geqslant 2$ (cf. decomposition (1.2)).

The greatest integer e for which a factorization as above exists is called the *multiplicity of the root* α for f. If $e = 1$, then α is said to be a *simple root* of f. If α is not a root of f, then one has $e = 0$. Note that e is the multiplicity of the root α if and only if one has $g(\alpha) \neq 0$ for a decomposition as in Definition 1.3.7; moreover, since $\deg(g) = \deg(f) - e$, one has $e \leqslant \deg(f)$, for any root α of f.

Corollary 1.3.8. *A non-constant polynomial $f \in D[x]$ of degree d has at most d roots in D (when the roots are counted with multiplicity).*

Proof. It directly follows by induction on d. $\qquad\qquad\qquad$ □

The notion of derivative of a polynomial can be given formally, with no use of infinitesimal calculus. This allows one to consider derivatives of polynomials in $D[x]$, for D any domain. Precisely, if

$$f := a_d x^d + a_{d-1} x^{d-1} + \cdots + a_1 x + a_0 \in D[x],$$

one defines the (*first*) *derivative* of f to be

$$\frac{df}{dx} = f' := d a_d x^{d-1} + (d-1) a_{d-1} x^{d-2} + \cdots + a_1 \qquad (1.3)$$

and recursively, the *kth derivatives*, for any $k \geqslant 2$, to be the polynomial given by $\frac{d^k f}{dx^k} = f^{[k]} := \frac{d(f^{[k-1]})}{dx}$, where $f^{[1]} := f'$.

Similarly, if $f \in D[x_1, \ldots, x_n]$, one can define the *partial derivative with respect to the indeterminate* x_j, $1 \leqslant j \leqslant n$, denoted by $\frac{\partial f}{\partial x_j}$, to be the (first) derivative of f considered as an element of $(D[x_1, \ldots, x_{j-1}, x_{j+1}, \ldots, x_n])[x_j]$. It is therefore clear that one can also define *higher order partial derivatives* and that the *Schwarz rule* holds, as it can be easily verified on monomials. We will not dwell on this.

Lemma 1.3.9. *$\alpha \in D$ is a multiple root of $f \in D[x]$ if and only if it is a common root of f and f'.*

Proof. Since $\alpha \in D$ is a root of f, by Lemma 1.3.6, $f = (x - \alpha)^e g$, for some $g \in D[x]$ s.t. $g(\alpha) \neq 0$ and for some positive integer e. One has $f' = e(x - \alpha)^{e-1} g + (x - \alpha)^e g'$. If $f' \neq 0$ (e.g. if D is of characteristic 0), then $e = 1$ if and only if $f'(\alpha) = g(\alpha) \neq 0$, proving the statement in this case. If otherwise $f' = 0$ (which can only occur in positive characteristic since f is non-constant by assumption), then $f(x) = (h(x))^p$, for some non-constant polynomial $h \in D[x]$ (cf. Lang, 2002, Proposition 1.12, p. 179), which implies that α is a root of multiplicity at least p.   □

1.3.2   *The case* $D = \mathbb{K}$ *a field*

Here we focus on the case $D = \mathbb{K}$ a field.

Proposition 1.3.10. *Let* $\mathbb{K}$ *be any field. Then* $\mathbb{K}[x]$ *contains infinite irreducible monic polynomials.*

Proof. $\mathbb{K}[x]$ contains irreducible polynomials as, e.g. $x - \alpha \in \mathbb{K}[x]$, for any $\alpha \in \mathbb{K}$. If $\mathbb{K}$ is an infinite field, we are done. If otherwise $\mathbb{K}$ is a finite field, assume by contradiction that $f_1, \dots, f_n \in \mathbb{K}[x]$ are all the monic, irreducible polynomials, for some positive integer n. The polynomial $f := 1 + (f_1 \cdots f_n) \in \mathbb{K}[x]$ is monic and it must irreducible, indeed none of the f_i's divides f, $1 \leqslant i \leqslant n$. This contradicts our assumption.   □

As a direct consequence of Lemma 1.3.6, by induction on the degree of the polynomial one obtains the following.

Theorem 1.3.11. *Let* $\mathbb{K}$ *be an algebraically closed field. Any non-constant polynomial* $f \in \mathbb{K}[x]$ *of degree* d *contains exactly* d *roots in* $\mathbb{K}$ *(when roots are counted with multiplicity). In particular,* f *factorizes in* $\mathbb{K}[x]$ *as*

$$f = a(x - \alpha_1) \cdots (x - \alpha_d),$$

where $a \in \mathbb{K} \setminus \{0\}$ *and* $\alpha_1, \dots, \alpha_d \in \mathbb{K}$ *are all the (not necessarily distinct) roots of* f.

Corollary 1.3.12. *If* $\mathbb{K}$ *is an algebraically closed field, then* $\mathbb{K}$ *contains infinitely many elements.*

Proof. It directly follows from Proposition 1.3.10 and Theorem 1.3.11. □

For what concerns polynomial rings $\mathbb{K}[x_1, \dots, x_n]$, a fundamental result is the following.

Theorem 1.3.13. *Let* $\mathbb{K}$ *be any infinite field and let* $f \in \mathbb{K}[x_1, \dots, x_n]$, $n \geqslant 1$, *be a non-constant polynomial. If there exists a subset* $J \subseteq \mathbb{K}$ *with infinitely many elements s.t.* $f(a_1, \dots, a_n) = 0$ *for any* $(a_1, \dots, a_n) \in J^n$, *where* $J^n := J \times \cdots \times J$ *the* n-*tuple product of* J, *then* $f = 0$ *in* $\mathbb{K}[x_1, \dots, x_n]$.

Proof. The proof is by induction on n. If $n = 1$, the statement directly follows from Corollary 1.3.8. Assume therefore $n \geqslant 2$ and that the statement holds true for all polynomials in $n - 1$ indeterminates with coefficients from $\mathbb{K}$. Take $f \in \mathbb{K}[x_1, \ldots, x_n]$ which, by the inductive hypothesis, can be considered to be non-constant with respect to all the indeterminates (in particular with respect to x_n). Let d be the degree of f with respect to x_n, i.e. when f is considered as an element of $(\mathbb{K}[x_1, \ldots, x_{n-1}])[x_n] \cong \mathbb{K}[x_1, \ldots, x_n]$. Thus, one can write

$$f = f_0 + x_n f_1 + \cdots + x_n^d f_d, \qquad (1.4)$$

with $f_j \in \mathbb{K}[x_1, \ldots, x_{n-1}]$, $0 \leqslant j \leqslant d$, and $f_d \neq 0$ by the assumption on the degree. With this set-up, assume by contradiction there exists an infinite subset $J \subseteq \mathbb{K}$ for which $f(a_1, \ldots, a_n) = 0$, $\forall \, (a_1, \ldots, a_n) \in J^n$. Since $f_d \neq 0$, by induction there exists $(a_1, \ldots, a_{n-1}) \in J^{n-1}$ s.t. $f_d(a_1, \ldots, a_{n-1}) \neq 0$. Thus, (1.4) gives $\phi_f(x_n) := f(a_1, \ldots, a_{n-1}, x_n) \in \mathbb{K}[x_n]$, where

$$\phi_f(x_n) = f_0(a_1, \ldots, a_{n-1}) + x_n f_1(a_1, \ldots, a_{n-1}) + \cdots + x_n^d f_d(a_1, \ldots, a_{n-1})$$

is a non-constant polynomial in $\mathbb{K}[x_n]$ since $f_d(a_1, \ldots, a_{n-1}) \neq 0$. This gives a contradiction since $\phi_f(x_n) \in \mathbb{K}[x_n]$ and $\phi_f(a) = f(a_1, \ldots, a_{n-1}, a) = 0$ for any $a \in J$ which is infinite. $\qquad \square$

Remark 1.3.14. Theorem 1.3.13 is false if $\mathbb{K}$ is a finite field; indeed, consider $f = x^p - x \in \mathbb{Z}_p[x]$, where $p \in \mathbb{Z}$ a prime; one has $f(\alpha) = 0$, for any $\alpha \in \mathbb{Z}_p$ (cf., e.g. Example 2.1.10-(iii) for details), even if f is not the zero-polynomial.

1.3.3 *Resultant of two polynomials in $D[x]$*

For the whole section, D will denote a UFD. Here we shall briefly recall some basic results which allow one to establish if two polynomials $f, g \in D[x]$ have a non-constant, common factor in $D[x]$ (recall Theorem 1.3.1). Let

$$f(x) = a_n x^n + \cdots + a_0, \quad g(x) = b_m x^m + \cdots + b_0, \quad \text{with } a_n b_m \neq 0 \quad (1.5)$$

be non-constant polynomials in $D[x]$.

Lemma 1.3.15 (Euler's lemma). *In the above assumptions, f and g have a non-constant common factor in $D[x]$ if and only if there exist non-zero polynomials $p(x), q(x) \in D[x]$, with $\deg(p) < m$, $\deg(q) < n$, such that*

$$pf = qg. \qquad (1.6)$$

Proof. If $\phi \in D[x]$ is a non-constant common factor of f and g, then $f = \phi q$ and $g = \phi p$, for some $p, q \in D[x]$, which verify the statement. Conversely, suppose (1.6) holds, with $\deg(p) < m$ and $\deg(q) < n$. In particular, any irreducible factor of g in $D[x]$ divides pf; since $\deg(p) < m = \deg(g)$, there exists at least a non-constant, irreducible factor ϕ of g which divides f, proving the statement. $\square$

With notation and assumptions as in (1.5), one defines the *Sylvester matrix* of the two polynomials $f, g \in D[x]$ to be the $(m + n) \times (m + n)$ matrix

$$S(f,g) := \begin{pmatrix} a_0 & a_1 & \cdots & a_n & 0 & 0 & \cdots & 0 \\ 0 & a_0 & a_1 & \cdots & a_n & 0 & \cdots & 0 \\ & & \cdots & & & & & \\ 0 & \cdots & 0 & 0 & a_0 & a_1 & \cdots & a_n \\ b_0 & b_1 & \cdots & b_m & 0 & 0 & \cdots & 0 \\ 0 & b_0 & b_1 & \cdots & b_m & 0 & \cdots & 0 \\ & & \cdots & & & & & \\ 0 & \cdots & 0 & 0 & b_0 & b_1 & \cdots & b_m \end{pmatrix}, \tag{1.7}$$

where the block containing $a_0, \ldots, a_n$ consists of $m = \deg(g)$ rows whereas the block containing $b_0, \ldots, b_m$ consists of $n = \deg(f)$ rows. One denotes by

$$R(f, g) := \det(S(f, g)) \in D, \tag{1.8}$$

which is called the *resultant* of the two polynomials. One has the following theorem.

Theorem 1.3.16. *Let f and g be polynomials as in (1.5). Then f and g have a non-constant common factor in $D[x]$ if and only if $R(f, g) = 0$.*

Proof. By Euler's lemma, the existence of polynomials $p, q \in D[x]$ satisfying (1.6), with $\deg(p) < m$ and $\deg(q) < n$, is equivalent to the existence of $n + m$ elements in D, say c_i, $0 \leqslant i \leqslant m - 1$, not all zero, and d_j, $0 \leqslant j \leqslant n - 1$, not all zero, such that

$$(c_{m-1}x^{m-1} + \cdots + c_0) f(x) = (-d_{n-1}x^{n-1} - \cdots - d_0) g(x), \tag{1.9}$$

holds true. This is equivalent to the fact that the c_i's and the d_j's give rise to a non-zero solution of the homogeneous linear system of

$n + m$ equations

$$\begin{cases} a_0 C_0 & = -b_0 D_0 \\ a_1 C_0 + a_0 C_1 = -b_1 D_0 - b_0 D_1 \\ \cdots \qquad\qquad \cdots \\ a_n C_{m-1} & = -b_m D_{n-1} \end{cases} \tag{1.10}$$

in the indeterminates C_i's and D_j's, respectively, $0 \leqslant i \leqslant m - 1$, $0 \leqslant j \leqslant n - 1$, whose coefficient matrix is exactly the transpose of the Sylvester matrix $S(f, g)$. One finally concludes by the well-known fact that the existence of a non-zero solution of (1.10) is equivalent to $R(f, g) = 0$ (by elementary transformations, this condition is equivalent to the existence of a non-zero solution of (1.10) in the field $\mathbb{K} := \mathcal{Q}(D)$; on the other hand, since (1.10) is homogeneous one has a non-zero solution in D). $\qquad\square$

Corollary 1.3.17. *If $D = \mathbb{K}$ is an algebraically closed field and f, g are polynomials as in (1.5), then $R(f, g) = 0$ if and only if f and g have a common root in $\mathbb{K}$.*

Proof. It is a direct consequence of Theorem 1.3.16 and of the fact that, if $\mathbb{K}$ is algebraically closed, then non-constant irreducible elements in $\mathbb{K}[x]$ are all of the form $(x - \alpha)$, $\alpha \in \mathbb{K}$. $\qquad\square$

When $f \in D[x]$ is such that $n = \deg(f) \geqslant 2$, one would like to find a condition similar to that given in Theorem 1.3.16 in order to establish if f has a *multiple factor* in $D[x]$, i.e. $f = g^e h$, for some integer $e \geqslant 2$ and some non-constant polynomial $g \in D[x]$. This question is obviously related to the existence of multiple roots of f, possibly in some field $\mathbb{K}$ containing D (cf. Definition 1.3.7 and Lemma 1.3.9). On the other hand, derivatives in positive characteristic not always well-behave (recall, e.g. the proof of Lemma 1.3.9). Indeed when D is a UFD of positive characteristic, there actually exist non-constant polynomials $f \in D[x]$ for which, either $f' = 0$ or f' constant even if $\deg(f) > 1$ (in both cases one cannot define $R(f, f')$). For the scope of this book, we can limit ourselves to the characteristic-zero case and refer the interested reader to, e.g. Lang (2002).

Thus, when D is a UFD of characteristic 0 and $f \in D[x]$ is a polynomial of degree $n \geqslant 2$, then $f' \in D[x]$ is a non-constant polynomial of degree $n-1$. As above, one can define the $(2n - 1) \times (2n - 1)$-Sylvester matrix $S(f, f')$

and consequently $R(f, f') \in D$, which is simply denoted by $\Delta(f)$ and called the *discriminant* of f.

Proposition 1.3.18. *Let D be a UFD of characteristic 0 and $f \in D[x]$ a polynomial of degree $n \geqslant 2$. Then f has a multiple factor in $D[x]$ if and only if $\Delta(f) = 0$.*

Proof. ($\Rightarrow$) By the assumptions, there exists a non-constant $g \in D[x]$ such that $f = g^2 k$, for some $k \in D[x]$. Then $f' = 2gg'k + g^2k'$, which shows that $g|f$ and $g|f'$. Since $\deg(f) \geqslant 2$ and since the characteristic is zero, $f' \in D[x]$ is a non-constant polynomial, so that $\Delta(f)$ makes sense and one concludes by Theorem 1.3.16.

($\Leftarrow$) Assume $\Delta(f) = 0$. From Theorem 1.3.16, it follows that f and f' have a non-constant common factor $g \in D[x]$ which we can assume to be irreducible. From $f = gh$, for some $h \in D[x]$, we deduce $f' = g'h + gh'$, where $g' \neq 0$. Since $g|f'$ by assumption, then $g|g'h$. Since $\deg(g) > \deg(g')$ and g is irreducible, then $g|h$ so $g^2|f$. $\qquad\square$

Corollary 1.3.19. *If $D = \mathbb{K}$ is an algebraically closed field of characteristic 0 and $f \in \mathbb{K}[x]$ is a polynomial of degree $n \geqslant 2$, then $\Delta(f) = 0$ if and only if f has a multiple root.*

1.3.4   *Resultant in $D[x_1, \ldots, x_n]$ and elimination*

In the previous section, we introduced the notion of resultant for two polynomials in $D_0[x]$, where D_0 any UFD. In particular, if D is a UFD then $D[x_1, \ldots, x_n] = (D[x_1, \ldots, x_{n-1}])[x_n]$, with $D_0 := D[x_1, \ldots, x_{n-1}]$ a UFD (cf. Theorem 1.3.1 and Remark 1.3.2). Thus, we can apply the machinery developed in Section 1.3.3 to any pair of non-constant polynomials $f, g \in D_0[x_n]$ to get $R_{x_n}(f, g) := R(f, g) \in D_0 = D[x_1, \ldots, x_{n-1}]$, which is called the *resultant polynomial of f and g with respect to the indeterminate* x_n. Note that $R_{x_n}(f, g)$ is a polynomial where the indeterminate x_n does not appear; this gives rise to the terminology *elimination theory*.

Proposition 1.3.20. *Let D be a UFD and let $f, g \in D[x_1, \ldots, x_n]$ be non-constant polynomials of the form*

$$f = a_d x_n^d + \cdots + a_0, \quad g(x) = b_m x_n^m + \cdots + b_0 \qquad (1.11)$$

with $a_d, b_m \in D[x_1, \ldots, x_{n-1}] \setminus \{0\}$. Then the resultant polynomial $R := R_{x_n}(f, g) \in D[x_1, \ldots, x_{n-1}]$ belongs to the ideal generated by f and g in

$D[x_1, \ldots, x_n]$. *More precisely there exist polynomials* $A, B \in D[x_1, \ldots, x_n]$, *with degrees w.r.t. the indeterminate* x_n *at most* $m - 1$ *and* $d - 1$, *respectively, such that* $Af + Bg = R$ *holds.*

Proof. One constructs the Sylvester matrix $S(f, g)$ as in (1.7). Then one finds the following relations:

$$\begin{cases} x^{m-1}f(x) = a_d x_n^{d+m-1} + \cdots + a_0 x_n^{m-1} \\ x^{m-2}f(x) = a_d x_n^{d+m-2} + \cdots + a_0 x_n^{m-2} \\ \cdots \cdots \\ f(x) = a_d x_n^d + \cdots + a_0 \\ x^{d-1}g(x) = b_m x_n^{d+m-1} + \cdots + b_0 x_n^{d-1} \\ x^{d-2}g(x) = b_m x_n^{d+m-2} + \cdots + b_0 x_n^{d-2} \\ \cdots \cdots \\ g(x) = b_m x_n^m + \cdots + b_0. \end{cases}$$

If we first multiply each relation by the co-factor of the corresponding element in the last column of $S(f, g)$ and then add-up all the relations obtained above, we get that a relation of the form $Af + Bg = R$ holds true. □

For more details, see, e.g. Hassett (2007, § 5.4, pp. 84–86).

1.4 Noetherian Rings and the Hilbert Basis Theorem

A ring R is said to be a *Noetherian ring* if any ideal I of R is finitely generated. $\mathbb{Z}$ and any field $\mathbb{K}$ are easy examples of Noetherian domains. Since $\mathbb{K}[x]$ is a Euclidean domain, it is a PID and so in particular it is a Noetherian ring. There actually exist rings which are not Noetherian: e.g. the ring $\mathbb{K}[x_1, x_2, x_3, \ldots]$ of polynomials with coefficients from a field $\mathbb{K}$ and with infinite indeterminates is an integral domain which cannot be Noetherian as the maximal ideal $(x_1, x_2, x_3, \ldots)$ cannot be finitely generated.

Note that $\mathbb{K}[x_1, \ldots, x_n]$, with $n \geqslant 2$, is not a PID: e.g. the maximal ideal $(x_1, \ldots, x_n)$ cannot be principal. On the other hand, when $n \geqslant 2$, $\mathbb{K}[x_1, \ldots, x_n]$ is actually Noetherian. This is a consequence of the following more general result.

Theorem 1.4.1 (Hilbert's basis theorem). *Let R be a Noetherian ring and let x be an indeterminate. Then the ring $R[x]$ is Noetherian.*

Remark 1.4.2. Let R be any ring and let $x_1, \ldots, x_n$ be indeterminates over R. One clearly has $R[x_1, \ldots, x_n] = \big(R[x_1, \ldots, x_{n-1}] \big)[x_n]$. Thus, as a consequence of Theorem 1.4.1, if R is Noetherian then also $R[x_1, \ldots, x_n]$ is, for any positive integer n. In particular, this holds for $R = \mathbb{K}$ any field.

Proof of Theorem 1.4.1. Let $f(x) \in R[x]$ be any polynomial. If $f(x)$ is not the zero-polynomial, then $f(x) := a_0 + a_1 x + a_2 x^2 + \cdots + a_i x^i \in R[x]$, where $a_j \in R$, $0 \leqslant j \leqslant i$, with $a_i \neq 0$, i.e. the integer $i \geqslant 0$ is the degree of $f(x)$ and $a_i := \mathrm{lc}(f(x)) \in R$ (recall Definition 1.3.4). If otherwise $f(x)$ is the zero-polynomial, its leading coefficient is $0 \in R$ by definition.

Let $I \subset R[x]$ be any non-trivial proper ideal (if I is either (0) or (1) there is nothing to prove). For any integer $i \geqslant 0$, denote by J_i the subset of R consisting of all leading coefficients of polynomials in I having degrees at most i. Similarly, denote by J the subset of R consisting of all leading coefficients of polynomials in I. It is clear that

$$J_0 \subseteq J_1 \subseteq J_2 \subseteq \cdots \subseteq J$$

and that $0 \in J_0$. By straightforward computations, one verifies that J and J_i, for any $i \geqslant 0$, are ideals of R. Being R Noetherian, all these ideals are finitely generated. Thus, for any $i \geqslant 0$, there exists a finite set $\{r_{ih}\}_{h \in H_i}$ of elements of R, with H_i a finite set of indexes, such that $J_i = (r_{ih})_{h \in H_i}$.

By the very definition of J_i, one can determine as many polynomials $\{f_{ih}(x)\}_{h \in H_i}$ s.t.

$$f_{ih}(x) \in I, \ \deg(f_{ih}(x)) \leqslant i \text{ and } \mathrm{lc}(f_{ih}(x)) = r_{ih}, \quad \forall\, h \in H_i.$$

Similarly, there exists a finite set $\{r_m\}_{m \in M}$ of elements of R, with M a finite set of indexes, such that $J = (r_m)_{m \in M}$, and as many polynomials $\{f_m(x)\}_{m \in M}$ s.t.

$$f_m(x) \in I \text{ and } \mathrm{lc}(f_m(x)) = r_m, \quad \forall\, m \in M.$$

Let $N := \mathrm{Max}_{m \in M}\{\deg(f_m(x))\}$ and consider the ideal I' generated by the polynomials $\{f_m(x)\}_{m \in M}$ and $\{f_{ih}(x)\}_{h \in H_i}$ for $i \leqslant N$, i.e.

$$I' := (\{f_m(x)\}_{m \in M}, \{f_{ih}(x)\}_{h \in H_i,\, i \leqslant N}).$$

One has $I' \subseteq I$ and I' is finitely generated.

To conclude the proof, it suffices to show that $I' = I$. Assume by contradiction that the inclusion $I' \subset I$ is strict. Thus, there exists an element $g(x) \in I \setminus I'$ of minimal degree with respect to this property.

If $\deg(g(x)) := d > N$, then $g(x) = b_0 + b_1 x + \cdots + b_d x^d$, with $0 \neq b_d \in J$, since $g(x) \in I$. By the assumption on J, we have $b_d = \Sigma_{m \in M} q_m r_m$, for some $q_m \in R$. Consider the polynomial

$$q(x) := \sum_{m \in M} q_m \, f_m(x) \, x^{d - \deg(f_m(x))} \in I'.$$

One has $\deg(q(x)) = \deg(g(x))$ and $\mathrm{lc}(q(x)) = \mathrm{lc}(g(x))$. Since $g(x) - q(x) \in I$ and $\deg(g(x) - q(x)) < d = \deg(g(x))$, by the minimality condition on $d = \deg(g(x))$, one must have $g(x) - q(x) \in I'$. This would imply $g(x) = (g(x) - q(x)) + q(x) \in I'$, a contradiction. If otherwise $\deg(g(x)) := i \leqslant N$, $g(x) = b_0 + b_1 x + \cdots + b_i x^i$, with $0 \neq b_i \in J_i$. As above, by the assumption on J_i, $b_i = \Sigma_{h \in H_i} q_{ih} r_{ih}$, for some $q_{ih} \in R$. Consider the polynomial

$$q_i(x) := \sum_{h \in H_i} q_{ih} \, f_{ih}(x) \, x^{i - \deg(f_{ih}(x))} \in I',$$

so that $\deg(q_i(x)) = \deg(g(x))$ and $\mathrm{lc}(q_i(x)) = \mathrm{lc}(g(x))$. Since $g(x) - q_i(x) \in I$ and $\deg(g(x) - q_i(x)) < i = \deg(g(x))$, one concludes as in the previous case. $\qquad \square$

Remark 1.4.3.

(1) If S is a subring of a Noetherian ring R, in general one cannot conclude that S is Noetherian too: consider, e.g. the ring $S := \mathbb{K}[x_1, x_2, x_3, \ldots]$ of polynomials in infinite indeterminates with coefficients from a field $\mathbb{K}$. From Remark 1.4.2 S is not Noetherian on the other hand, since it is an integral domain, it admits a quotient field $\mathcal{Q}(S) := R$. Thus, S is a subring of a Noetherian ring R.

(2) On the contrary, any quotient R/I of a Noetherian ring R is Noetherian too, as it easily follows from the bijective correspondence between ideals of R/I and ideals of R containing I (cf. also Atiyah and McDonald, 1969, Proposition 7.1).

Proposition 1.4.4. *A ring R is Noetherian if and only if every ascending chain*

$$I_1 \subseteq I_2 \subseteq I_3 \subseteq \cdots \subseteq I_n \subseteq I_{n+1} \subseteq \cdots \subset R \qquad (1.12)$$

of proper ideals is stationary, i.e. there exists an integer n_0 among the indexes s.t. $I_{n_0} = I_{n_0+h}$, for any integer $h \geqslant 1$.

Proof. ($\Rightarrow$) Suppose that R is Noetherian and that (1.12) is an ascending chain of proper ideals. Recalling what discussed in Section 1.1, $I := \Sigma_n I_n$ is a proper ideal of R; indeed if $1 \in I$, there should exist an index n_1 such that $1 \in I_{n_1}$, which contradicts the assumptions on the chain (1.12). Since R is Noetherian, there exist $r_1, \ldots, r_m \in I$ s.t. $I = (r_1, \ldots, r_m)$. Let $r_i \in I_{n(i)}$ and let $n_0 := \text{Max}_{1 \leqslant i \leqslant m}\{n(i)\}$. Then, by the ascending condition, $I_{n(i)} \subseteq I_{n_0}$ so, in particular, $r_i \in I_{n_0}$, for any $1 \leqslant i \leqslant m$. Thus, $I \subseteq I_{n_0}$, i.e. $I_{n_0} = I_{n_0+h}$, for any integer $h \geqslant 1$.

($\Leftarrow$) Suppose that the second part of the statement holds true and let I be any proper ideal of R. If I were not finitely generated, one could find an infinite sequence $a_1, a_2, a_3, \ldots$ of distinct generators of I giving rise to an ascending sequence of proper ideals $(a_1) \subset (a_1, a_2) \subset (a_1, a_2, a_3) \subset \cdots$ which could not be stationary, a contradiction.

1.5 R-Modules, R-Algebras and Finiteness Conditions

Let R be a ring. An abelian group $(M, +)$ is called a R-*module* if there exists a R-multiplication map $R \times M \to M$, $(a, m) \to am$, such that for any $a, b \in R$ and any $m, n \in M$ the following hold:

$$(a + b)m = am + bm, \; a(m + n) = am + an, \; (ab)m = a(bm), \; 1m = m.$$

Easy examples of R-modules are, e.g. $M = R$, $M = I$ as well as $M = R/I$ for any ideal $I \subset R$, $M = R^n := R \oplus \cdots \oplus R$ where the R-multiplication is considered componentwise. If $R = \mathbb{K}$ is a field, a $\mathbb{K}$-module is nothing but a $\mathbb{K}$-vector space.

If $N \subset M$ is a subgroup s.t. for any $n \in N$ and any $r \in R$ one has $rn \in N$, then N is said to be a R-*submodule* of M. If $T \subset M$ is a subset of a R-module M, we denote by $\langle T \rangle := \{\Sigma_i a_i t_i \,|\, t_i \in T, \, a_i \in R\}$ the subset of elements in M of the form $\Sigma a_i t_i$, where in the sums only finitely many a_i's are non-zero. The set $\langle T \rangle$ is a submodule of M which is called the R-*module generated by T* and T is said to be a *set of generators* of the R-module $\langle T \rangle$. A R-module M is said to be *finitely generated*, if $M = \langle T \rangle$ for some finite subset $T \subset M$.

Let M and N be two R-modules. A map $\varphi : M \to N$ is a R-*module homomorphism* if it is a group homomorphism which is also R-linear, namely

$$\varphi(x + y) = \varphi(x) + \varphi(y), \;\; \varphi(r\,x) = r\,\varphi(x), \;\; \forall\, x, y \in M, \;\; \forall\, r \in R.$$

A *R-algebra* is a ring S together with a ring-homomorphism $\varphi : R \to S$ (recall that by assumption $\varphi(1_R) = 1_S$, so $\mathrm{Ker}(\varphi) \subsetneq R$ is always a proper ideal). $S' \subset S$ is a *R-subalgebra* of S, if S' is a subring of S which is a R-algebra.

Remark 1.5.1.

(i) Note that a R-algebra is also a R-module by $r \cdot s := \varphi(r) s$, for any $r \in R$ and any $s \in S$, where the two structure are compatible.

(ii) Easy examples of R-algebras are, e.g. the ring of polynomials $R[x_1, \ldots, x_n]$ for any $n \geqslant 1$, the quotient ring R/I for any ideal $I \subsetneq R$ with $\varphi = \pi_I$ the canonical projection.

(iii) If $R = \mathbb{K}$ is a field and $S \neq 0$, then φ necessarily is injective, i.e. $\mathbb{K}$ can be identified with its image in S. Thus, a $\mathbb{K}$-algebra is nothing but a ring containing the field $\mathbb{K}$ as a subring. If in particular $S = \mathbb{F}$ is also a field, then $\mathbb{K} \hookrightarrow \mathbb{F}$ is said to be a *field extension*. The *degree of the field extension*, denoted by $[\mathbb{F} : \mathbb{K}]$, is $\dim_{\mathbb{K}}(\mathbb{F})$ as a $\mathbb{K}$-vector space.

(iv) Since R is commutative and with identity, there exits a unique ring homomorphism $\mathbb{Z} \to R$, $n \to n \cdot 1_R$. In other words, any (commutative and with identity) ring R is automatically a $\mathbb{Z}$-algebra.

(v) Let $\varphi : R \to S$ and $\psi : R \to T$ be ring homomorphisms. A *R-algebra homomorphism*, say $\eta : S \to T$, is a ring homomorphism which is also a homomorphism of R-modules. This occurs if and only if $\eta \circ \varphi = \psi$.

Definition 1.5.2. In the above notation, S is said to be a *finite R-algebra* if S is finitely generated as a R-module.

When $R = \mathbb{K}$ is a field, easy examples of finite $\mathbb{K}$-algebras are, e.g. $\mathbb{K}[x]/(x^n)$ as well as any (algebraic) field extension of finite degree.

Let S be a R-algebra, with $R \xrightarrow{\varphi} S$ as above. Consider $s_1, \ldots, s_n$ elements in S and $x_1, \ldots, x_n$ indeterminates over R. One can define a ring-homomorphism

$$\Phi := \Phi_{s_1, \ldots, s_n} : R[x_1, \ldots, x_n] \to S \tag{1.13}$$

by the rules $\Phi(x_i) = s_i$, $1 \leqslant i \leqslant n$, and $\Phi(a) = \varphi(a)$, $\forall\, a \in R$. Then $\mathrm{Im}(\Phi) \subseteq S$ is the smallest subalgebra of S containing all polynomial expressions in the elements $s_1, \ldots, s_n$ with coefficient from the subring $\varphi(R)$. The subalgebra $\mathrm{Im}(\Phi)$ is simply denoted by $R[s_1, \ldots, s_n]$.

Definition 1.5.3. A R-algebra S is said to be a *R-algebra of finite type* if $S = R[s_1, \ldots s_n]$, for some finitely many elements $s_1, \ldots, s_n \in S$.

In particular, a ring R is said to be *finitely generated* if it is a $\mathbb{Z}$-algebra of finite type.

Example 1.5.4. If S is a finite R-algebra, then it is also a R-algebra of finite type. The converse does not hold in general; indeed, $R[x_1, \ldots, x_n]$, where $n \geqslant 1$ an integer and x_i indeterminates over R, $1 \leqslant i \leqslant n$, is a R-algebra of finite type which is not a finite R-algebra.

Let $\mathbb{K}$ and $\mathbb{F}$ be fields such that $\mathbb{F}$ is a $\mathbb{K}$-algebra. Let $s_1, \ldots, s_n \in \mathbb{F} \setminus \mathbb{K}$. One denotes by $\mathbb{K}(s_1, \ldots, s_n)$ the quotient field $Q(\mathbb{K}[s_1, \ldots, s_n])$, which is the smallest subfield of $\mathbb{F}$ containing $\mathbb{K}$ and $s_1, \ldots, s_n \in \mathbb{F}$.

Definition 1.5.5. A field extension $\mathbb{K} \subset \mathbb{F}$ is called a *finitely generated field extension* if $\mathbb{F} = \mathbb{K}(s_1, \ldots, s_n)$, for some finitely many elements $s_1, \ldots, s_n$ in $\mathbb{F}$. When $n = 1$, the field extension is called *simple*.

Remark 1.5.6. Note that if $\mathbb{K}$ and $\mathbb{F}$ are fields such that $\mathbb{F}$ is a $\mathbb{K}$-algebra of finite type, then $\mathbb{K} \subseteq \mathbb{F}$ is a finitely generated field extension. In particular, for a field $\mathbb{F}$, we have

$$(*) \ \mathbb{F} \text{ is a finite } \mathbb{K} - \text{algebra} \ \Rightarrow \ (**) \ \mathbb{F} \text{ is a } \mathbb{K} - \text{algebra of finite type}$$

$$\Rightarrow (***) \ \mathbb{K} \subseteq \mathbb{F} \text{ is a finitely generated field extension.}$$

The opposite implications do not hold in general: to show that in general "$(**)$ *does not imply* $(*)$", recall, e.g. Example 1.5.4 whereas, to note that in general "$(***)$ *does not imply* $(**)$", we can deduce it from the next result.

Proposition 1.5.7. *Let $\mathbb{K}$ be any field. Let $x_1, \ldots, x_n$ be indeterminates over $\mathbb{K}$, for any integer $n \geqslant 1$, and let $\mathbb{K}(x_1, \ldots, x_n) := Q(\mathbb{K}[x_1, \ldots, x_n])$ be the* field of rational functions *with coefficients from $\mathbb{K}$, which is a finitely generated field extension of $\mathbb{K}$. Then $\mathbb{K}(x_1, \ldots, x_n)$ is not a $\mathbb{K}$-algebra of finite type.*

Proof. For simplicity, we consider the case $n = 1$, the general case being similar. Therefore, in what follows we will simply put $x_1 = x$. Assume by contradiction there exist finitely many $s_1, \ldots, s_t \in \mathbb{K}(x)$ such that $\mathbb{K}(x) = \mathbb{K}[s_1, \ldots, s_t]$, and let $s \in \mathbb{K}[x]$ be the product of all denominators of $s_1, \ldots, s_t \in \mathbb{K}(x)$. In such a case, for any $z \in \mathbb{K}(x)$ there would exist a positive integer m (depending on z) such that $s^m z \in \mathbb{K}[x]$. This is a contradiction; indeed, it suffices to take any irreducible non-constant polynomial $c \in \mathbb{K}[x]$ which does not divide s in $\mathbb{K}[x]$, and consider $z := \frac{1}{c} \in \mathbb{K}(x)$. $\qquad \square$

1.6 Integrality

Let R and S be rings such that $R \subseteq S$.

Definition 1.6.1. An element $v \in S$ is said to be *integral over the subring* R if there exists a monic polynomial $f(t) \in R[t]$, where t is an indeterminate over R, s.t. $f(v) = 0$.

Remark 1.6.2.

(i) If $v \in R$, then v is integral over R.

(ii) If $R \subseteq S$ is a field extension, $v \in S$ is integral over R if and only if it is algebraic over R.

(iii) $v \in \mathbb{Q}$ *is integral over* $\mathbb{Z}$ *if and only if* $v \in \mathbb{Z}$.

Proof. Indeed, if $v = \frac{r}{s} \in \mathbb{Q}$, with r and s coprime, then an expression of the form $v^n + a_{n-1}v^{n-1} + \cdots + a_0 = 0$, $a_i \in \mathbb{Z}$, $0 \leqslant i \leqslant n-1$, gives $r^n + a_{n-1}r^{n-1}s + \cdots + a_0 s^n = 0$, i.e. s divides r^n. This implies that s divides r and so $s = \pm 1$. $\qquad\square$

(iv) *Let R be any UFD and let $\mathcal{Q}(R)$ be its quotient field. Then $v \in \mathcal{Q}(R)$ is integral over R if and only if $v \in R$.*

The proof is similar to that of (iii) above. Note that the previous statement, in particular, applies to the polynomial ring $R = \mathbb{K}[t]$, where $\mathbb{K}$ any field and t an indeterminate over $\mathbb{K}$.

Proposition 1.6.3. *Let R and S be rings s.t. $R \subseteq S$. The following conditions are equivalent:*

(i) *$v \in S$ is integral over R;*

(ii) *$R[v]$ is a finite R-algebra, i.e. it is a finitely generated R-module;*

(iii) *there exists a subring B of S, containing $R[v]$, which is a finitely generated R-module.*

Proof. (i) $\Rightarrow$ (ii): from (i), there exist a positive integer n and elements $a_i \in R$, $0 \leqslant i \leqslant n-1$, such that for any non-negative integer r one has

$$v^{n+r} = -(a_{n-1}v^{n+r-1} + \cdots + a_0 v^r).$$

Thus, one gets that, for any integer $m \geqslant 0$, v^m belongs to the R-submodule of S generated by the elements $1, v, v^2, \ldots, v^{n-1}$, which implies that $R[v]$ is a finitely generated R-module.

(ii) $\Rightarrow$ (iii): it suffices to consider $B = R[v]$.

(iii) $\Rightarrow$ (i): let $R[v] \subseteq B \subseteq S$ be ring inclusions, such that B is a finitely generated R-module. Let $v_1, \ldots, v_n \in B$ be generators of B as a R-module. Since $v, v_i \in B$ then $vv_i \in B$, for any $1 \leqslant i \leqslant n$, as B is a ring. By the assumptions on B, there exist elements $a_{ij} \in R$, $1 \leqslant i, j \leqslant n$, such that

$$vv_i = \sum_{j=1}^{n} a_{ij} v_i, \quad 1 \leqslant i \leqslant n.$$

Denoting by δ_{ij} the Kronecker symbol, the previous equality gives

$$\det(\delta_{ij} v - a_{ij}) \, v_i = 0, \quad 1 \leqslant i \leqslant n.$$

Since $v_1, \ldots, v_n$ generates B, this implies $\det(\delta_{ij} v - a_{ij}) = 0$, i.e. v is a root of the characteristic polynomial of the matrix (a_{ij}), $1 \leqslant i, j \leqslant n$, which is a monic polynomial with coefficients from R. $\qquad\square$

Corollary 1.6.4. *Let R and S be rings s.t. $R \subseteq S$. The set $C := \{v \in S \mid v \text{ integral over } R\}$ is a subring of S such that $R \subseteq C \subseteq S$.*

Proof. It is clear that as sets $R \subseteq C$. Take now $v, w \in C$ any two elements. Since $R \subseteq R[v]$ and since w is integral over R, then w is also integral over $R[v]$. From Proposition 1.6.3, $R[v, w]$ is a finitely generated $R[v]$-module. By transitivity, $R[v, w]$ is a finitely generated R-module. Therefore, from Proposition 1.6.3, $R[v, w] \subseteq C$; in particular, vw, $v \pm w \in C$, which proves that C is a subring of S containing R as a subring. $\qquad\square$

C as in Corollary 1.6.4 is called the *integral closure of R in S*.

Definition 1.6.5. An integral domain R is said to be *integrally closed* if it coincides with its integral closure in $\mathcal{Q}(R)$, i.e. $v \in \mathcal{Q}(R)$ and v integral over R implies $v \in R$.

From Remark 1.6.2-(iii) and (iv), $\mathbb{Z}$ and any UFD are integrally closed.

Proposition 1.6.6 (cf. Milne (2017, Proposition 1.18, p. 19)). *Let R be any integral domain and $\mathbb{F} := \mathcal{Q}(R)$ be its quotient field. Let $\mathbb{K}$ be any field such that $\mathbb{K} \supseteq \mathbb{F}$. If $\alpha \in \mathbb{K}$ is algebraic over $\mathbb{F}$, then there exists $d \in R$ such that $d\alpha \in \mathbb{K}$ is integral over R.*

Proof. By assumptions, there exists a relation of the form

$$\alpha^n + a_{n-1}\alpha^{n-1} + \cdots + a_0 = 0, \ a_i \in \mathbb{F}, \ 0 \leqslant i \leqslant n - 1.$$

Let $d \in R$ be the product of all denominators of the a_i's. Then $da_i \in R$, for any $0 \leqslant i \leqslant n - 1$. If we multiply the previous relation by d^n, we get

$$(d\alpha)^n + da_{n-1}(d\alpha)^{n-1} + \cdots + d^n a_0 = 0,$$

i.e. $d\alpha$ is a root of a monic polynomial in $R[x]$. $\qquad\square$

Corollary 1.6.7 (cf. Milne (2017, Corollary 1.19, p. 20)). *Let R be an integral domain and $\mathbb{F} := \mathcal{Q}(R)$ be its quotient field. For any algebraic field extension $\mathbb{F} \subseteq \mathbb{K}$, consider C the integral closure of R in $\mathbb{K}$. Then $\mathbb{K} = \mathcal{Q}(C)$.*

Proof. The proof of Proposition 1.6.6 shows that any element $\alpha \in \mathbb{K}$ can be written as $\alpha = \frac{\beta}{d}$, where β integral over R and $d \in R$. $\qquad\square$

1.7 Zariski's Lemma

In Section 1.5, we introduced finiteness conditions for $\mathbb{K}$-algebras, where $\mathbb{K}$ is a field.

Remark 1.7.1. Recall that for a field $\mathbb{F}$ which is a simple extension of $\mathbb{K}$, i.e. $\mathbb{F} = \mathbb{K}(v)$ for some $v \in \mathbb{F} \setminus \mathbb{K}$, two cases occur:

(i) if $v \in \mathbb{F}$ is *transcendental* over $\mathbb{K}$, then $\mathbb{F} \cong \mathbb{K}(x)$, with x an indeterminate, and so Proposition 1.5.7 implies that $\mathbb{K} \subset \mathbb{F}$ is a finitely generated field extension which is neither a $\mathbb{K}$-algebra of finite type nor a finite $\mathbb{K}$-algebra;

(ii) if $v \in \mathbb{F}$ is *algebraic* over $\mathbb{K}$, then $\mathbb{F} = \mathbb{K}[v]$ is a finite $\mathbb{K}$-algebra (so in particular also a $\mathbb{K}$-algebra of finite type) and the field extension $\mathbb{K} \subset \mathbb{F}$ is finite (*a fortiori* finitely generated), i.e. $[\mathbb{F} : \mathbb{K}] = \deg(f_v(x)) < +\infty$ where $f_v(x) \in \mathbb{K}[x]$ is the minimal polynomial of $v \in \mathbb{F}$. In other words, when v is algebraic the three conditions $(*)$, $(**)$ and $(***)$ in Remark 1.5.6 coincide.

The next fundamental result shows that the same occurs for any finitely generated field extension.

Lemma 1.7.2 (Zariski's lemma). *Let $\mathbb{K} \subset \mathbb{F}$ be a field extension. Then $\mathbb{F}$ is a $\mathbb{K}$-algebra of finite type if and only if $[\mathbb{F} : \mathbb{K}] < +\infty$.*

Remark 1.7.3.

(i) Note that, in the assumptions of Zariski's Lemma, $\mathbb{K} \subset \mathbb{F}$ is in any case a finitely generated field extension (recall Remark 1.5.6).

(ii) There are field extensions $\mathbb{K} \subset \mathbb{F}$ which do not satisfy any of the finiteness conditions introduced in Section 1.5. Take, e.g. the field extension $\mathbb{Q} \subset \overline{\mathbb{Q}}$ (respectively, $\mathbb{Z}_p \subset \overline{\mathbb{Z}_p}$, where $p \in \mathbb{Z}$ a prime) given by algebraic closure. This cannot be a finitely generated field extension otherwise, being an algebraic extension, it would be of finite degree over $\mathbb{Q}$ (cf. Lang, 2002, Proposition 1.6, for details), which is a contradiction since $\overline{\mathbb{Q}}$ has infinite dimension as a $\mathbb{Q}$-vector space. In particular, it is neither a finite $\mathbb{Q}$-algebra nor a $\mathbb{Q}$-algebra of finite type, as it easily follows from Zariski's Lemma. The same discussion holds verbatim for the field extension $\mathbb{Z}_p \subset \overline{\mathbb{Z}_p}$.

Proof of Zariski's Lemma. First of all, we can assume $\mathbb{F}$ with infinite elements, otherwise there is nothing else to prove.

($\Leftarrow$) This implication is obvious (cf. Example 1.5.4).

($\Rightarrow$) By assumption we have $\mathbb{F} = \mathbb{K}[v_1, \ldots, v_n]$, for some integer n. The proof proceeds by induction on n. The case $n = 1$ has already been discussed in Remark 1.7.1; therefore assume $n \geqslant 2$ and use inductive hypothesis. We have $\mathbb{K} \subset \mathbb{K}_1 := \mathbb{K}(v_1) \subset \mathbb{F}$. Since $\mathbb{F} = \mathbb{K}[v_1, \ldots, v_n]$, one also has $\mathbb{F} = (\mathbb{K}(v_1))[v_2, \ldots, v_n] = \mathbb{K}_1[v_2, \ldots, v_n]$, i.e. $\mathbb{F}$ is a $\mathbb{K}_1$-algebra of finite type. By induction, the extension $\mathbb{K}_1 \subset \mathbb{F}$ is algebraic of finite degree. By transitivity of degree extension, one has $[\mathbb{F} : \mathbb{K}] = [\mathbb{F} : \mathbb{K}_1] \cdot [\mathbb{K}_1 : \mathbb{K}]$ so, to deduce the finiteness of $[\mathbb{F} : \mathbb{K}]$ it suffices to show that v_1 is algebraic over $\mathbb{K}$. Assume by contradiction that v_1 is transcendental over $\mathbb{K}$. Since $\mathbb{F} = \mathbb{K}_1[v_2, \ldots, v_n]$, with $v_2, \ldots, v_n$ algebraic over $\mathbb{K}_1$, from Proposition 1.6.6 there exists $d \in \mathbb{K}[v_1]$ s.t. dv_i is integral over $\mathbb{K}[v_1]$, for any $2 \leqslant i \leqslant n$. Let $f \in \mathbb{K}_1$ be any element.

Claim 1.7.4. For any integer $N >> 0$, $d^N f \in \mathbb{K}[v_1, dv_2, \ldots, dv_n]$.

Proof. Since $f \in \mathbb{K}_1 \subset \mathbb{F}$, then

$$f = \sum_{j_1 \cdots j_n} \alpha_{j_1 \ldots j_n} \, v_1^{j_1} \, v_2^{j_2} \, \ldots \, v_n^{j_n}.$$

Thus, for any N sufficiently large, any monomial of $d^N f$ is of the form

$$d^N \alpha_{j_1 \ldots j_n} v_1^{j_1} v_2^{j_2} \cdots v_n^{j_n} = d^{N-(j_2+\cdots+j_n)} \alpha_{j_1 \ldots j_n} v_1^{j_1} (dv_2)^{j_2} \cdots (dv_n)^{j_n},$$

where $d^{N-(j_2+\cdots+j_n)} \alpha_{j_1 \ldots j_n} v_1^{j_1} \in \mathbb{K}[v_1]$. □

From Corollary 1.6.4, $d^N f$ is therefore integral over $\mathbb{K}[v_1]$. Since by assumption v_1 is transcendental over $\mathbb{K}$, then $\mathbb{K}[v_1]$ is a PID (in particular, a UFD). From Remark 1.6.2-(iv), $\mathbb{K}[v_1]$ is therefore integrally closed, i.e. $d^N f \in \mathbb{K}[v_1]$. If we take $c \in \mathbb{K}[v_1]$ irreducible, not dividing d, and if we consider $f := c^{-1}$, from the previous discussion, for any $N >> 0$ we would get $d^N f \in \mathbb{K}[v_1]$, a contradiction. Thus, v_1 is algebraic and the theorem is proved. □

1.8 Transcendence Degree

Let $\mathbb{K} \subset \mathbb{F}$ be a field extension and let $s_1, \ldots, s_n \in \mathbb{F}$. Similarly to (1.13), consider the $\mathbb{K}$-algebra homomorphism

$$\Phi : \mathbb{K}[x_1, \ldots, x_n] \to \mathbb{F}, \quad \Phi(x_j) = s_j, \quad 1 \leqslant j \leqslant n, \qquad (1.14)$$

where $x_1, \ldots, x_n$ are indeterminates over $\mathbb{K}$.

Definition 1.8.1. The elements $s_1, \ldots, s_n \in \mathbb{F}$ are said to be *algebraically independent over* $\mathbb{K}$ if $\mathrm{Ker}(\Phi) = (0)$. Otherwise, they are said to be *algebraically dependent over* $\mathbb{K}$ and any non-zero $f \in \mathrm{Ker}(\Phi)$ is said to be a *relation of algebraic dependence over* $\mathbb{K}$ among the s_i's.

If $s_1, \ldots, s_n$ are algebraically independent over $\mathbb{K}$, the homomorphism Φ above extends to an isomorphism $\mathcal{Q}^{(n)} := \mathbb{K}(x_1, \ldots x_n) \cong \mathbb{K}(s_1, \ldots, s_n)$.

Definition 1.8.2. Let $s_1, \ldots, s_n \in \mathbb{F}$ be elements which are algebraically independent over $\mathbb{K}$. The set $\{s_1, \ldots, s_n\}$ is said to be a (finite) *transcendence basis* of $\mathbb{F}$ over $\mathbb{K}$ if $\mathbb{F}$ is an algebraic extension of $\mathbb{K}(s_1, \ldots, s_n)$.

Lemma 1.8.3. *Let* $\{s_1, \ldots, s_n\}$ *be a transcendence basis of* $\mathbb{F}$ *over* $\mathbb{K}$ *and let* $t_1, \ldots, t_h \in \mathbb{F}$ *be algebraically independent over* $\mathbb{K}$. *Then* $h \leqslant n$.

Proof. Since $\mathbb{F}$ is an algebraic extension of $\mathbb{K}(s_1, \ldots, s_n)$, there exists a non-zero polynomial $f \in \mathbb{K}[y_1, x_1, \ldots, x_n] = A^{(n+1)}$ s.t. $f(t_1, s_1, \ldots, s_n) = 0$. Since $s_1, \ldots, s_n$ are algebraically independent, the polynomial f is non-constant with respect to the indeterminate y_1. Since moreover $t_1 \in \mathbb{F}$ is transcendent over $\mathbb{K}$, the polynomial f has to be non-constant with respect to at least one of the indeterminates x_i, $1 \leqslant i \leqslant n$; up to a re-labeling of the indeterminates, we can assume this occurs for x_1. In such a case s_1 is algebraic over $\mathbb{K}(t_1, s_2, \ldots, s_n)$.

By transitivity of algebraic extensions, $\mathbb{F}$ is an algebraic extension of $\mathbb{K}(t_1, s_2, \ldots, s_n)$. Moreover $t_1, s_2, \ldots, s_n$ are algebraically independent over $\mathbb{K}$ (otherwise s_1 would be algebraically dependent to $s_2, \ldots, s_n$). Thus, $\{t_1, s_2, \ldots, s_n\}$ is a transcendence basis of $\mathbb{F}$ over $\mathbb{K}$. Consider the following statement, for some $1 \leqslant i \leqslant h - 1$:

(S_i) *Up to a possible permutation of* $s_2, \ldots, s_n$, *the set* $\{t_1, t_2, \ldots, t_i, s_{i+1}, \ldots s_n\}$ *is a transcendence basis of* $\mathbb{F}$ *over* $\mathbb{K}$.

(S_1) has already been proved above. We proceed by induction on i; assume therefore we have proved (S_i), for some $1 \leqslant i \leqslant h - 1$, we want to prove (S_{i+1}); by the inductive hypothesis, there exists a non-zero polynomial $g \in \mathbb{K}[y_{i+1}, y_1, \ldots, y_i, x_{i+1}, \ldots, x_n]$ such that

$$g(t_{i+1}, t_1, \ldots, t_i, s_{i+1}, \ldots, s_n) = 0.$$

The polynomial g is non-constant with respect to the indeterminate y_{i+1} (since $t_1, \ldots, t_i, s_{i+1}, \ldots, s_n$ are algebraically independent) as well as with respect to at least one indeterminates x_j, e.g. x_{i+1} (since $t_1, \ldots, t_i, t_{i+1}$ are algebraically independent). Thus, s_{i+1} is algebraic over $\mathbb{K}(t_1, \ldots, t_{i+1}, s_{i+2}, \ldots, s_n)$. Moreover $t_1, \ldots, t_{i+1}, s_{i+2}, \ldots, s_n$ are algebraically independent: indeed if one had

$$h(t_1, \ldots, t_{i+1}, s_{i+2}, \ldots, s_n) = 0$$

for some non-zero polynomial $h \in \mathbb{K}[y_1, \ldots, y_{i+1}, x_{i+2}, \ldots, x_n]$, by the inductive hypothesis the polynomial h could not be constant with respect to the indeterminate y_{i+1}, so t_{i+1} would be algebraic over $\mathbb{K}(t_1, \ldots, t_i, s_{i+2}, \ldots, s_n)$; in such a case s_{i+1} would be algebraic over $\mathbb{K}(t_1, \ldots, t_i, s_{i+2}, \ldots, s_n)$, contradicting the inductive hypothesis. It follows that $\{t_1, \ldots, t_{i+1}, s_{i+2}, \ldots, s_n\}$ is a transcendence basis of $\mathbb{F}$ over $\mathbb{K}$, i.e. (S_{i+1}) has been proved. By induction, (S_h) holds true; in particular $h \leqslant n$. $\qquad\square$

If $\mathbb{F}$ admits a finite transcendence basis over $\mathbb{K}$, we will say that $\mathbb{F}$ has *finite transcendence degree over* $\mathbb{K}$. In such a case, from Lemma 1.8.3, any two transcendence bases have the same cardinality. This non-negative integer is called *transcendence degree of* $\mathbb{F}$ *over* $\mathbb{K}$ and it is denoted by

$$\operatorname{trdeg}_{\mathbb{K}}(\mathbb{F}). \tag{1.15}$$

Remark 1.8.4.

(i) If $\mathbb{K} \subseteq \mathbb{F}$ is an algebraic extension, $\mathrm{trdeg}_{\mathbb{K}}(\mathbb{F}) = 0$.

(ii) If $\mathbb{F}$ does not admit a finite transcendence basis over $\mathbb{K}$ (e.g. the quotient field of the polynomial ring $\mathbb{K}[x_1, x_2, \ldots]$ with infinite indeterminates), one poses $\mathrm{trdeg}_{\mathbb{K}}(\mathbb{F}) = +\infty$.

(iii) If $x_1, \ldots, x_n$ are indeterminates over $\mathbb{K}$, then for $\mathcal{Q}^{(n)} := \mathbb{K}(x_1, \ldots, x_n)$ one has $\mathrm{trdeg}_{\mathbb{K}}(\mathcal{Q}^{(n)}) = n$ and $\{x_1, \ldots, x_n\}$ is a transcendence basis of $\mathcal{Q}^{(n)}$ over $\mathbb{K}$.

(iv) If $\mathbb{F} = \mathbb{K}(s_1, \ldots, s_n)$ and $s_1, \ldots, s_n \in \mathbb{F}$ are algebraically independent, then $\mathrm{trdeg}_{\mathbb{K}}(\mathbb{F}) = n$ and $F \cong \mathcal{Q}^{(n)}$. In such a case, $\mathbb{F}$ is said to be a *purely transcendental extension* of $\mathbb{K}$.

(v) For a finitely generated field extension $\mathbb{K} \subset \mathbb{F} = \mathbb{K}(s_1, \ldots, s_n)$ with $\mathrm{trdeg}_{\mathbb{K}}(\mathbb{F}) = n$, the field $\mathbb{F}$ is called a *field of algebraic functions in the s_i's*.

Proposition 1.8.5. *Let $\mathbb{K} \subset \mathbb{L} \subset \mathbb{F}$ be field extensions and assume* $\mathrm{trdeg}_{\mathbb{K}}(\mathbb{L}) < +\infty$ *and* $\mathrm{trdeg}_{\mathbb{L}}(\mathbb{F}) < +\infty$. *Then also* $\mathrm{trdeg}_{\mathbb{K}}(\mathbb{F}) < +\infty$ *and* $\mathrm{trdeg}_{\mathbb{K}}(\mathbb{F}) = \mathrm{trdeg}_{\mathbb{K}}(\mathbb{L}) + \mathrm{trdeg}_{\mathbb{L}}(\mathbb{F})$. *More precisely, if $s_1, \ldots, s_n \in \mathbb{L}$ is a transcendence basis of $\mathbb{L}$ over $\mathbb{K}$ and $t_1, \ldots, t_m \in \mathbb{F}$ is a transcendence basis of $\mathbb{F}$ over $\mathbb{L}$, then $s_1, \ldots, s_n, t_1, \ldots, t_m$ is a transcendence basis of $\mathbb{F}$ over $\mathbb{K}$.*

Proof. Assume by contradiction there exists a non-zero polynomial $g \in K[x_1, \ldots, x_n, y_1, \ldots, y_m]$, where x_i's and y_j's are indeterminates over $\mathbb{K}$. Consider the polynomial $h_s(y_1, \ldots, y_n) := g(s_1, \ldots, s_n, y_1, \ldots, y_m) \in \mathbb{L}[y_1, \ldots, y_m]$. From the fact that $s_1, \ldots, s_n$ are algebraically independent over $\mathbb{K}$, this polynomial is non-zero. In this case, $h_s(y_1, \ldots, y_n)$ would give a non-zero algebraic relation over $\mathbb{L}$ among the elements $t_1, \ldots, t_m$, contradicting the assumption. Thus, $s_1, \ldots, s_n, t_1, \ldots, t_m$ are algebraically independent over $\mathbb{K}$. We are left to showing that $\mathbb{K}(s_1, \ldots, s_n, t_1, \ldots, t_m) \subseteq \mathbb{F}$ is an algebraic extension. Note that $\mathbb{L}(t_1, \ldots, t_m) \subseteq \mathbb{F}$ is an algebraic extension, since $\{t_1, \ldots, t_m\}$ is a transcendence basis of $\mathbb{F}$ over $\mathbb{L}$. The same occurs to the extension $\mathbb{K}(s_1, \ldots, s_n, t_1, \ldots, t_m) \subseteq \mathbb{L}(t_1, \ldots, t_m)$, since $\{s_1, \ldots, s_n\}$ is a transcendence basis of $\mathbb{L}$ over $\mathbb{K}$. Therefore, the composition $\mathbb{K}(s_1, \ldots, s_n, t_1, \ldots, t_m) \subseteq \mathbb{L}(t_1, \ldots, t_m) \subseteq \mathbb{F}$ is an algebraic extension and we are done.

Proposition 1.8.6. *Let $\mathbb{K}$ be a field and let R be any integral $\mathbb{K}$-algebra, with quotient field $\mathcal{Q}(R)$. If $\mathrm{trdeg}_{\mathbb{K}}(\mathcal{Q}(R)) := n < +\infty$, then the non-negative integer n equals the maximum number of elements in R which are algebraically independent over $\mathbb{K}$.*

Proof. If $A \subset R$ is any set of elements of R which are algebraically independent over $\mathbb{K}$ then, from the fact that $A \subset R \subseteq \mathcal{Q}(R)$ and from Lemma 1.8.3, one has that $|A| \leqslant n$. We want to show that there actually exist subsets $A \subset R$ for which their cardinality reaches the maximum n.

Take $s_1, \ldots, s_n \in \mathcal{Q}(R)$ any transcendence basis of $\mathcal{Q}(R)$ over $\mathbb{K}$. Let $z \in R \setminus \{0\}$ s.t. $zs_1, \ldots, zs_n \in R$. We want to show that $zs_1, \ldots, zs_n$ are algebraically independent over $\mathbb{K}$. Assume by contradiction there exists a non-zero polynomial

$$f := \sum_{i_1, \ldots, i_n} a_{i_1, \ldots, i_n} x_1^{i_1} \cdots x_n^{i_n} \in \mathbb{K}[x_1, \ldots, x_n],$$

where $a_{i_1, \ldots, i_n} \in \mathbb{K}$ and $x_1, \ldots, x_n$ indeterminates over $\mathbb{K}$, such that

$$0 = f(zs_1, \ldots, zs_n) = \sum_{i_1, \ldots, i_n} a_{i_1, \ldots, i_n} z^{i_1 + \cdots + i_n} s_1^{i_1} \cdots s_n^{i_n}.$$

If we expand the previous expression in powers w.r.t. z, this reads

$$(*) \quad 0 = c_r(s_1, \ldots, s_n) z^r + c_{r-1}(s_1, \ldots, s_n) z^{r-1} + \cdots + c_0(s_1, \ldots, s_n),$$

where $c_j(x_1, \ldots, x_n) \in \mathbb{K}[x_1, \ldots, x_n]$ for any $0 \leqslant j \leqslant r$.

Assume first $c_0(x_1, \ldots, x_n) \in \mathbb{K}[x_1, \ldots, x_n] \setminus \{0\}$. In such a case equality $(*)$ reads $c_0(s_1, \ldots, s_n) = -\Sigma_{j=1}^r c_j(s_1, \ldots, s_n) z^j$. Since $z \in R$ divides the right-side-member of the previous equality whereas the left-side-member is a polynomial expression in $s_1, \ldots, s_n \in \mathcal{Q}(R)$ with coefficients from $\mathbb{K}$, it follows that $c_0(s_1, \ldots, s_n) = 0$. This contradicts the fact that $s_1, \ldots, s_n$ in $\mathcal{Q}(R)$ are algebraically independent over $\mathbb{K}$.

If otherwise $c_j(x_1, \ldots, x_n) = 0$, for any $0 \leqslant j \leqslant \ell - 1 < r$, where $\ell \leqslant r$ is the minimum integer for which $c_\ell(x_1, \ldots, x_n) \in \mathbb{K}[x_1, \ldots, x_n] \setminus \{0\}$, equality $(*)$ reads in this case $0 = -z^\ell \cdot (\Sigma_{j=\ell}^r c_j(y_1, \ldots, y_n) z^{j-\ell})$. Since $\mathcal{Q}(R)$ is a field and since $z \neq 0$, the previous equality implies therefore $0 = \Sigma_{j=\ell}^r c_j(y_1, \ldots, y_n) z^{j-\ell}$, where $c_\ell(x_1, \ldots, x_n) \neq 0$, and one can conclude as in the previous case. $\qquad \square$

Definition 1.8.7. Let $\mathbb{K}$ be a field and let R be an integral $\mathbb{K}$-algebra with quotient field $\mathcal{Q}(R)$. By small abuse of terminology, one defines the *transcendence degree of R over $\mathbb{K}$*, denoted by $\mathrm{trdeg}_\mathbb{K}(R)$, to be the transcendence degree of $\mathcal{Q}(R)$ over $\mathbb{K}$, namely

$$\mathrm{trdeg}_\mathbb{K}(R) := \mathrm{trdeg}_\mathbb{K}(\mathcal{Q}(R)). \tag{1.16}$$

Remark 1.8.8.

(i) If $x_1, \ldots, x_n$ are indeterminates over $\mathbb{K}$, then $\operatorname{trdeg}_{\mathbb{K}}(\mathbb{K}[x_1, \ldots, x_n]) = n$ in the sense of Definition 1.8.7.

(ii) If $\Phi : R \twoheadrightarrow R'$ is a surjective homomorphism of integral $\mathbb{K}$-algebras, then

$$\operatorname{trdeg}_{\mathbb{K}}(R) \geqslant \operatorname{trdeg}_{\mathbb{K}}(R'), \tag{1.17}$$

in the sense of Definition 1.8.7. Indeed, set $n := \operatorname{trdeg}_{\mathbb{K}}(R)$. In view of Proposition 1.8.6, for any choice of $s_1, \ldots, s_{n+1} \in R$, these elements are algebraically dependent over $\mathbb{K}$, i.e. there exists $g \in \mathbb{K}[x_1, \ldots, x_{n+1}] \setminus \{0\}$, where $x_1, \ldots, x_{n+1}$ indeterminates over $\mathbb{K}$, such that $g(s_1, \ldots, s_{n+1}) = 0$. Set $f_i := \Phi(s_i) \in R'$, for any $1 \leqslant i \leqslant n + 1$; one has therefore

$$g(f_1, \ldots, f_{n+1}) = \Phi(g(s_1, \ldots, s_{n+1})) = \Phi(0) = 0,$$

the first equality following from the fact that Φ is a $\mathbb{K}$-algebra homomorphism. Since Φ is surjective, any element of R' is the image via Φ of (at least) an element of R, so this implies that the maximal number of algebraically independent elements of R' over $\mathbb{K}$ is at most n. From Proposition 1.8.6, it follows that $\operatorname{trdeg}_{\mathbb{K}}(R') \leqslant n = \operatorname{trdeg}_{\mathbb{K}}(R)$, in the sense of Definition 1.8.7.

(iii) If $\mathbb{F} = \mathbb{K}(s_1, \ldots, s_n)$ is a finitely generated field extension of $\mathbb{K}$, then $\operatorname{trdeg}_{\mathbb{K}}(\mathbb{F}) \leqslant n$ as it follows from (i), (ii) and the fact that one has a surjective $\mathbb{K}$-algebra homomorphism $\Phi : \mathbb{K}[x_1, \ldots, x_n] \twoheadrightarrow \mathbb{K}[s_1, \ldots, s_n]$.

1.9 Tensor Products of R-Modules and of R-Algebras

Let M, N and P be R-modules. A map $f : M \times N \to P$ is said to be *R-bilinear* if, for any $x \in M$ and any $y \in N$, the maps

$$f(x, -) : N \to P, \quad y \to f(x, y) \quad \text{and} \quad f(-, y) : M \to P, \quad x \to f(x, y)$$

are R-linear.

In what follows we shall construct a R-module T, called the *tensor product of the modules M and N*, in such a way that, for any R-module P, R-bilinear maps $M \times N \to P$ turn out to be in bijective correspondence with R-linear maps $T \to P$. Indeed, one has

Proposition 1.9.1. *Let M and N be R-modules. There exists a pair (T, g), where T a R-module and $g : M \times N \to T$ a bilinear map, satisfying the following property:*

$(*)$ *given any R-module P and any R-bilinear map $f : M \times N \to P$, there exists a unique linear map $f' : T \to P$ such that $f = f' \circ g$.*

Moreover, if (T', g') is another pair satisfying property $(*)$, then there exists a unique isomorphism of R-modules $j : T \to T'$ such that $j \circ g = g'$.

Proof. Let C be the free R-module whose elements are given by (finite) linear combinations of elements of $M \times N$ with coefficients from R, i.e. of the form $\Sigma_{i=1}^{n} r_i (x_i, y_i)$, where $n \in \mathbb{N}$, $x_i \in M$, $y_i \in N$. Let D be the R-submodule of C generated by elements of the form

$$(x + x', y) - (x, y) - (x', y), \quad (x, y + y') - (x, y) - (x, y'),$$

$$(r\,x, y) - r\,(x, y), \quad (x, r\,y) - r\,(x, y)$$

and put $T := C/D$. For any element (x, y) of the basis of C, we denote with the symbol $x \otimes y$ its image in T. Thus, T is generated by elements of the form $x \otimes y$ and, from above, one has

$$(x + x') \otimes y = x \otimes y + x' \otimes y, \ x \otimes (y + y') = x \otimes y + x \otimes y', \ (r\,x) \otimes y = x \otimes r\,y = r\,x \otimes y.$$

Similarly, the quotient map $g : M \times N \longrightarrow T$, defined as $g(x, y) := x \otimes y$, is R-bilinear. Let P be any R-module; any map $f : M \times N \to P$ extends by linearity to a R-module homomorphism $\overline{f} : C \to P$. If in particular f is R-bilinear, from the above relations, $\overline{f}$ vanishes on all generators of D and so on D. Thus, it determines a well-defined homomorphism $f' : T \to P$ of R-modules such that $f'(x \otimes y) = f(x, y)$. The map f' is uniquely determined by this condition, so the pair (T, g) as above satisfies condition $(*)$.

Let (T', g') be another pair satisfying $(*)$; replacing (P, f) with (T', g') in the statement of $(*)$, one obtains a unique R-linear map $j : T \to T'$ such that $g' = j \circ g$. Changing role between T and T', one gets a R-linear map $j' : T' \to T$ such that $g = j' \circ g'$. Since $j \circ j'$ and $j' \circ j$ are both the identities, it follows that j is an isomorphism. $\square$

Definition 1.9.2. The R-module T in Proposition 1.9.1 is called the *tensor product* of M and N and it is denoted by $M \otimes_R N$ (or simply by $M \otimes N$, if no confusion arises).

By its definition, if $(x_i)_{i \in I}$ and $(y_j)_{j \in J}$ are families of generators of M and N, respectively, then $M \otimes N$ is generated by the elements $x_i \otimes y_j$, $i \in I$, $j \in J$; in particular, if M and N are finitely generated R-modules, the same occurs for $M \otimes N$.

Example 1.9.3. One word of warning; the element $x \otimes y$ strictly depends on the modules M and N. Take, e.g. $R = \mathbb{Z}$ and let $M = \mathbb{Z}$, $N = \mathbb{Z}_2$ and

$M' = 2\mathbb{Z}$ be the submodule of M generated by 2. Let $\bar{1} \in \mathbb{Z}_2$ and consider $z := 2 \otimes \bar{1}$. Viewed as an element of $M \otimes N = \mathbb{Z} \otimes \mathbb{Z}_2$, z is the zero element indeed $2 \otimes \bar{1} = 2 \cdot 1 \otimes \bar{1} = 1 \otimes 2 \cdot \bar{1} = 1 \otimes \bar{0} = 0_{M \otimes N}$. On the other hand, as an element of $M' \otimes N = 2\mathbb{Z} \otimes \mathbb{Z}_2$, z is non-zero.

The proof of Proposition 1.9.1 can be extended to multilinear maps $f : M_1 \times \cdots \times M_r \to P$ of R-modules, giving rise to *multi-tensor product*

$$M_1 \otimes \cdots \otimes M_r \tag{1.18}$$

of modules. We left to the reader to check straightforward computations.

Proposition 1.9.4. *Let M, N and P be R-modules. The following are uniquely determined isomorphisms*:

(i) $M \otimes N \cong N \otimes M$, *defined by* $x \otimes y \to y \otimes x$,
(ii) $(M \otimes N) \otimes P \cong M \otimes (N \otimes P) \cong M \otimes N \otimes P$, *defined by* $(x \otimes y) \otimes z \to x \otimes (y \otimes z) \to x \otimes y \otimes z$,
(iii) $(M \oplus N) \otimes P \cong (M \otimes P) \oplus (N \otimes P)$, *defined by* $(x, y) \otimes z \to (x \otimes z, y \otimes z)$,
(iv) $R \otimes M \cong M$, *defined by* $r \otimes x \to r\,x$.

Proof. See Atiyah and McDonald (1969, Proposition 2.14). $\square$

1.9.1 *Restriction and extension of scalars*

Let $R \xrightarrow{f} S$ be a ring homomorphism and let N be a S-module. Then N has a natural structure of R-module defined by:

$$r \cdot x := f(r)\,x, \quad \text{for any } r \in R \text{ and any } x \in N.$$

This structure of R-module on N is said to be obtained by *restriction of scalars*. In particular, the homomorphism f naturally defines a structure of R-module on the ring S (cf. Section 1.5).

Proposition 1.9.5. *With notation as above, assume that N is finitely generated as a S-module and that S is finitely generated as a R-module. Then N is finitely generated as a R-module.*

Proof. Let $y_1, \ldots, y_n$ be a system of generators of N as a S-module and $x_1, \ldots, x_m$ a system of generators of S as a R-module. The set $\{x_i y_j\}$, $1 \leqslant i \leqslant n$, $1 \leqslant j \leqslant m$, is a system of mn generators of N as a R-module. $\square$

Let M be a R-module; from above one can consider the R-module $S \otimes_R M$. This R-module has also a structure of S-module given by:

$$s \cdot (s' \otimes x) = (ss') \otimes x, \quad \text{for any } s, s' \in S \text{ and any } x \in M.$$

The S-module $S \otimes_R M$ is said to be obtained from M by *extension of scalars*.

Proposition 1.9.6. *Assume that M is a finitely generated R-module. Then $S \otimes_R M$ is a finitely generated S-module.*

Proof. If $x_1, \ldots, x_m$ is a system of generators of M as a R-module, the elements $1 \otimes x_i$, $1 \leqslant i \leqslant m$, generate $S \otimes_R M$ as a S-module. $\qquad \square$

Let now M, N and P be R-modules and $f : M \times N \to P$ be a R-bilinear map. Since, for any $x \in M$, the induced map $f(x, -) : N \to P$, $y \to f(x, y)$ is R-linear, then f induces a map $M \longrightarrow \text{Hom}(N, P)$ which is R-linear, since f is R-linear with respect to the variable $x \in M$. Conversely, any R-homomorphism $M \xrightarrow{\phi} \text{Hom}_R(N, P)$ defines a R-bilinear map $f_\phi : M \times N \to P$, $(x, y) \longrightarrow \phi(x)(y)$.

This shows there exists a bijective correspondence between the set of R-bilinear maps $M \times N \to P$ and $\text{Hom}(M, \text{Hom}(N, P))$. At the same time, by definition of tensor product, the set of R-bilinear maps $M \times N \to P$ bijectively corresponds to $\text{Hom}(M \otimes_R N, P)$. Thus, one has a canonical isomorphism of R-modules $\text{Hom}(M \otimes_R N, P) \cong \text{Hom}(M, \text{Hom}(N, P))$.

1.9.2   *Tensor product of algebras*

Let S and T be two R-algebras, with $R \xrightarrow{\varphi} S$ and $R \xrightarrow{\psi} T$ the corresponding structural morphisms. From Remark 1.5.1-(i), we can consider the R-module $U := S \otimes_R T$. We will endow U with a R-algebra structure.

Consider the map $S \times T \times S \times T \longrightarrow U$, $(s, t, s', t') \to ss' \otimes tt'$; it is R-linear in each factor. By definition of multi-tensor product (1.18), the previous map induces a homomorphism of R-modules $S \otimes T \otimes S \otimes T \to U$. From Proposition 1.9.4 this homomorphism corresponds to a homomorphism of R-modules $U \otimes U \to U$ which, by Proposition 1.9.1, corresponds to a R-bilinear map:

$$\mu : U \times U \longrightarrow U, \quad \mu(s \otimes t, \, s' \otimes t') := s\,s' \otimes t\,t'. \tag{1.19}$$

By linearity one has $\mu(\Sigma_i(s_i \otimes t_i), \Sigma_j(s'_j \otimes t'_j)) = \Sigma_{i,j}(s_i s'_j \otimes t_i t'_j)$. In other words, the map μ defines a product on U.

Proposition 1.9.7. *The product μ endows the R-module U with a struc-ture of commutative ring, with identity $1 \otimes 1$. Furthermore, U is a R-algebra.*

Proof. The first part of the statement is straightforward. For the last part, note that the map $R \longrightarrow U$, $r \to \varphi(r) \otimes 1 = 1 \otimes \psi(r)$ is a ring homomorphism. Indeed one has a commutative diagram of ring homomorphisms

$$
\begin{array}{ccc}
 & S & \\
{}^{\varphi}\nearrow & & \searrow^{f} \\
R & & U\,, \\
\searrow_{\psi} & & \nearrow_{g} \\
 & T &
\end{array}
$$

where f and g are, respectively, defined by $f(s) = s \otimes 1$ and $g(t) = 1 \otimes t$.

$\square$

1.10 Graded Rings and Modules, Homogeneous Ideals

Let S be a ring (commutative and with identity, as always) and let moreover $G(+)$ be an abelian group.

Definition 1.10.1. S is said to be a *G-graded ring* (equivalently, *endowed with a G-graduation*), if S has a decomposition

$$
S = \bigoplus_{g \in G} S_g, \tag{1.20}
$$

where each S_g is a sub-group of the abelian group $S(+)$ such that:

(i) $1 \in S_0$, and
(ii) for any $(g, h) \in G \times G$ one has $S_g \cdot S_h \subseteq S_{g+h}$, where for any subsets $\mathcal{A}, \mathcal{B} \subseteq S$ one denotes

$$
\mathcal{A} \cdot \mathcal{B} := \{ab \,|\, a \in \mathcal{A}, b \in \mathcal{B}\} \quad \text{and} \quad \mathcal{A} + \mathcal{B} := \{a + b \,|\, a \in \mathcal{A}, b \in \mathcal{B}\}.
$$

The group S_g is called the *degree-g graded part* of S whereas its elements are called *homogeneous elements of degree g* of S. When in particular $G = \mathbb{Z}$, then S is simply called a *graded ring*.

For any non-empty subset $\mathcal{F} \subseteq S$ of a G-graded ring, one poses

$$
\mathcal{F}_g := \mathcal{F} \cap S_g, \quad \forall \, g \in G \tag{1.21}
$$

and

$$H(\mathcal{F}) := \bigcup_{g \in G} \mathcal{F}_g, \tag{1.22}$$

where the latter is called the *set of homogeneous elements* of $\mathcal{F}$.

Notation 1.10.2. From now on in this book, to distinguish homogeneous elements from non-homogeneous ones, capital letters A, B, etcetera will be used to denote homogeneous elements of a graded ring $\mathcal{S}$. When, moreover, we want more precisely specify the degree of a given homogeneous element, we will sometimes use the symbol A_g to stress that it is an element in $\mathcal{S}_g$. On the contrary, elements in $\mathcal{S}$ which are not homogeneous will be simply denoted by small letters like a, b, etcetera.

Remark 1.10.3. From the previous definitions and remarks, one has that:

(a) any $f \in \mathcal{S}$ can be uniquely written as a finite sum

$$f = F_{g_1} + \cdots + F_{g_n}, \tag{1.23}$$

where $F_{g_i} \in \mathcal{S}_{g_i}$, for distinct integers $1 \leqslant i \leqslant n$ and distinct elements $g_1, \ldots, g_n \in G$. The elements $F_{g_1}, \ldots, F_{g_n}$ are called the *homogeneous components* of f and (1.23) is called the *decomposition of f into its homogeneous components*;

(b) $\mathcal{S}_0$ is a subring of $\mathcal{S}$, whereas $\mathcal{S}_g$ is a $\mathcal{S}_0$-submodule of $\mathcal{S}$, for any $g \in G$. In particular, if $\mathcal{S}_0 \cong \mathbb{K}$ is a field, then $\mathcal{S}_g$ is a $\mathbb{K}$-vector space;

(c) if $G = \mathbb{Z}$, for any integer n we set

$$\mathcal{S}_{>n} := \bigoplus_{d > n} \mathcal{S}_d.$$

If $\mathcal{S}_d = \{0\}$ for any $d < 0$, then $\mathcal{S}$ is more precisely a *non-negatively* graded ring and $\mathcal{S}_{>n}$ is an ideal of $\mathcal{S}$, for any integer n. In particular, one has $\mathcal{S} = \mathcal{S}_{>-1}$.

Definition 1.10.4. The ideal $\mathcal{S}_{>0}$ is simply denoted by $\mathcal{S}_+$ and is called the *irrelevant ideal* of $\mathcal{S}$.

(cf. Chapter 3 for geometric motivations of this terminology).

Definition 1.10.5. Let $\mathcal{I}$ be an ideal of a graded ring $\mathcal{S}$. Then $\mathcal{I}$ is said to be a *homogeneous ideal* if $\mathcal{I} = \bigoplus_{g \in G} \mathcal{I}_g$, i.e. $f \in \mathcal{I}$ if and only if all the homogeneous components of f belong to $\mathcal{I}$.

Example 1.10.6. It is easy to see that, e.g. the polynomial ring $S :=$ $\mathbb{K}[x_1, x_2]$ is a (non-negatively) graded ring, where the graduation is induced by the degree (cf. Section 1.10.1 below for a more general treatment). Then one easily verifies that $S_+ = (x_1, x_2)$ is a homogeneous ideal. On the other hand, $\mathcal{I} := (x_1 + x_2^2, x_1 x_2)$ is not homogeneous. To see this, consider, e.g.

$$\mathcal{I} \ni f = (x_1 + x_2^2)(1 + 2x_2) + (x_1 x_2)x_1^2$$
$$= (x_1) + (2x_1 x_2 + x_2^2) + (2x_2^3) + (x_1^3 x_2),$$

where the right-hand side of the previous equality is the decomposition of f into its homogeneous parts; then $F_1 := x_1 \in S_1$ and $F_2 := (2x_1 x_2 + x_2^2) \in S_2$ do not belong to $\mathcal{I}$, whereas $F_3 = 2x_2^3 = 2x_2(x_1 + x_2^2) - 2(x_1 x_2) \in \mathcal{I}$ and $F_4 = x_1^3 x_2 \in \mathcal{I}$.

Proposition 1.10.7. *Let S be a G-graded ring and let $\mathcal{I}$ be an ideal. Then $\mathcal{I}$ is a homogeneous ideal if and only if $\mathcal{I}$ can be generated by a family of homogeneous elements.*

Proof. ($\Rightarrow$) Let $\mathcal{I} := (f_\ell)_{\ell \in L}$. By Remark 1.10.3-(a), any generator f_ℓ of $\mathcal{I}$ can be uniquely decomposed by means of its homogeneous components:

$$f_\ell = F_{\ell, g_{t_\ell}} + F_{\ell, g_{t_\ell + 1}} + \cdots + F_{\ell, g_{t_\ell + k_\ell}},$$

for suitable integers t_ℓ, k_ℓ depending on the choice of $\ell \in L$. Since $\mathcal{I}$ is homogeneous, for any $\ell \in L$ these components belong to $\mathcal{I}$ so that $\mathcal{I} :=$ $\left(F_{\ell, g_j}\right)_{\ell \in L, t_\ell \leqslant j \leqslant t_\ell + k_\ell}$, as desired.

($\Leftarrow$) Let $\mathcal{I} := (F_\ell)_{\ell \in L}$, where $\{F_\ell\}_{\ell \in L}$ is a family of homogeneous generators of $\mathcal{I}$. Then, any $a \in \mathcal{I}$ is of the form $a = \Sigma_{\ell \in L} a_\ell F_\ell$, for some $a_\ell \in S$. Decomposing any a_ℓ as in (1.23), one easily sees that any homogeneous component of a is in $\mathcal{I}$, i.e. $\mathcal{I}$ is homogeneous. $\quad\square$

Proposition 1.10.8.

(a) *Let S be a G-graded ring and let $\mathcal{I}_1$, $\mathcal{I}_2$ be homogeneous ideals. Then $\mathcal{I}_1 \cdot \mathcal{I}_2$, $\mathcal{I}_1 \cap \mathcal{I}_2$, $\mathcal{I}_1 + \mathcal{I}_2$ are homogeneous ideals.*

(b) *If S is a non-negatively graded ring and $\mathcal{I}$ is a homogeneous ideal, then:*

   (i) *$\sqrt{\mathcal{I}}$ is homogeneous;*

   (ii) *$\mathcal{I}$ is prime if and only if, for any pair $(F, G) \in H(S) \times H(S)$ s.t. $FG \in \mathcal{I}$, then either $F \in \mathcal{I}$ or $G \in \mathcal{I}$.*

Proof. (a) is a straightforward consequence of the definitions.

To prove $(b) - (i)$, for any $f \in \sqrt{\mathcal{I}}$ consider its decomposition into its homogeneous components

$$f = F_\ell + F_{\ell+1} + \cdots + F_{\ell+m}, \tag{1.24}$$

for some integers $m, \ell \geqslant 0$. By definition of $\sqrt{\mathcal{I}}$, there exists a positive integer $r \geqslant 1$ s.t. $f^r \in \mathcal{I}$. From (1.24), the decomposition of f^r into homogeneous components is of the form $f^r = F_\ell^r + o(r\ell)$, where $o(r\ell) \in (\mathcal{S}_+)_{r\ell}$. Since $\mathcal{I}$ is homogeneous then $F_\ell^r \in \mathcal{I}$, i.e. $F_\ell \in \sqrt{\mathcal{I}}$. Thus, also $f - F_\ell \in \sqrt{\mathcal{I}}$. Recursively applying the same reasoning, one can conclude.

For $(b) - (ii)$, the implication $(\Rightarrow)$ directly follows from the definition of prime and homogeneous ideal. Let us prove the other implication. Take elements $a, b \in \mathcal{S}$ such that $ab \in \mathcal{I}$. Consider the decompositions of a and b in $\mathcal{S}$ into their homogeneous components

$$a := A_s + A_{s+1} + \cdots + A_{s+t} \text{ and } b := B_n + B_{n+1} + \cdots + B_{n+m}$$

for some integers $t, s \geqslant 0$, $m, n \geqslant 0$. Then $ab = \sum_i H_i \in \mathcal{I}$, where for any index i s.t. $n + s \leqslant i \leqslant n + m + s + t$, H_i is the degree-i homogeneous component of ab, with $H_i := \Sigma A_{s+j} B_{n+k}$, where summation is on indices $0 \leqslant j \leqslant t$ and $0 \leqslant k \leqslant m$ such that $s + j + m + k = i$.

Since $\mathcal{I}$ is homogeneous, then $H_i \in \mathcal{I}$ for any $n+s \leqslant i \leqslant n+m+s+t$. In particular, this holds for $H_{s+t+n+m} = A_{s+t} B_{n+m} \in \mathcal{I}$. By the assumption on $\mathcal{I}$, either $A_{s+t} \in \mathcal{I}$ or $B_{n+m} \in \mathcal{I}$. We can reduce to the case that e.g. $A_{s+t} \notin \mathcal{I}$ and $B_{n+m} \in \mathcal{I}$ (if indeed both of them are in $\mathcal{I}$, one replaces a and b with $a - A_{s+t}$ and $b - B_{n+m}$ respectively and proceeds).

By induction assume one has proved $B_{n+m}, B_{n+m-1}, \ldots, B_{n+m-r} \in \mathcal{I}$, for some $r \geqslant 0$. Consider $H_{m+n+s+t-r-1} \in \mathcal{I}$, where

$$H_{m+n+s+t-r-1} = A_{s+t} B_{n+m-r-1} + A_{s+t-1} B_{n+m-r} + \cdots$$
$$+ A_{s+t-r-1} B_{n+m}.$$

By inductive hypothesis and $A_{s+t} \notin \mathcal{I}$, one has $B_{n+m-r-1} \in \mathcal{I}$. By recursive application of the same strategy, one proves that all homogeneous components of b are in $\mathcal{I}$, i.e. $\mathcal{I}$ is prime. $\qquad\square$

By using decomposition in homogeneous components, one can easily prove the following easy result.

Proposition 1.10.9. *Let $\mathcal{S}$ be a G-graded ring and let $\mathcal{I} \subseteq \mathcal{S}$ be an ideal. Consider the canonical projection $\pi : \mathcal{S} \twoheadrightarrow \mathcal{S}/\mathcal{I}$.*

(i) *For any $g \in G$ one has $\pi(S_g) \cong S_g/\mathcal{I}_g$; in particular*

$$S/\mathcal{I} \cong \sum_{g \in G} S_g/\mathcal{I}_g. \tag{1.25}$$

(ii) *Moreover, (1.25) is a direct sum if and only if $\mathcal{I}$ is homogeneous.*

From the previous result, when $\mathcal{I}$ is a homogeneous ideal, if one poses

$$(S/\mathcal{I})_g := S_g/\mathcal{I}_g, \text{ for any } g \in G,$$

then the ring $S/\mathcal{I}$ is a endowed with a *G-graduation induced* by the G-graduation of S.

Definition 1.10.10. If S is a graded, integral domain, the field $\mathcal{Q}(S)$ contains as a sub-field the set consisting of all fractions $\frac{G_1}{G_2}$ such that $G_1, G_2 \in H(S)$ are of the same degree, with $G_2 \neq 0$. This sub-field will be denoted by $\mathcal{Q}_0(S)$ and called the *sub-field of degree-zero, homogeneous fractions* of S.

1.10.1 *Homogeneous polynomials*

We now discuss in more details a concrete example of non-negatively graded ring. Let $\mathbb{K}$ be any field and let $X_0, \ldots, X_n$ be indeterminates over $\mathbb{K}$. A non-zero polynomial $F \in \mathbb{K}[X_0, \ldots, X_n]$ is said to be *homogeneous* if all its monomials have the same degree.

Definition 1.10.11. For any integer $d \geqslant 0$, we will denote by $\mathbb{K}[X_0, \ldots, X_n]_d$ the set consisting of the zero-polynomial together with all degree-d, homogeneous polynomials in $\mathbb{K}[X_0, \ldots, X_n]$. This is a $\mathbb{K}$-vector space, whose *canonical basis* is given by monic monomials of degree d in the indeterminates $X_0, \ldots, X_n$.

Lemma 1.10.12. *For any integer $d \geqslant 0$, one has*

$$\mathsf{b}(n, d) := \dim_{\mathbb{K}}(\mathbb{K}[X_0, \ldots, X_n]_d) = \binom{n+d}{d}, \tag{1.26}$$

where $\mathsf{b}(n, d)$ stands for the binomial coefficient of indexes n and d, with the convention $\mathsf{b}(n, 0) := \binom{n}{0} = 1$.

Proof. If $d = 0$, for any n one has $\mathbb{K}[X_0, \ldots, X_n]_0 = \mathbb{K}$ and we are done in this case. Therefore, we may assume $d \geqslant 1$.

If $n = 1$, for any integer d, (1.26) holds as the canonical basis of $\mathbb{K}[X_0, X_1]_d$ is given by X_0^d, $X_0^{d-1}X_1$, $X_0^{d-2}X_1^2, \ldots, X_1^d$, consisting of $d + 1 = \binom{d+1}{d}$ monomials.

We can therefore assume $n \geqslant 2$ and proceed by double induction on n and d. Let $\mathtt{M}_n^d$ denote the canonical basis of $\mathbb{K}[X_0, \ldots, X_n]_d$; its cardinality $|\mathtt{M}_n^d|$ equals $\dim_{\mathbb{K}}(\mathbb{K}[X_0, \ldots, X_n]_d)$. The set $\mathtt{M}_n^d$ is a disjoint union of the two non-empty subsets $\mathtt{M}_n^d = \mathtt{M}' \cup \mathtt{M}''$, where $\mathtt{M}'$ is the set of monomials in $\mathtt{M}_n^d$ containing X_n whereas $\mathtt{M}'' = \mathtt{M}_n^d \setminus \mathtt{M}'$. On the one hand, $\mathtt{M}''$ is the canonical basis of $\mathbb{K}[X_0, \ldots, X_{n-1}]_d$; by the inductive hypothesis on n, $|\mathtt{M}''| = \binom{n-1+d}{d}$. On the other, if we divide each monomial appearing in $\mathtt{M}'$ by X_n, we get the canonical basis of $\mathbb{K}[X_0, \ldots, X_n]_{d-1}$ which by construction bijectively corresponds to $\mathtt{M}'$ and which, by the inductive hypothesis on d, consists of $\binom{n+d-1}{d-1}$ elements. One concludes by observing that

$$|\mathtt{M}_n^d| = |\mathtt{M}'| + |\mathtt{M}'| = \binom{n+d-1}{d-1} + \binom{n-1+d}{d} = \binom{n+d}{d}$$

and the proof is complete.   □

Remark 1.10.13.

(i) Previous results show that the polynomial ring

$$\mathcal{S}^{(n)} := \mathbb{K}[X_0, \ldots, X_n] = \oplus_{d \geqslant 0} \mathcal{S}_d^{(n)} \tag{1.27}$$

is a non-negatively graded ring. In particular, $\mathcal{S}_0^{(n)} = \mathbb{K}$ and each graded piece $\mathcal{S}_d^{(n)}$, for any $d \geqslant 1$, is a (free) $\mathcal{S}_0^{(n)}$-module, i.e. $\mathbb{K}$-vector space of finite dimension given by (1.26). Graduation of $\mathcal{S}^{(n)}$ is given by the *degree* of the homogeneous components of a polynomial.

(ii) One comment for the reader; we use here the symbol $\mathcal{S}^{(n)}$ to denote $\mathbb{K}[X_0, \ldots, X_n]$ (and not other notation like $A^{(n+1)}$ which instead will be used later on, cf., e.g. Chapter 2) since we want to shed light onto the graded structure of this ring. Moreover, the index (n) is used instead of $(n + 1)$, the correct number of indeterminates, since as customary $\mathcal{S}^{(n)}$ will be identified with the ring of *homogeneous coordinates* of the projective space $\mathbb{P}^n$ in Chapter 3.

Notation 1.10.14. Following Notation 1.10.2, homogeneous polynomials in $\mathcal{S}^{(n)}$ will be denoted by capital letters like F, G, etcetera. When, in particular, we consider a polynomial in $\mathcal{S}_d^{(n)}$, we will sometimes more precisely denote it by F_d, to explicitly remind its degree. On the contrary, polynomials which are not homogeneous will be denoted by small letters f, g, etc., as done in the previous sections. If $f \in \mathcal{S}^{(n)}$ is of degree m, then f can be uniquely decomposed as a linear combination of its degree-d homogeneous parts, for $0 \leqslant d \leqslant m$, which we write

$$f = \sum_{d=0}^{m} F_d, \tag{1.28}$$

where not necessarily all $F_d \neq 0$.

Proposition 1.10.15.

(i) *F is homogeneous of degree d if and only if the following identity between polynomials in $\mathbb{K}[X_0, \ldots, X_n, t]$ holds:*

$$F(tX_0, \ldots, tX_n) = t^d F(X_0, \ldots, X_n). \tag{1.29}$$

(ii) *If F is homogeneous of degree d, then the* Euler's identity

$$\sum_{i=0}^{n} X_i \frac{\partial F}{\partial X_i} = d\, F \tag{1.30}$$

holds true.

(iii) *Let $F, g \in \mathcal{S}^{(n)}$ be non-constant polynomials, where F is homogeneous and g divides F. Then $g = G$ is also homogeneous. In particular, irreducible factors of a homogeneous polynomial are all homogeneous.*

Proof.

(i) Condition (1.29) is obviously necessary. The fact that it is also sufficient directly follows from the decomposition as in (1.28). Indeed, for $f \in \mathcal{S}^{(n)}$, condition $f(tX_0, \ldots, tX_n) = t^d f(X_0, \ldots, X_n)$ forces f to be an element of $\mathcal{S}_d^{(n)}$.

(ii) Taking the partial derivative with respect to the indeterminate t of (1.29) and then evaluating for $t = 1$ gives Euler's identity.

(iii) One has $F = gh$, for some $h \in \mathcal{S}^{(n)}$. Assume by contradiction that g is not homogeneous. From (1.28), there exist integers $k > 0$, $j \geqslant 0$ such that

$$g = G_j + G_{j+1} + \cdots + G_{j+k}, \quad \text{with } G_j, G_{j+k} \neq 0.$$

Since F is homogeneous, then h has to be non-homogeneous, i.e. there exist integers $r > 0$, $i \geqslant 0$ such that

$$h = H_i + H_{i+1} + \cdots + H_{i+r}, \quad \text{with } H_i, H_{i+r} \neq 0.$$

Now, $F = gh$ would give

$$F = G_j H_i + (G_j H_{i+1} + G_{j+1} H_i) + \cdots + G_{j+r} H_{i+k},$$

where $G_j H_i \neq 0$, $G_{j+r} H_{i+k} \neq 0$, since $\mathcal{S}^{(n)}$ is an integral domain. Thus, $i + j = \deg(G_j H_i) < \deg(G_{j+r} H_{i+k}) = i + j + r + k$ would contradict the hypothesis on F. □

We now define important operators on homogeneous and non-homogeneous polynomials, which will be frequently used in the next chapters (cf., e.g. Chapter 3). In what follows, $x_1, \ldots, x_n$ and $X_0, \ldots, X_n$ will denote indeterminates over $\mathbb{K}$.

Definition 1.10.16. For any integer $i \in \{0, \ldots, n\}$, let

$$\delta_i : \mathbb{K}[X_0, \ldots, X_n] \to \mathbb{K}[X_0, \ldots, X_{i-1}, X_{i+1}, \ldots, X_n] \qquad (1.31)$$

be the map defined by

$$\delta_i(f(X_0, \ldots, X_{i-1}, X_i, X_{i+1}, \ldots, X_n)) := f(X_0, \ldots, X_{i-1}, 1, X_{i+1}, \ldots, X_n).$$

Viceversa, let

$$\mathfrak{h}_i : \mathbb{K}[x_1, \ldots, x_n] \to \mathbb{K}[X_0, \ldots, X_n] \qquad (1.32)$$

be the map defined in such a way that, if $f \in \mathbb{K}[x_1, \ldots, x_n]$ is of degree d, then

$$\mathfrak{h}_i(f(x_1, \ldots, x_n)) := X_i^d \, f\left(\frac{X_0}{X_i}, \frac{X_1}{X_i}, \ldots, \frac{X_n}{X_i}\right),$$

where in the above definition one has replaced x_h, with $\frac{X_{h-1}}{X_i}$, for $1 \leqslant h \leqslant i$, and with $\frac{X_h}{X_i}$ for $i + 1 \leqslant h \leqslant n$.

Lemma 1.10.17. *For any integer $i \in \{0, \ldots, n\}$:*

(i) *the map δ_i is a $\mathbb{K}$-algebra homomorphism;*

(ii) *the map $\mathfrak{h}_i$ is not a $\mathbb{K}$-algebra homomorphism; on the other hand it is multiplicative, i.e.*

$$\mathfrak{h}_i(fg) = \mathfrak{h}_i(f)\mathfrak{h}_i(g)$$

and, when $\deg(f) = \deg(g) = \deg(f + g)$, it is also additive, i.e.

$$\mathfrak{h}_i(f + g) = \mathfrak{h}_i(f) + \mathfrak{h}_i(g);$$

(iii) *for any $f \in \mathbb{K}[x_1, \ldots, x_n]$ of degree d, $\mathfrak{h}_i(f) \in \mathbb{K}[X_0, \ldots, X_n]_d$;*

(iv) *$\delta_i \circ \mathfrak{h}_i = \mathrm{id}_{\mathbb{K}[x_1, \ldots, x_n]}$;*

(v) *X_i does not divide f if and only if $\delta_i(f)$ has the same degree of f;*

(vi) *let $F \in \mathcal{S}^{(n)}$ be any homogeneous polynomial and let $m \geqslant 0$ be the multiplicity of X_i as a factor of F (recall (1.2)), then $\mathfrak{h}_i(\delta_i(F)) = \frac{F}{X_i^m}$.*

Proof. Properties (i), (ii), (iv) and (v) are straightforward, whereas (iii) follows by applying Proposition 1.10.15-(i). To prove (vi), it suffices to consider the case where X_i does not divide F and verifying in this case that $\mathfrak{h}_i(\delta_i(F)) = F$. To prove this, taking into account (v), it suffices to prove it on monomials of degree d and then use linearity on each monomial appearing in F. $\qquad\square$

Note that, if $F \in \mathcal{S}^{(n)}$ is homogeneous, in general $\delta_i(F)$ does not remain homogeneous (it remains homogeneous only if F is constant w.r.t. X_i). For this reason δ_i is called the *dehomogenizing operator w.r.t. X_i* and, correspondingly, $\delta_i(F)$ is called the *dehomogenized polynomial of F w.r.t. X_i*. On the contrary, by Lemma 1.10.17-(ii), $\mathfrak{h}_i(f)$ is called the *homogenized polynomial of f w.r.t. X_i* and, consequently, $\mathfrak{h}_i$ is called the *homogenizing operator w.r.t. X_i*.

As in Section 1.3.4, one can consider resultant of two homogeneous polynomials with respect to a given indeterminate. One has:

Theorem 1.10.18. *Let $F, G \in \mathcal{S}^{(n)}$ be non-constant, homogeneous polynomials, with $\deg(F) = d$ and $\deg(G) = m$. Write*

$$F = A_d + A_{d-1}X_n + \cdots + A_0 X_n^d \text{ and } G = B_m + A_{m-1}X_n + \cdots + B_0 X_n^m,$$

where $A_j \in \mathbb{K}[x_0, \ldots, X_{n-1}]_j$, $B_k \in \mathbb{K}[x_0, \ldots, X_{n-1}]_k$, with $0 \leqslant j \leqslant d$, $0 \leqslant k \leqslant m$ and $A_0 B_0 \neq 0$. Then the resultant polynomial $R_{X_n}(F, G)$ is either 0 or it is homogeneous of degree dm.

Proof. If $F, G \in (\mathbb{K}[X_0, \ldots, X_{n-1}])[X_n] =: D[X_n]$ have a non-constant common factor then, by Theorem 1.3.16, one concludes $R_{X_n}(F, G) = 0$.

Assume therefore that F and G have no non-constant common factor, i.e. $R_{X_n}(F, G) \in \mathbb{K}[X_0, \ldots, X_{n-1}] \setminus \{0\}$. For simplicity, let $R(X_0, \ldots, X_{n-1}) := R_{X_n}(F, G)$. By (1.7) and the fact that the polynomials A_j and B_i are homogeneous, for any $t \in \mathbb{K}$ one has that $R(tX_0, \ldots, tX_{n-1})$

is given by the determinant of the $(m + d) \times (m + d)$ matrix:

$$C := \begin{pmatrix} t^d A_d & t^{d-1} A_{d-1} & \cdots & & \cdots & A_0 & 0 & 0 & \cdots & 0 \\ 0 & t^d A_d & t^{d-1} A_{d-1} & \cdots & \cdots & & A_0 & 0 & \cdots & 0 \\ \cdots & \cdots & \cdots & & \cdots & \cdots & & & \cdots & \cdots \\ \cdots & \cdots & \cdots & & \cdots & t^d A_d & t^{d-1} A_{d-1} & \cdots & \cdots & A_0 \\ t^m B_m & t^{m-1} B_{m-1} & \cdots & & \cdot\cdot & B_0 & 0 & & \cdots & 0 \\ 0 & t^m B_m & t^{m-1} B_{m-1} & \cdots & \cdots & & \cdot\cdot & B_0 & \cdots & 0 \\ \cdots & \cdots & \cdots & & \cdots & \cdots & & \cdots & \cdots & \cdots \\ \cdots & \cdots & & \cdots & t^m B_m & \cdots & & \cdots & \cdots & B_0 \end{pmatrix}.$$

If we multiply the ith-row of C by t^{m-i+1}, $1 \leqslant i \leqslant m$, and the $(m+j)$th-row of C by t^{d-j+1}, $1 \leqslant j \leqslant d$, we get the following matrix

$$\widetilde{C} := \begin{pmatrix} t^{m+d} A_d & t^{m+d-1} A_{d-1} & \cdots & \cdots & & \cdots & & \cdots & t^m A_0 & 0 & \cdots & 0 \\ 0 & t^{m+d-1} A_d & \cdots & \cdots & & \cdots & & \cdots & t^m A_1 & t^{m-1} A_0 & \cdots & 0 \\ \cdots & \cdots & \cdots & \cdots & & \cdots & & \cdots & \cdots & \cdots & \cdot\cdot & \cdot\cdot \\ 0 & 0 & \cdots & t^{d+1} A_d & \cdots & & \cdots & & \cdots & \cdots & \cdot\cdot & t A_0 \\ t^{m+d} B_m & t^{m+d-1} B_{m-1} & \cdots & t^{d+1} B_1 & t^d B_0 & 0 & & 0 & & \cdots & \cdot\cdot & 0 \\ 0 & t^{m+d-1} B_m & \cdots & t^{d+1} B_2 & t^d B_1 & & \cdot\cdot & & 0 & & \cdots & \cdot\cdot & 0 \\ \cdots & \cdots & \cdots & \cdots & & \cdots & & \cdots & \cdots & \cdots & \cdot\cdot & \cdot\cdot \\ 0 & 0 & \cdots & \cdots & 0 & t^{m+1} B_m & t^m B_{m-1} & & \cdots & & \cdot\cdot & t B_0 \end{pmatrix}.$$

Using linearity for $\det(\widetilde{C})$, once with respect to rows and then with respect to columns, we get the following relation:

$$t^p R(tX_0, \ldots, tX_n) = \det(\widetilde{C}) = t^q R(X_0, \ldots, X_n),$$

where $p := m + (m-1) + \cdots + 1 + d + (d-1) + \cdots 1 = \binom{m+1}{2} + \binom{d+1}{2}$ whereas $q := (m+d) + (m+d-1) + \cdots + 1 = \binom{m+d+1}{2}$. Thus, $R(tX_0, \ldots, tX_n) = t^{q-p} R(X_0, \ldots, X_n)$ and one then concludes by Proposition 1.10.15–(i) and by the fact that $q - p = md$. $\qquad\square$

To conclude this section, we note that for homogeneous polynomials in two variables, we have the following useful result.

Proposition 1.10.19. *Let $\mathbb{K}$ be algebraically closed, d a positive integer and $F \in \mathbb{K}[X_0, X_1]_d$. Then there exist $\lambda \in \mathbb{K} \setminus \{0\}$ and d pairs $(a_i, b_i) \in \mathbb{K}^2$, $1 \leqslant i \leqslant d$ (not necessarily all distinct pairs but each of them different from $(0, 0)$) such that*

$$F(X_0, X_1) = \lambda(a_1 X_1 - b_1 X_0) \cdots (a_d X_1 - b_d X_0).$$

The pairs (a_i, b_i) are uniquely determined up to order whereas each pair is determined up to proportionality. These pairs are called roots of the homogeneous polynomial F.

Proof. Assume that r is the multiplicity of X_0 as a factor of F, with $0 \leqslant r \leqslant d$. Thus, $F = X_0^r G$, where $G \in \mathbb{K}[X_0, X_1]_{d-r}$ is not divisible by X_0. If $r = d$, then $G = \lambda$ is a constant and the pairs in the statement all coincide with $(0, 1)$, which is therefore counted with multiplicity d. Let us assume therefore that $r < d$. Then $G \in \mathbb{K}[X_0, X_1]_{d-r}$ is a non-constant, homogeneous polynomial (cf. Proposition 1.10.15-(iii)), which is not divisible by X_0; thus $\delta_0(G) \in \mathbb{K}[X_1]$. One concludes by applying Theorem 1.3.11 to the polynomial $\delta_0(G)$ and then by homogenizing via $\mathfrak{h}_0$ each linear factor. $\qquad\square$

1.10.2 *Graded modules and graded morphisms*

If $\mathcal{S}$ is a G-graded ring and $\mathcal{S}'$ is a H-graded ring, for $G(+)$ and $H(+)$ abelian groups, a ring homomorphism $f : S \to S'$ is said to be *homogeneous* if there exists a group homomorphism $\phi : G \to H$ s.t., for any $g \in G$ one has $f(\mathcal{S}_g) \subseteq \mathcal{S}'_{\phi(g)}$. If f and ϕ are both isomorphisms, then f is an *isomorphism of ϕ-graded rings*. If moreover $G = H$, a homomorphism $f : \mathcal{S} \to \mathcal{S}'$ is said to be *homogeneous of degree* 0 if $\phi = \mathrm{id}_G$. An isomorphism of degree 0 is simply called an *isomorphism* of graded rings. If $G = H = \mathbb{Z}$ and if $f : \mathcal{S} \to \mathcal{S}'$ is homogeneous, then $\phi : \mathbb{Z} \to \mathbb{Z}$ is given by the multiplication with an integer d, so $f(\mathcal{S}_g) \subseteq \mathcal{S}'_{dg}$, for any $g \in \mathbb{Z}$. In this case, f is said to be *homogeneous of degree d*.

If $\mathcal{S}$ is a G-graded ring, a $\mathcal{S}$-modulo $\mathcal{M}$ is called a *G-graded module* if $\mathcal{M}$ admits a decomposition $\mathcal{M} = \bigoplus_{g \in G} \mathcal{M}_g$, where each $\mathcal{M}_g$ is an abelian sub-group of the group $\mathcal{M}(+)$, s.t., for any $(g, h) \in G \times G$, $\mathcal{S}_g \cdot \mathcal{M}_h \subseteq \mathcal{M}_{g+h}$, where we used notation as above. If $G = \mathbb{Z}$, $\mathcal{M}$ is simply said to be a *graded module*.

As an easy example, any homogeneous ideal of a G-graded ring $\mathcal{S}$ is a G-graded $\mathcal{S}$-module. Same terminology and definitions introduced for graded rings can be extended to graded modules, in particular one has the notion of *homogeneous homomorphisms* of graded modules.

Given a G-graded $\mathcal{S}$-module $\mathcal{M}$, one can change its graduation by using the following procedure: fix $h \in G$ and define

$$\mathcal{M}(h) := \bigoplus_{g \in G} \mathcal{M}(h)_g \quad \text{where } \mathcal{M}(h)_g := \mathcal{M}_{h+g}. \tag{1.33}$$

Then, $M(h)$ is a G-graded S-module which is isomorphic to $\mathcal{M}$ as an S-module but in general not as a G-graded S-module (cf. Exercise 1.5).

Example 1.10.20. Let $\mathbb{K}$ be a field and V be a $\mathbb{K}$-vector space of dimension $n + 1 \geqslant 1$. The *symmetric algebra* over V is $\operatorname{Sym}(V) := \bigoplus_{d \in \mathbb{N}} \operatorname{Sym}^d(V)$, which is a non-negatively graded ring, simply denoted by $S(V)$. Its degree d homogeneous part will be therefore denoted by $S(V)_d$. A similar proof as that for (1.26) shows that $\dim(S(V)_d) = \mathsf{b}(n, d) = \binom{n+d}{d}$. Moreover, $S(V)_0 \cong \mathbb{K}$ and $S(V)$ is generated as a $\mathbb{K}$-algebra by $S(V)_1$. As usual, for $F \in S(V)_d$, we write $d = \deg(F)$.

The choice of an (ordered) basis $(\underline{e}_0, \dots, \underline{e}_n)$ for V induces the *dual basis* $(\underline{e}^0, \dots, \underline{e}^n)$ on $V^* \cong \operatorname{Hom}(V, \mathbb{K})$, which is defined by $\underline{e}^i(\underline{e}_j) = \underline{e}_j(\underline{e}^i) = \delta_{ij}$, where δ_{ij} the *Kronecker's symbol*. One usually poses

$$\underline{e}^i := X_i \text{ and } \underline{e}_i := \partial_i, \quad 0 \leqslant i \leqslant n.$$

In this way, $S(V^*)$ is naturally identified with $\mathcal{S}^{(n)} = \mathbb{K}[X_0, \dots, X_n]$ as in Section 1.10.1. Dually, the ring $S(V)$ is identified with the *ring of differential operators*, which is denoted by $\mathcal{D}_n := \mathbb{K}[\partial_0, \dots, \partial_n]$ and which is (abstractly) isomorphic to $\mathcal{S}^{(n)}$ as a graded ring. The degree-d homogeneous part $\mathcal{D}_{n,d}$ of the ring $\mathcal{D}_n$ is called the $\mathbb{K}$-vector space of *degree-d homogeneous differential operators*, i.e. differential operators where only degree-d monomials in $\partial_0, \dots, \partial_n$ appear (the zero-differential operator is considered to be homogeneous of any degree).

Let $\underline{X} := (X_0, \dots, X_n)$ be indeterminates and, for any *multi-index* $\mathtt{I} := (i_0, \dots, i_n) \in \mathbb{N}^{n+1}$, let $|\mathtt{I}| := i_0 + \cdots + i_n$ be the *length* of the multi-index. We will pose

$$\underline{X}^{\mathtt{I}} := X_0^{i_0} \dots X_n^{i_n}. \tag{1.34}$$

Thus, any degree-d homogeneous polynomial in $\mathcal{S}_d^{(n)}$ can be written also as $F = \Sigma_{|\mathtt{I}|=d} F_{\mathtt{I}} \, \underline{X}^{\mathtt{I}}$, (similar notation can be used for differential operators). In the above notation, from (1.29) F is homogeneous of degree d if and only if $F(t\underline{X}) = t^d F(\underline{X})$, $\forall t \in \mathbb{K}$ and Euler's identity (cf. (1.30)) reads $d\, F(\underline{X}) = \Sigma_{i=0}^n \partial_i F(\underline{X})$.

1.11   Localization

Let R be a ring and let $S \subset R$ be a subset. S is said to be a *multiplicative system* if

(i) $0 \notin S$,

(ii) $1 \in S$,

(iii) for any $s, t \in S$, one has $st \in S$.

In the cartesian product $R \times S$ one poses the following relation:

$$(a, s) \equiv (b, t) \iff \exists\, u \in S \text{ s.t. } (at - bs)u = 0. \tag{1.35}$$

It is easy to see that $\equiv$ is an equivalence relation.

Remark 1.11.1. When in particular R is an integral domain and $S = R \setminus \{0\}$, then $\equiv$ coincides with the equivalence relation used to construct the field of fractions $\mathcal{Q}(R)$, i.e. $(a, s) \equiv (b, t)$ if and only if $(at - bs) = 0$.

Definition 1.11.2. The quotient set $\frac{R \times S}{\equiv}$ will be denoted by R_S and called *localization of R with respect to S*. The $\equiv$–equivalence class of a pair $(a, s) \in R \times S$ will be denoted by $\frac{a}{s}$. Operations on R endow R_S with a structure of a (commutative and with identity) ring by the following rules:

$$\frac{a}{s} + \frac{b}{t} := \frac{at + bs}{st} \quad \text{and} \quad \frac{a}{s}\frac{b}{t} := \frac{ab}{st}.$$

One has a natural ring homomorphism

$$j : R \to R_S, \quad \text{defined by } a \xrightarrow{\ j\ } \frac{a}{1},$$

which is called the *localization homomorphism*.

Proposition 1.11.3.

(i) *With notation as above, one has that* $\mathrm{Ker}(j) = \{a \in R \mid sa = 0$ *for some* $s \in S\}$.

(ii) *For any* $s \in S$, $j(s) = \frac{s}{1} \in R_S$ *is invertible.*

Proof.

(i) Take $a \in R$ and assume that $as = 0$ in R, for some $s \in S$. Then $j(a) = \frac{a}{1}$ is such that $(a, 1) \equiv (as, s) = (0, s)$, i.e. $j(a) = 0 \in R_S$. Conversely, for any $a \in \mathrm{Ker}(j)$, one has $(a, 1) \equiv (0, s)$, for some $s \in S$. This means there exists $u \in S$ s.t. $0 = u(as - 0) = a(us)$ in R, i.e. a is a zero-divisor in R.

(ii) $j(s)^{-1} := \frac{1}{s}$. $\qquad\qquad\qquad\qquad\qquad\qquad\qquad\qquad\qquad\square$

Remark 1.11.4.

(i) From Proposition 1.11.3, if R is an integral domain then j is injective, in which case R can be identified with its image $j(R)$ and so considered as a subring of R_S. Moreover, for any choice of multiplicative system $S \subset R$, R_S is identified with a subring of $\mathcal{Q}(R)$, the field of fractions of R.

(ii) When otherwise R is an arbitrary ring and S coincides with the set of all non-zero divisors in R, R_S is called the *ring of total fractions of R*.

Definition 1.11.5. If R is a graded ring and S is a multiplicative system of R, the elements of R_S are endowed with a graduation induced by that of R: $\deg\left(\frac{a}{s}\right) := \deg(a) - \deg(s)$. The subset

$$R_{(S)} := \left\{ \frac{a}{s} \in R_S \mid a, s \in H(R),\ s \in S,\ \deg(a) = \deg(s) \right\}$$

is a subring of R_S which is called the *homogeneous localization of R with respect to S*. $R_{(S)}$ is therefore the subring of homogeneous, degree-zero elements in R_S.

Recalling Definition 1.1.3, one has:

Proposition 1.11.6. *Let R be a ring, $S \subset R$ a multiplicative system and j the associated localization homomorphism.*

(i) *Any ideal of R_S is an extended ideal.*
(ii) *Let $J \subset R$ be a proper ideal. $J^e \subseteq R_S$ is a proper ideal of R_S if and only if $J \cap S = \emptyset$.*
(iii) *Let I_1 and I_2 be ideals in R_S. Then $I_1 = I_2$ if and only if $I_1^c = I_2^c$.*
(iv) *If R is Noetherian, then R_S is Noetherian.*

Proof.

(i) Let $I \subseteq R_S$ be any ideal. For any $\frac{a}{s} \in I$, one has $j(s)\frac{a}{s} = \frac{s}{1}\frac{a}{s} = a \in I$ and $a \in I^c$. Thus, $j(a)\frac{1}{s} = \frac{a}{s} \in (I^c)^e$, i.e. $I \subseteq (I^c)^e$. By Remark 1.1.4-(ii) one concludes that $I = (I^c)^e$.

(ii) $1 \in J^e$ if and only if there exist $a \in R$ and $s \in S$ s.t. $1 = \frac{a}{s} \in J^e$, i.e. if and only if $(a, s) \equiv (d, d)$, for some $d \in S$. By (1.35), this is equivalent to the existence of $u \in S$ s.t. $uad = uds \in S$, since S is a multiplicative system, i.e. if and only if $a(ud) \in J \cap S$.

(iii) The statement follows from (i) as, for any ideal $I \subseteq R_S$, one has $I = (I^c)^e$.

(iv) Let $I_1 \subseteq I_2 \subseteq \cdots \subseteq R_S$ be any ascending chain of ideals of R_S. Since R is noetherian, the ascending chain $I_1^c \subseteq I_2^c \subseteq \cdots \subseteq R$ is stationary (cf. Proposition 1.4.4). This means there exists a positive integer n_0 s.t. $I_{n_0}^c = I_{n_0+k}^c$ for any integer $k \geqslant 1$. From (i) and (iii) we get $I_{n_0} = (I_{n_0}^c)^e = (I_{n_0+k}^c)^e = I_{n_0+k}$, for any $k \geqslant 1$, i.e. the original ascending chain of ideals of R_S is stationary too. $\qquad\square$

Remark 1.11.7. From Proposition 1.11.6-(ii) and (iii), the correspondence

$$\{\text{Ideals of } R_S\} \to \{\text{Ideals of } R\}, \text{ defined by } I \to I^c,$$

is injective and it restricts to a bijective correspondence

$$\{\text{Proper ideals of } R_S\} \to \{\text{Ideals of } R \text{ not intersecting } S\}.$$

As a matter of notation, if $J \subset R$ is any ideal such that $J \cap S = \emptyset$, then the ideal J^e of R_S is usually denoted by

$$JR_S \tag{1.36}$$

(cf., e.g. Atiyah and McDonald, 1969).

Corollary 1.11.8.

(i) *If $J \subset S$ is an ideal such that $J \cap S = \emptyset$, then one has $(R_S/JR_S) \cong (R/J)_S$.*

(ii) *If $\mathfrak{p} \subset R$ is a prime ideal s.t. $\mathfrak{p} \cap S = \emptyset$, then $\mathfrak{p}R_S$ is prime, such that $(\mathfrak{p}R_S)^c = \mathfrak{p}$.*

(iii) *Any prime ideal of R_S is of the form $\mathfrak{p}R_S$, for some prime ideal $\mathfrak{p}$ of R s.t. $\mathfrak{p} \cap S = \emptyset$.*

Proof.

(i) By assumption, the sequence

$$0 \to J \overset{\iota}{\to} R \overset{\pi}{\to} R/J \to 0$$

is *exact*, i.e. ι is injective, $\mathrm{Im}(\iota) = \mathrm{Ker}(\pi)$ and π is surjective. From the fact that localization is an exact operation (cf. for details Atiyah and McDonald, 1969, Proposition 3.3), the previous exact sequence gives rise to the exact sequence

$$0 \to JR_S \to R_S \to (R/J)_S \to 0,$$

which proves the statement.

(ii) Assume first that $\mathfrak{p}$ is a prime ideal in R. If $\frac{a}{s}\frac{b}{t} \in \mathfrak{p}R_S$, there exists $u \in S$ s.t. $uab = stu \in \mathfrak{p}$. Since $u \in S$ and $S \cap \mathfrak{p} = \emptyset$, then $ab \in \mathfrak{p}$, which implies either $a \in \mathfrak{p}$ or $b \in \mathfrak{p}$. Thus, either $\frac{a}{s} \in \mathfrak{p}R_S$ or $\frac{b}{t} \in \mathfrak{p}R_S$. Let now $b \in (\mathfrak{p}R_S)^c$ be any element. By definition of $\mathfrak{p}R_S$, this means there exist $a \in \mathfrak{p}$ and $s,t \in S$ s.t. $\frac{b}{t} = \frac{a}{s} \in \mathfrak{p}R_S$. By (1.35), there exists $u \in S$ s.t. $u(at - bs) = 0$, i.e. $ubs = uat \in \mathfrak{p}$. On the other hand, since $us \in S$ and $\mathfrak{p} \cap S = \emptyset$, then $b \in \mathfrak{p}$ because $\mathfrak{p}$ is prime. Thus $(\mathfrak{p}R_S)^c \subseteq \mathfrak{p}$. By Remark 1.1.4-(ii), one deduces $(\mathfrak{p}R_S)^c = \mathfrak{p}$.

(iii) Let $\mathfrak{q} \subset R_S$ be any prime ideal. From Remark 1.1.4-(iii) and Proposition 1.11.6-(ii), $\mathfrak{p} := \mathfrak{q}^c$ is a prime ideal of R s.t. $\mathfrak{p} \cap S = \emptyset$. From Proposition 1.11.6-(i) and (iii), $\mathfrak{q} = \mathfrak{p}\,R_S$. $\qquad\square$

Remark 1.11.9. Previous results show, in particular, that there exists a bijective correspondence between prime ideals of R_S and prime ideals of R not intersecting S.

1.11.1   *Local rings and localization*

A ring R is said to be a *local ring* if it contains a unique maximal ideal $\mathfrak{m}$. The field $R/\mathfrak{m}$ is called the *residue field* of the local ring $(R, \mathfrak{m})$. Any field $\mathbb{F}$ is a local ring, of maximal ideal (0) and residue field $\mathbb{F}$; for any prime $p \in \mathbb{Z}$ and any positive integer n, $\mathbb{Z}_{p^n}$ is a local ring, of maximal ideal (p) and residue field $\mathbb{Z}_p$.

Proposition 1.11.10. *Let R be a ring and $\mathfrak{m}$ be a maximal ideal of R. Then $(R, \mathfrak{m})$ is local if and only if $U(R) = R \setminus \mathfrak{m}$.*

Proof. If $(R, \mathfrak{m})$ is local, then every non-invertible element is contained in $\mathfrak{m}$. Conversely, assume that $U(R) = R \setminus \mathfrak{m}$; for any proper ideal $I \subset R$ one necessarily has $I \subseteq \mathfrak{m}$, i.e. all proper ideals of R are contained in $\mathfrak{m}$ which is therefore the only maximal ideal of R. $\qquad\square$

We will now briefly discuss how localization in some cases can produce local rings from a given ring R.

Remark 1.11.11.

(i) If $\mathfrak{p}$ is any prime ideal of a ring R, the set $S := R \setminus \mathfrak{p}$ is a multiplicative system. In this case, one usually denotes the localization R_S by

$$R_{\mathfrak{p}} \qquad\qquad (1.37)$$

which, by abuse of terminology, is called the *localization of R w.r.t.* $\mathfrak{p}$ (cf., e.g. Atiyah and McDonald, 1969). From Remark 1.11.9, there is therefore a bijective correspondence between prime ideals of $R_{\mathfrak{p}}$ and prime ideals $\mathfrak{q}$ of R contained in $\mathfrak{p}$. If moreover R is a graded ring, and $\mathfrak{p}$ is a homogeneous prime ideal, the homogeneous localization $R_{(S)}$ (cf. Definition 1.11.5) with $S := R \setminus \mathfrak{p}$ is denoted by

$$R_{(\mathfrak{p})}. \tag{1.38}$$

(ii) If $f \in R$ is any non-nilpotent element, the set $S := \{1, f, f^2, f^3, \ldots\}$ is a multiplicative system and the localization R_S will be denoted by

$$R_f. \tag{1.39}$$

When moreover R is a graded ring and f is a homogeneous element in R, the homogeneous localization $R_{(S)}$ will be denoted by

$$R_{([f])}. \tag{1.40}$$

(iii) If R is an integral domain, the ideal (0) is prime and one has $R_{(0)} = \mathcal{Q}(R)$. If moreover R is a graded ring, then $R_{((0))} = \mathcal{Q}_0(R)$, where $\mathcal{Q}_0(R)$ as in Definition 1.10.10.

In the above notation, one has the following.

Proposition 1.11.12.

(i) *For any prime ideal $\mathfrak{p}$ in R, $(R_{\mathfrak{p}}, \mathfrak{p}R_{\mathfrak{p}})$ is a local ring with residue field isomorphic to $\mathcal{Q}(R/\mathfrak{p})$.*

(ii) *For any graded ring R and any homogeneous, prime ideal $\mathfrak{p}$, $(R_{(\mathfrak{p})}, \mathfrak{p}R_{(\mathfrak{p})})$ is a local ring with residue field isomorphic to $\mathcal{Q}_0(R/\mathfrak{p})$.*

Proof.

(i) From Corollary 1.11.8–(ii), $\mathfrak{p}R_{\mathfrak{p}}$ is a prime ideal of $R_{\mathfrak{p}}$; in particular all its elements are not invertible in $R_{\mathfrak{p}}$. Consider any $\frac{a}{s} \in R_{\mathfrak{p}} \setminus \mathfrak{p}R_{\mathfrak{p}}$; then $a \notin \mathfrak{p}$. Thus, $j(a) = \frac{a}{1}$ is invertible in $R_{\mathfrak{p}}$. One concludes that $(R_{\mathfrak{p}}, \mathfrak{p}R_{\mathfrak{p}})$ is local by Proposition 1.11.10. At last, by definition of localization, it is clear that $R_{\mathfrak{p}}/\mathfrak{p}R_{\mathfrak{p}} \cong \mathcal{Q}(R/\mathfrak{p})$.

(ii) It follows from (i) and the fact that $R_{(\mathfrak{p})} \subset R_{\mathfrak{p}}$ as a subring. $\qquad\square$

1.12 Krull-Dimension of a Ring

Let R be a ring. A *chain of prime ideals* of R is a finite sequence of inclusions

$$\mathfrak{p}_0 \subset \mathfrak{p}_1 \subset \mathfrak{p}_2 \subset \cdots \mathfrak{p}_m, \qquad (1.41)$$

where $\mathfrak{p}_i$ is a prime ideal of R, for any $0 \leqslant i \leqslant m$, and where the inclusions are strict. The *length of the chain* (1.41) is set to be m by definition. The chain (1.41) is said to be *maximal* if it cannot be part of a chain of prime ideals of length higher than m.

Given a prime ideal $\mathfrak{p}$ in R, one defines the *height of* $\mathfrak{p}$, denoted by $\mathrm{ht}(\mathfrak{p})$, as follows:

$$\mathrm{ht}(\mathfrak{p}) := \mathrm{Sup}_{m \in \mathbb{Z}} \{ m \mid \mathfrak{p}_0 \subset \mathfrak{p}_1 \subset \cdots \subset \mathfrak{p}_m = \mathfrak{p} \}. \qquad (1.42)$$

Set

$$\mathrm{Spec}(R) := \{ \mathfrak{p} \mid \mathfrak{p} \text{ prime ideal of } R \}. \qquad (1.43)$$

Definition 1.12.1. Given R a ring we define the *Krull-dimension* of R, denoted by $K \dim(R)$, as $K \dim(R) := \mathrm{Sup}_{\mathfrak{p} \in \mathrm{Spec}(R)} \{ \mathrm{ht}(\mathfrak{p}) \}$.

Example 1.12.2.

(i) If $R = \mathbb{K}$ is a field, then $K \dim(\mathbb{K}) = 0$.

(ii) If R is any PID, e.g. $R = \mathbb{Z}$ or $R = \mathbb{K}[x]$, where $\mathbb{K}$ is a field and x an indeterminate over $\mathbb{K}$, then $K \dim(R) = 1$. Indeed, being a PID, any ideal in R is principal; thus any prime ideal is generated by a prime element of R which is also irreducible. Therefore any (maximal) chain of prime ideals in R is of the form $(0) \subset (p)$, where $p \in R$ irreducible.

(ii) Let $\mathbb{K}$ be a field, $n \geqslant 2$ be a positive integer and $R := \mathbb{K}[x_1, \ldots, x_n]$, where x_i indeterminates over $\mathbb{K}$, for $1 \leqslant i \leqslant n$. The ideal $(x_1) \subset R$ is a prime ideal, since $\frac{R}{(x_1)} \cong \mathbb{K}[x_2, \ldots, x_n]$ is an integral domain. Moreover, $(0) \subset (x_1)$ is a chain of prime ideals in R, thus $\mathrm{ht}((x_1)) \geqslant 1$. Note moreover that the previous chain of prime ideals is also maximal. Indeed, if by contradiction there were a prime ideal $\mathfrak{p} \neq (0), (x_1)$ such that

$$0 \subset \mathfrak{p} \subset (x_1),$$

then $x_1 \notin \mathfrak{p}$ but any element $p \in \mathfrak{p}$ should be divisible by x_1 in R, namely of the form $p = x_1^k \, f$, with k a positive integer and $f \in R$, where f not divisible by x_1 in R. This is a contradiction; indeed, since

$\mathfrak{p}$ is a prime ideal and since by assumption $x_1 \notin \mathfrak{p}$, then one must have $f \in \mathfrak{p}$ which would imply that f is divisible by x_1. Therefore one more precisely has $\operatorname{ht}((x_1)) = 1$.

Using similar arguments as above, one shows that the chain of prime ideals

$$(0) \subset (x_1) \subset (x_1, x_2) \subset (x_1, x_2, x_3) \subset \cdots \subset (x_1, x_2, \ldots, x_n)$$

is maximal, which implies $K \dim(R) \geqslant n$. From Corollary 1.12.4-(i) below, we will see that this is actually an equality.

(iii) If otherwise $\mathbb{K}$ is a field and $x_1, x_2, x_3, \ldots$ is a sequence of infinite indeterminates over $\mathbb{K}$, the polynomial ring $R := \mathbb{K}[x_1, x_2, x_3, \ldots]$ of polynomials in infinite indeterminates and with coefficients from $\mathbb{K}$ is such that $K \dim(R) = +\infty$. Indeed, for any positive integer n, there exists in R the chain of prime ideals

$$(0) \subset (x_1) \subset (x_1, x_2) \subset (x_1, x_2, x_3) \subset \cdots \subset (x_1, x_2, \ldots, x_n),$$

which is not maximal; in other words R contains chains of prime ideals of arbitrarily large length.

The next result shows that Krull-dimension extends the definition of transcendence degree for integral $\mathbb{K}$-algebras of finite type given in Definition 1.8.7 to more general rings which are not necessarily integral $\mathbb{K}$-algebras of finite type. Precisely, recalling Definition 1.8.7, one has the following.

Theorem 1.12.3. *Let $\mathbb{K}$ be a field and let R be an integral $\mathbb{K}$-algebra of finite type. Then $K \dim(R) = \operatorname{trdeg}_{\mathbb{K}}(R)$. In particular, any integral $\mathbb{K}$-algebra of finite type has finite Krull-dimension.*

Corollary 1.12.4.

(i) *If $\mathbb{K}$ is a field and if $x_1, \ldots, x_n$ are indeterminates over $\mathbb{K}$, then $K \dim(\mathbb{K}[x_1, \ldots, x_n]) = n$.*

(ii) *If $\mathbb{K}$ is a field and if $A \subset B$ is an integral extension of integral $\mathbb{K}$-algebras, then $K \dim(A) = K \dim(B)$.*

Proof. (i) follows from the fact that $\operatorname{trdeg}_{\mathbb{K}}(\mathbb{K}[x_1, \ldots, x_n]) = n$ as in Remark 1.8.8, whereas (ii) follows from the fact that $\operatorname{trdeg}_{\mathbb{K}}(A) = \operatorname{trdeg}_{\mathbb{K}}(B)$. $\qquad\square$

Remark 1.12.5. One warning for the reader. Theorem 1.12.3 does not hold for more general integral $\mathbb{K}$-algebras. Indeed, let $\mathbb{K}$ be a field, $n \geqslant 1$ be an integer and $x_1, \ldots, x_n$ be indeterminates over $\mathbb{K}$. The field of rational functions $\mathbb{K}(x_1, \ldots, x_n)$ is a $\mathbb{K}$-algebra which is not of finite type over $\mathbb{K}$ (cf. Proposition 1.5.7) and, being a field, it is such that $K \dim(\mathbb{K}(x_1, \ldots, x_n)) = 0$; on the other hand, being a purely transcendental extension of $\mathbb{K}$, one has $\mathrm{trdeg}_{\mathbb{K}}(\mathbb{K}(x_1, \ldots, x_n)) = n$ (cf. Remark 1.8.4-(iii) and (iv)). Thus, $R := \mathbb{K}(x_1, \ldots, x_n)$ is an integral $\mathbb{K}$-algebra for which $K \dim(R) < \mathrm{trdeg}_{\mathbb{K}}(R)$.

Proof of Theorem 1.12.3. Set $r := \mathrm{trdeg}_{\mathbb{K}}(R)$. Since R is an integral $\mathbb{K}$-algebra of finite type, then $r < +\infty$.

We first show that $r \geqslant K \dim(R)$. By Definition 1.8.7, $r = \mathrm{trdeg}_{\mathbb{K}}(\mathcal{Q}(R))$ and this integer equals the maximum number of elements in R which are algebraically independent over $\mathbb{K}$ (cf. Proposition 1.8.6).

First consider the case $r = 0$, i.e. $\mathbb{K} \subseteq \mathcal{Q}(R)$ is an algebraic field extension. Since $\mathbb{K} \subseteq R \subseteq \mathcal{Q}(R)$, all elements of R are algebraic over $\mathbb{K}$. Since R is an integral domain, we claim that R is a field; if the claim is true, then equality $K \dim(R) = \mathrm{trdeg}_{\mathbb{K}}(R) = r = 0$ holds. We therefore need to showing that R is a field, namely that $U(R) = R \setminus \{0\}$; it suffices to showing that, for any $a \in R \setminus \{0\}$, $\mathbb{K}[a]$ is a field. We may therefore assume that $R = \mathbb{K}[a]$. Since R is an integral $\mathbb{K}$-algebra of finite type and $a \in R$ is algebraic over $\mathbb{K}$, then $\mathbb{K} \subseteq R$ is an algebraic simple field extension (cf. Remark 1.7.1) and we are done.

Assume now $r \geqslant 1$. For any $\mathfrak{p} \in \mathrm{Spec}(R)$, $\frac{R}{\mathfrak{p}}$ is an integral $\mathbb{K}$-algebra of finite type, as it follows from the fact that R is and from the existence of the canonical epimorphism $\pi : R \twoheadrightarrow \frac{R}{\mathfrak{p}}$. For any length-$m$ prime ideal chain

$$\mathfrak{p}_0 = (0) \subset \mathfrak{p}_1 \subset \cdots \subset \mathfrak{p}_m$$

in R, for some $m \geqslant 0$, factoring by $\mathfrak{p}_1$ yields a prime ideal chain in $\frac{R}{\mathfrak{p}_1}$

$$(0) \subset \frac{\mathfrak{p}_2}{\mathfrak{p}_1} \subset \cdots \subset \frac{\mathfrak{p}_m}{\mathfrak{p}_1},$$

which is of length $m - 1$. If we can show that $\mathrm{trdeg}_{\mathbb{K}}\left(\frac{R}{\mathfrak{p}_1}\right) < r$ then, by induction on r, we can deduce that $K \dim\left(\frac{R}{\mathfrak{p}_1}\right) < r$ so that $K \dim(R) \leqslant r$. Since we have the canonical surjection $\pi : R \twoheadrightarrow \frac{R}{\mathfrak{p}_1}$, by Remark 1.8.8-(ii), we have that $\mathrm{trdeg}_{\mathbb{K}}\left(\frac{R}{\mathfrak{p}_1}\right) \leqslant \mathrm{trdeg}_{\mathbb{K}}(R) = r$. We need to showing that equality does not hold. Assume by contradiction that equality holds. Therefore, there exist $y_1, \ldots, y_r \in R$ such that their images $\pi(y_1), \ldots, \pi(y_r) \in \frac{R}{\mathfrak{p}_1}$ are algebraically independent over $\mathbb{K}$. Take any element $z \in \mathfrak{p}_1 \setminus \{0\}$; thus, the elements $y_1, \ldots, y_r, z \in R$ are algebraically dependent over $\mathbb{K}$ (cf. Proposition 1.8.6) so there exists a polynomial $f(x_1, x_2, \ldots, x_{r+1})$ in $\mathbb{K}[x_1, x_2, \ldots, x_{r+1}]$, where $x_1, x_2, \ldots, x_{r+1}$ indeterminates over $\mathbb{K}$, such that $f(y_1, \ldots, y_r, z) = 0$. Since R is an integral domain, we may therefore assume that $f := f(x_1, x_2, \ldots, x_{r+1})$ is irreducible and that f is not a polynomial in the only indeterminate x_{r+1}, otherwise $z \in \mathfrak{p}_1 \subset R$ would be algebraic over $\mathbb{K}$ so one would have $\mathbb{K}[z] = \mathbb{K}(z)$ (cf. Remark 1.7.1) and so $z \in R$ would be invertible, which contradicts $z \in \mathfrak{p}_1$. Thus $\overline{f}(x_1, \ldots, x_r) := f(x_1, x_2, \ldots, x_r, 0) \in \mathbb{K}[x_1, \ldots, x_r]$ is a non-constant polynomial for which $\overline{f}(\pi(y_1), \pi(y_2), \ldots, \pi(y_r)) = 0$ in $\frac{R}{\mathfrak{p}_1}$, contradicting the algebraic independence of $\pi(y_1), \pi(y_2), \ldots, \pi(y_r)$ over $\mathbb{K}$. This completes the proof of $r \geqslant K \dim(R)$.

Now, we show that $r \leqslant K \dim(R)$; we will use induction on r. If $r = 0$, by the previous step, we already know that in this case $K \dim(R) = \mathrm{trdeg}_{\mathbb{K}}(R) = r = 0$ holds true. Let us assume $r \geqslant 1$ and let $R = \mathbb{K}[s_1, \ldots, s_n]$, for some $n > 0$ and some $s_i \in R$, $1 \leqslant i \leqslant n$. Up to a possible permutation, we may assume that s_1 is transcendental over $\mathbb{K}$. Let us consider the multiplicative system $S := \mathbb{K}[s_1] \setminus \{0\} \subset R$. Localizing R by S gives $R_S \cong \mathbb{K}(s_1)[s_2, \ldots, s_n]$, i.e. R_S is an integral $\mathbb{K}(s_1)$-algebra of finite type. Since we have the field extensions

$$\mathbb{K} \subset \mathbb{K}(s_1) \subset \mathcal{Q}(R) \cong \mathbb{K}(s_1, \ldots, s_n)$$

with $r = \mathrm{trdeg}_{\mathbb{K}}(R)$, by Proposition 1.8.5 we have $r = \mathrm{trdeg}_{\mathbb{K}}(R) = \mathrm{trdeg}_{\mathbb{K}}(\mathbb{K}(s_1)) + \mathrm{trdeg}_{\mathbb{K}(s_1)}(\mathcal{Q}(R))$. Therefore, as in Definition 1.8.7, one gets

$$\mathrm{trdeg}_{\mathbb{K}(s_1)}(\mathcal{Q}(R)) = \mathrm{trdeg}_{\mathbb{K}(s_1)}(R_S) = r - 1.$$

By induction, it follows that there exists a maximal chain of prime ideals in R_S of length $r - 1$, namely

$$(*) \quad \mathfrak{q}_0 \subset \mathfrak{q}_1 \subset \cdots \subset \mathfrak{q}_{r-1}.$$

For any $0 \leqslant j \leqslant r - 1$, consider the contracted ideal (cf. Definition 1.1.3) $\mathfrak{p}_j := \mathfrak{q}_j^c = \mathfrak{q}_j \cap R$. Any $\mathfrak{p}_j$ is a prime ideal in R (cf. Corollary 1.11.8-(iii))such that $\mathfrak{p}_j \cap S = \emptyset$. The induced chain of prime ideals in R

$$(**) \quad \mathfrak{p}_0 \subset \mathfrak{p}_1 \subset \cdots \subset \mathfrak{p}_{r-1}$$

is proper, i.e. the inclusions are strict, otherwise $\mathfrak{p}_i = \mathfrak{p}_j$, for some $0 \leqslant i \neq j \leqslant r - 1$, would give $\mathfrak{p}_i^e = \mathfrak{p}_i R_S = \mathfrak{q}_i = \mathfrak{p}_j^e = \mathfrak{p}_j R_S = \mathfrak{q}_j$, contradicting $(*)$. Since $\mathfrak{p}_{r-1} \cap S = \emptyset$, then $\mathfrak{p}_{r-1} \cap \mathbb{K}[s_1] = \{0\}$. Thus the canonical epimorphism $\pi : R \twoheadrightarrow \frac{R}{\mathfrak{p}_{r-1}}$ injectively maps $\mathbb{K}[s_1]$ into the quotient ring $\frac{R}{\mathfrak{p}_{r-1}}$, since $\mathrm{Ker}(\pi) = \mathfrak{p}_{r-1}$. Note that $\mathfrak{p}_{r-1} \in \mathrm{Spec}(R)$ cannot be a maximal ideal of R; otherwise $\frac{R}{\mathfrak{p}_{r-1}}$ would be a field; being moreover a $\mathbb{K}$-algebra of finite type, since R is, then by Zariski's lemma (cf. Lemma 1.7.2) we would have $\left[\frac{R}{\mathfrak{p}_{r-1}} : \mathbb{K} \right] < +\infty$, so any element of $\frac{R}{\mathfrak{p}_{r-1}}$ would be algebraic over $\mathbb{K}$, which is forbidden by the facts that $\mathbb{K}[s_1] \subset \frac{R}{\mathfrak{p}_{r-1}}$ and that s_1 is transcendental over $\mathbb{K}$. Therefore, there exists a maximal ideal $\mathfrak{p}_r \subset R$ such that $\mathfrak{p}_{r-1} \subset \mathfrak{p}_r$. Using $(**)$ we get therefore a chain

$$\mathfrak{p}_0 \subset \mathfrak{p}_1 \subset \cdots \subset \mathfrak{p}_{r-1} \subset \mathfrak{p}_r$$

of prime ideals of R which shows that $K \dim(R) \geqslant r$, as wanted. $\qquad\square$

Exercises

Exercise 1.1. Let R be a ring (commutative and with identity, as always in this book).

 (i) Prove that if I is a prime ideal in R then, for any expression $I = I' \cap I''$, with I', I'' ideals in R, then either $I = I'$ or $I = I''$.
 (ii) Let $\sqrt{(0)} := \{a \in R \,|\, \exists\, n \geqslant 1 \text{ s.t. } a^n = 0\}$ be the set of nilpotents elements of R, which is called the *nilradical of R*. Prove that $\sqrt{(0)}$ is an ideal in R such that $\sqrt{(0)} = \cap_{\mathfrak{p} \in \mathrm{Spec}(R)} \mathfrak{p}$.
 (iii) Let $a \in R$ be a nilpotent element and let u be invertible in R. Show that $a + u$ is invertible in R.

Exercise 1.2. Let R be a ring and let $I, J \subset R$ be ideals. Show that the set

$$(I : J) := \{r \in R \,|\, r J \subseteq I\}$$

is an ideal, which is called the *quotient ideal of I by J*. Show moreover that, for any ideals I and J in R, one has (cf. Atiyah and McDonald, 1969, p. 26):

$$I \subseteq (I : J), \ (I : J) J \subseteq I, \ \sqrt{I \cap J} = \sqrt{I} \cap \sqrt{J}.$$

Exercise 1.3. Let R be ring and let x be an indeterminate over R. Let

$$f(x) := a_0 + a_1 x + a_2 x^2 + \cdots + a_n x^n \in R[x].$$

Show that $f(x) \in R[x]$ is:

(i) invertible if and only if a_0 is invertible in R and $a_1, \ldots, a_n$ are nilpotent elements in R;

(ii) nilpotent if and only if $a_0, a_1, \ldots, a_n$ are nilpotent elements;

(iii) a zero-divisor if and only if there exists $r \in R^* = R \setminus \{0\}$ such that $r \cdot f(x) = 0$.

Exercise 1.4. Let $\mathbb{K}$ be a field and let $R \subset \mathbb{K}$ be a subring such that $\mathbb{K}$ is a finite R-algebra. Show that R is also a field (cf. Hulek, 2003, Lemma 1.22).

Exercise 1.5. Let $\mathcal{S}$ be a non-negatively graded ring and let $\mathcal{M}$ be a non-negatively graded $\mathcal{S}$-module. Let $h \in \mathbb{N}$ and let $\mathcal{M}(h)$ be the $\mathcal{S}$-module defined as in (1.33). Show that $\mathcal{M}(h)$ is a graded $\mathcal{S}$-module which is isomorphic to $\mathcal{M}$ as a $\mathcal{S}$-module. Give an example where $\mathcal{M}(h)$ is not isomorphic to $\mathcal{M}$ as a graded $\mathcal{S}$-module.

Chapter 2

Algebraic Affine Sets

From now on $\mathbb{K}$ will always denote a field. To start with, we consider basic facts with no further assumptions on $\mathbb{K}$. We will then discuss some examples showing where difficulties come out in this general setting. Thus, we will focus once and for all on the case of $\mathbb{K}$ algebraically closed (unless otherwise explicitly stated), in particular containing infinitely many elements (cf. Corollary 1.3.12).

2.1 Algebraic Affine Sets and Ideals

For any field $\mathbb{K}$, let $\mathbb{A}^n_{\mathbb{K}}$ (or simply $\mathbb{A}^n$, when the field $\mathbb{K}$ is understood), the *n–dimensional numerical (standard) affine space* over $\mathbb{K}$. This is the set $\mathbb{K}^n$ of ordered n-tuples $(p_1, \ldots, p_n)$, where $p_i \in \mathbb{K}$, $1 \leqslant i \leqslant n$. The use of the symbol $\mathbb{A}^n$ (instead of $\mathbb{K}^n$) resides on the fact that we want to take into account the *geometric nature* of $\mathbb{K}^n$, i.e. considering its elements as *points* instead as vectors. We will indeed endowed $\mathbb{A}^n_{\mathbb{K}}$ with a structure of topological space (cf. Definition 2.1.13), and this structure we want to be kept in mind to be different from that of $\mathbb{K}^n$ as a numerical n-dimensional vector space over $\mathbb{K}$.

An element $P = (p_1, \ldots, p_n)$ in $\mathbb{A}^n$ will be called a *point* of the affine space and $p_1, \ldots, p_n$ will be its *coordinates* in $\mathbb{A}^n$. Correspondingly, the *(numerical) vector* $(p_1, \ldots, p_n) \in \mathbb{K}^n$, i.e. the vector having those components with respect to the canonical basis of $\mathbb{K}^n$, will be simply denoted by $\underline{P}$ and called *coordinate vector* of the point P. The point $O \in \mathbb{A}^n$, whose coordinate vector is $\underline{O} = (0, 0, \ldots, 0) \in \mathbb{K}^n$, is the *origin* of $\mathbb{A}^n$.

Let $\underline{x} := (x_1, \ldots, x_n)$ be a n–tuple of indeterminates over $\mathbb{K}$. In this chapter, we will denote by $A_{\mathbb{K}}^{(n)}$, or simply by $A^{(n)}$, the polynomial ring $\mathbb{K}[\underline{x}] = \mathbb{K}[x_1, \ldots, x_n]$. An element $f(\underline{x}) = f(x_1, \ldots, x_n) \in A^{(n)}$ will be simply denoted by f, if no confusion arises. It is then clear that any $f \in A^{(n)}$ can be regarded as a map

$$f : \mathbb{A}^n \to \mathbb{K}, \quad P \to f(P) = ev_P(f), \tag{2.1}$$

where $ev_P(f)$ denotes the *evaluation* of $f \in A^{(n)}$ at the point $P \in \mathbb{A}^n$.

Definition 2.1.1. The subset of $\mathbb{A}^n$

$$Z_a(f) := f^{-1}(0) = \{P \in \mathbb{A}^n \mid f(P) = 0\},$$

will be called the *set of zeroes* (or simply the *zero-set*) of f.

The subscript a in $Z_a(-)$ stands for the word *affine*, to distinguish from the *projective case* which will be considered in Chapter 3.

Remark 2.1.2. From Definition 2.1.1 it is already clear the reason why, for geometric objects, one is mostly concerned with algebraically closed fields. Indeed, consider the polynomials 1, $x_1^2 + 1 \in A^{(1)}$. Whatever $\mathbb{K}$ is, $Z_a(1) = \emptyset$. On the other hand, if, e.g. $\mathbb{K} = \mathbb{R}$ then $Z_a(1) = \emptyset = Z_a(x_1^2 + 1)$, since $x_1^2 + 1$ is irreducible in $A^{(1)}$. If otherwise $\mathbb{K} = \mathbb{C}$, in this case $Z_a(1) = \emptyset \subset Z_a(x_1^2 + 1) = \{i, -i\}$ and the cardinality of the zero-set of the polynomial $x_1^2 + 1 \in A^{(1)}$ equals its degree.

More generally, one poses

Definition 2.1.3. For any $F \subset A^{(n)}$ finite subset of polynomials,

$$Z_a(F) := \bigcap_{f \in F} Z_a(f)$$

is called the *zero-set* of F. More precisely, if $F := \{f_1, \ldots, f_t\}$, then

$$Z_a(F) = Z_a(f_1, \ldots, f_t) = \bigcap_{i=1}^{t} Z_a(f_i) = \{P \in \mathbb{A}^n \mid f_i(P) = 0, \ \forall \, i = 1, \ldots, n\}.$$

Definition 2.1.4. A subset $Y \subseteq \mathbb{A}^n$ is called an *Algebraic Affine Set* (*AAS*, for short), if $Y = Z_a(F)$, for some finite subset $F \subset A^{(n)}$. When F is explicitly given as $F = \{f_1, \ldots, f_t\}$, the polynomials $f_i \in A^{(n)}$, $1 \leqslant i \leqslant t$, are said to determine a *system of equations defining* Y.

Given $F = \{f_1, \ldots, f_t\}$ as above, consider the ideal these polynomials generate in $A^{(n)}$. Thus, $I := (f_1, \ldots, f_t)$ is a finitely generated ideal.

Proposition 2.1.5. *For any field $\mathbb{K}$, one has*

$$Z_a(F) = Z_a(I) := \{P \in \mathbb{A}^n \mid f(P) = 0, \; \forall \, f \in I\}.$$

Proof. It is clear that $Z_a(I) \subseteq Z_a(f_1, \ldots, f_t)$. On the other hand, since any $f \in I$ is $f = \Sigma_{i=1}^{t} g_i f_i$, for suitable $g_i \in A^{(n)}$, $1 \leqslant i \leqslant t$, it is clear that $P \in Z_a(f_1, \ldots, f_t)$ implies $P \in Z_a(I)$. $\qquad\square$

Remark 2.1.6. From Proposition 2.1.5, an AAS Y does not depend on the system of equations defining it. In particular any finite set of generators of the same ideal I defines the same algebraic affine set Y in $\mathbb{A}^n$.

Example 2.1.7. As above, let $\mathbb{K}$ be any field.

(i) In $\mathbb{A}^1$, one has $\{0\} = Z_a(x_1) = Z_a((x_1)) = Z_a(\{x_1^n\}_{n \in F})$, for $F \subset \mathbb{N}$ any (possibly infinite) subset.

(ii) In $A^{(2)}$ consider the three ideals $I = (x_1, x_2)$, $J = (x_1, x_2 - x_1^2)$ and $K = (x_2, x_2 - x_1^2)$. One has $I = J$; indeed J is a proper ideal in $A^{(1)}$, $I \subseteq J$ (since x_1, $x_2 = 1(x_2 - x_1^2) + x_1(x_1) \in J$) and I is maximal (since $A^{(2)}/I \cong \mathbb{K}$ is a field). On the other hand, $K \subsetneq I$, as it easily follows from the fact that $K = (x_2, x_1^2)$. However, $Z_a(I) = Z_a(J) = Z_a(K) = O = (0,0) \in \mathbb{A}^2$.

To understand the geometric counterpart of the three ideals, take for simplicity $\mathbb{K} = \mathbb{R}$. The origin O is the AAS cut out by the generators of the three ideals I, J and K above. But one has main differences among the three ideals as it follows.

The generators of the ideals I give rise to affine lines, which are the two coordinate axes of the Cartesian plane, which transversally intersect along the origin O (cf. Figure 2.1).

The ideal J gives rise to the intersection between the vertical axis $Z_a(x_1)$ and the parabola $Z_a(x_2 - x_1^2)$ with vertex at O (cf. Figure 2.2); once again this intersection is transversal.

For what concerns the ideal K, its two generators give rise respectively to the parabola $Z_a(x_2 - x_1^2)$ and to its tangent line $Z_a(x_2)$ at the origin O;

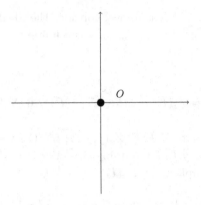

Fig. 2.1   Origin as $Z_a(I)$.

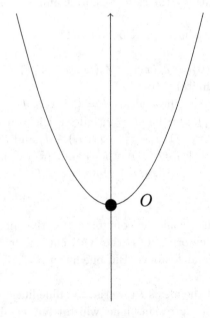

Fig. 2.2   Origin as $Z_a(J)$.

thus the line $Z_a(x_2)$ has intersection multiplicity 2 with the parabola at the origin O (cf. Figure 2.3).

In all the three cases, passing to $Z_a(-)$ discards multiplicities from the picture. We will give more precise algebraic motivations for this phenomenon in Remark 2.1.14-(iv).

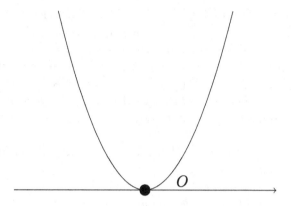

Fig. 2.3   Origin as $Z_a(K)$.

A direct consequence of Theorem 1.4.1 is the following.

Corollary 2.1.8. *Let $\mathbb{K}$ be any field. Any AAS in $\mathbb{A}^n$ is of the form $Z_a(I)$, for some ideal $I \subseteq A^{(n)}$.*

Remark 2.1.9. In particular, Definitions 2.1.3 and 2.1.4 make sense for any subset $F \subseteq A^{(n)}$; no matter F is finite or not, the ideal $I := (F)$ is always finitely generated since $A^{(n)}$ is Noetherian (cf. Remark 1.4.2).

We discuss easy examples, some of which show basic differences between the cases with either $\mathbb{K}$ algebraically closed or not, and either infinite or not.

Example 2.1.10.

(i) In $A^{(1)}$, consider the subset $F := \{x_1^n\}_{n \in \mathbb{N}}$. The ideal I generated by F is $I = (F) = (x_1)$, since $\mathbb{K}[x_1]$ is a PID. Moreover $Z_a(F) = Z_a((F)) = Z_a((x_1)) = \{0\}$.

(ii) Consider $f \in A^{(1)}$ a non-constant polynomial. If $\mathbb{K}$ is not algebraically closed, there exist polynomials f with no root in $\mathbb{K}$ (e.g. when f is irreducible of $\deg(f) \geqslant 2$). In such a case, $Z_a((f)) = \emptyset = Z_a((1))$ even if the inclusion $(f) \subset (1) = A^{(1)}$ is strict (being (f) a proper maximal ideal in $A^{(1)}$). When otherwise $\mathbb{K}$ is algebraically closed, for any $f \in A^{(1)}$, one has $\emptyset \neq Z_a((f))$; more precisely, $Z_a((f)) = \{\alpha_1, \ldots, \alpha_k\}$, where $k \leqslant \deg(f)$ and α_i are all the distinct roots of f, the case $k = n$ occurring if and only if f has simple roots.

(iii) Conversely to (ii), taking $\alpha_1, \ldots, \alpha_k \in \mathbb{K}$ distinct elements and posing $f := \Pi_{i=1}^k (x_1 - \alpha_i) \in \mathbb{K}[x_1]$, one has $\{\alpha_1, \ldots, \alpha_k\} = Z_a((f)) \subset \mathbb{A}_{\mathbb{K}}^1$.

(iv) If $\mathbb{K}$ is a finite field, there do exist non-zero polynomials $f \in A^{(1)}$ for which $Z_a(f) = \mathbb{A}^1$. Take, e.g. $\mathbb{K} = \mathbb{Z}_p$, $p \in \mathbb{Z}$ a prime, and consider $f := x_1^p - x_1 \in \mathbb{Z}_p[x]$. Then $Z_a(f) = \mathbb{A}^1 = Z_a(0)$; indeed, as a consequence of Lagrange's theorem and the fact that $U(\mathbb{Z}_p)$ is cyclic of order $p-1$, f totally splits in $\mathbb{Z}_p[x_1]$ as $f = \Pi_{j \in \mathbb{Z}_p}(x_1 - j)$.

Recalling operations on ideals as in Section 1.1, one can discuss first properties of AAS's in $\mathbb{A}^n$.

Proposition 2.1.11. *Let $\mathbb{K}$ be any field. Then:*

(i) *for any subsets F and G of $A^{(n)}$ one has*

$$F \subseteq G \Longrightarrow Z_a(F) \supseteq Z_a(G). \tag{2.2}$$

In particular, for any ideals $I_1 \subseteq I_2$ in $A^{(n)}$, one has $Z_a(I_1) \supseteq Z_a(I_2)$;

(ii) *for any $I_1, I_2 \subseteq A^{(n)}$ ideals, $Z_a(I_1 \cap I_2) = Z_a(I_1) \cup Z_a(I_2)$;*

(iii) *for any family $\{I_\alpha\}_{\alpha \in A}$ of ideals in $A^{(n)}$, $Z_a\left(\Sigma_{\alpha \in A} I_\alpha\right) = \cap_{\alpha \in A} Z_a(I_\alpha)$;*

(iv) *for any $P \in \mathbb{A}^n$, whose coordinates are $(p_1, \ldots, p_n)$, the ideal*

$$\mathfrak{m}_P := (x_1 - p_1, \ldots, x_n - p_n) \tag{2.3}$$

is maximal in $A^{(n)}$ and $Z_a(\mathfrak{m}_P) = \{P\}$.

(v) *for any ideal $I \subseteq A^{(n)}$, $Z_a(I) = Z_a(\sqrt{I})$.*

Proof.

(i) For any $P \in Z_a(G)$ and for any $f \in G$, one has $f(P) = 0$. Since $F \subseteq G$, then $P \in Z_a(F)$ also. The rest of the statement follows from Proposition 2.1.5, and from Remark 2.1.9.

(ii) Since $I_1 \cap I_2 \subseteq I_1, I_2$, by (i) we have $Z_a(I_1) \cup Z_a(I_2) \subseteq Z_a(I_1 \cap I_2)$. On the other hand, if $P \notin Z_a(I_1) \cup Z_a(I_2)$, there exist $f_1 \in I_1$ and $f_2 \in I_2$ s.t. $f_1(P) \neq 0$ and $f_2(P) \neq 0$. Thus, $f_1 f_2 \in I_1 \cap I_2$ is such that $(f_1 f_2)(P) := f_1(P) f_2(P) \neq 0$, since $\mathbb{K}$ is a field, which shows that $P \notin Z_a(I_1 \cap I_2)$.

(iii) One has $I_\alpha \subseteq \Sigma_{\alpha \in A} I_\alpha$, for any $\alpha \in A$. Therefore, by (i), $Z_a(\Sigma_{\alpha \in A} I_\alpha) \subseteq Z_a(I_\alpha)$, for any $\alpha \in A$, which implies that $Z_a(\Sigma_{\alpha \in A} I_\alpha) \subseteq \cap_{\alpha \in A} Z_a(I_\alpha)$. On the other hand, for any $P \in \cap_{\alpha \in A} Z_a(I_\alpha)$, one has that $f_\alpha(P) = 0$, for any $f_\alpha \in I_\alpha$ and for any $\alpha \in A$. By definition of $\Sigma_{\alpha \in A} I_\alpha$, this implies that $P \in Z_a(\Sigma_{\alpha \in A} I_\alpha)$.

(iv) For any $P \in \mathbb{A}^n$, we have the $\mathbb{K}$-algebra homomorphism

$$\Phi_P : A^{(n)} \to \mathbb{K}$$

defined by the rules $\Phi_P(k) = k$, $\forall\, k \in \mathbb{K}$, and $\Phi_P(x_i) = p_i$, $1 \leqslant i \leqslant n$. In particular, for any $f \in A^{(n)}$, $\Phi_P(f) = ev_P(f)$, where $ev_P(-)$ is the evaluation at the point P as above.

Clearly, the morphism Φ_P is surjective. Thus, $\mathrm{Ker}(\Phi_P)$ is a maximal ideal of $A^{(n)}$ and, by the definition of $\mathfrak{m}_P$ as in (2.3), one has $\mathfrak{m}_P \subseteq \mathrm{Ker}(\Phi_P)$. To prove the maximality of $\mathfrak{m}_P$ it therefore suffices to show that equality holds. To do this, denote by $\underline{y} := (y_1, \ldots, y_n)$ a set of new indeterminates in such a way that $\underline{y} := \underline{x} - \underline{P}$, where $\underline{P}$ is the coordinate vector associated to the point $P \in \mathbb{A}^n$. For any $f = f(\underline{x}) \in \mathrm{Ker}(\Phi_P)$, one has

$$f(\underline{x}) = f(\underline{y} + \underline{P}) =: g(\underline{y}) \in \mathbb{K}[\underline{y}] \cong A^{(n)}.$$

By the assumptions on $f(\underline{x})$, one has $g(\underline{y}) \in (y_1, \ldots, y_n) \subset \mathbb{K}[\underline{y}]$, i.e.

$$g(\underline{y}) = y_1 h_1(\underline{y}) + \cdots + y_n h_n(\underline{y}), \text{ for some } h_i(\underline{y}) \in \mathbb{K}[\underline{y}], \ 1 \leqslant i \leqslant n.$$

This implies that

$$f(\underline{x}) = g(\underline{x} - \underline{P}) = (x_1 - p_1) h_1(\underline{x} - \underline{P}) + \cdots + (x_n - p_n) h_1(\underline{x} - \underline{P}) \in \mathfrak{m}_P,$$

which shows that $\mathrm{Ker}(\Phi_P) \subseteq \mathfrak{m}_P$. The last part of the assertion is trivial.

(v) The inclusion $Z_a(\sqrt{I}) \subseteq Z_a(I)$ directly follows from (i) and the fact that, for any ideal I one has $I \subseteq \sqrt{I}$ (recall Section 1.1). On the other hand, for any $g \in \sqrt{I}$, there exists a positive integer n such that $g^n \in I$ so that, for any $P \in Z_a(I)$, one has $(g^n)(P) = 0$. Since $(g^n)(P) = (g(P))^n$, as a power in $\mathbb{K}$, this forces $g(P) = 0$, i.e. $P \in Z_a(\sqrt{I})$, which proves the other inclusion. $\qquad\square$

Note that, for any field $\mathbb{K}$, one has $\emptyset = Z_a((1))$ and $\mathbb{A}^n = Z_a((0))$, i.e. they are AAS's. Therefore, using Proposition 2.1.11-(ii) and (iii), one easily observes:

Proposition 2.1.12. *The set*

$$\mathcal{C}_a^n := \{AAS's\ in\ \mathbb{A}^n\} \tag{2.4}$$

is the set of closed subsets *of a topology on* $\mathbb{A}^n$.

Definition 2.1.13. The topology on $\mathbb{A}^n$ having $\mathcal{C}_a^n$ as the set of closed subsets is called the *Zariski topology* of $\mathbb{A}^n$ and will be denoted by $\mathfrak{Zar}_{\mathbb{A}^n}$, or simply by $\mathfrak{Zar}_a^n$.

Remark 2.1.14.

(i) From now on, unless otherwise specified, $\mathbb{A}^n$ will be always endowed with the topology $\mathfrak{Zar}_a^n$. The AAS's will be called also (*Zariski*) *closed subsets of* $\mathbb{A}^n$ (for brevity, the term Zariski will be omitted in the sequel). An *open subset* in $\mathfrak{Zar}_a^n$ will be of the form $Z_a(I)^c$, for some ideal I in $A^{(n)}$, i.e. the complement of a closed subset of $\mathbb{A}^n$. Any subset $Y \subseteq \mathbb{A}^n$ will be endowed from now on with the topology induced by $\mathfrak{Zar}_a^n$ on Y. This topology will be called the *Zariski topology of* Y and denoted by $\mathfrak{Zar}_{a,Y}^n$ (or simply $\mathfrak{Zar}_Y$, when the inclusion $Y \subseteq \mathbb{A}^n$ is clearly understood).

(ii) From Proposition 2.1.11-(iv), it follows that $\mathfrak{Zar}_a^n$ is T_1 and that $\mathcal{C}_a^n$ contains all finite subsets of $\mathbb{A}^n$. It is clear that for $n = 1$ these, together with the empty-set, are the only proper closed subsets of $\mathbb{A}^1$; if moreover $\mathbb{K}$ is finite, these are the only closed subsets at all. If otherwise $\mathbb{K}$ is infinite (e.g. when $\mathbb{K}$ is algebraically closed, cf. Corollary 1.3.12), any non-empty open subset of $\mathbb{A}^1$ contains infinite elements and any two non-empty open subsets intersect, i.e. $\mathfrak{Zar}_1^a$ is not T_2. We will show that this more generally holds for any $n \geqslant 1$ (cf. Remark 4.1.1). This is false when $\mathbb{K}$ is finite: take, e.g. $\mathbb{K} = \mathbb{Z}_2$, $U_1 = Z_a(x)^c$ and $U_2 = Z_a(x-1)^c$.

(iii) It is then clear that, when $\mathbb{K}$ is either $\mathbb{R}$ or $\mathbb{C}$, the euclidean topology of $\mathbb{A}_{\mathbb{K}}^n$ is finer than $\mathfrak{Zar}_a^n$.

(iv) Proposition 2.1.11-(v) shows that we do not have a bijective correspondence between ideals in $A^{(n)}$ and elements in $\mathcal{C}_a^n$, since any ideal I which is not radical, defines the same closed subset of its radical ideal $\sqrt{I}$. On the other hand, passing to radical ideals allows one to discard *non-reduced* pieces from the scene. To understand why, consider the following examples.

Example 2.1.15. Let $\mathbb{K}$ be any field. Consider, e.g. $x_2, x_2^2 \in A^{(2)}$; the associated closed sets in $\mathbb{A}^2$ is the same $Z_a(x_2) = Z_a(x_2^2)$, given by the x_1-axis. Algebraically speaking, this is due to the fact that $\sqrt{(x_2^2)} = (x_2)$. The point set given by $Z_a(x_2) = Z_a(x_2^2)$ does not take into account that the x_1-axis in the second case should have to be considered endowed with *multiplicity 2 at any of its points*. In terms of ideals, the situation is completely different: (x_2) is prime, being $A^{(2)}/(x_2)$ an integral domain, whereas (x_2^2) is not prime, since $A^{(2)}/(x_2^2)$ is not a reduced ring (recall Definition 1.1.1). This phenomenon occurred also in Example 2.1.7-(ii): the ideal $I = J$ is radical, since maximal (cf. Lemma 1.1.2-(ii)) whereas K is not radical since $K = (x^2, y)$ but $x \notin K$. On the other hand, $\sqrt{K} = I$, which motivates $O = Z_a(I) = Z_a(K)$.

When $\mathbb{K}$ is not algebraically closed, the failure of the bijective correspondence between ideals of $A^{(n)}$ and elements in $\mathcal{C}_a^n$ is even worse: indeed it fails even if we restrict to radical ideals of $A^{(n)}$, as some of the examples discussed in the following remark show.

Remark 2.1.16.

(i) Let $\mathbb{K}$ be a non-algebraically closed field. Let $f \in A^{(1)}$ be a non-constant polynomial with no roots in $\mathbb{K}$ (e.g. when f is irreducible). Then $Z_a((f)) = \emptyset = Z_a((1))$, even if (f) is a proper ideal. If we moreover assume that $f \in A^{(1)}$ is irreducible, then (f) is a maximal ideal (since $A^{(1)}$ is a PID), and so radical (as it has been proved in Lemma 1.1.2 (ii)). Thus, $Z_a((f)) = \emptyset$, even if $(f) = \sqrt{(f)} \subset (1) = A^{(1)}$.

(ii) Consider once again Example 2.1.10-(iv), where we found $Z_a(x^p - x) = Z_a(0) = \mathbb{Z}_p$, even if $(0) \subset (x^p - x)$. Note moreover that the ideal $(x^p - x)$ is radical: indeed $(x^p - x) = \cap_{a \in \mathbb{Z}_p}(x - a)$, where each $(x - a)$ is a maximal (so prime ideal) in $\mathbb{Z}_p[x]$ (cf. the *primary decomposition of an ideal* Atiyah and McDonald, 1969, Chapter 4).

(iii) Let $f := x_1^2 + x_2^2 + 1 \in \mathbb{K}[x_1, x_2]$, where we consider $\mathbb{K}$ to be either $\mathbb{R}$ or $\mathbb{C}$. In both cases, f is irreducible. Since $\mathbb{K}[x_1, x_2]$ is a UFD (but not a PID), the ideal (f) is prime (so radical) but not maximal (since $\mathbb{K}[x_1, x_2]/(f)$ is not a field). Thus, from the algebraic point of view, properties of the ideal (f) remain unchanged, no matter if $\mathbb{K}$ is either $\mathbb{R}$ or $\mathbb{C}$.

On the contrary the geometric situation is completely different. When $\mathbb{K} = \mathbb{R}$, one has $Z_a(f) = \emptyset = Z_a((1))$ even if $\sqrt{(f)} = (f)$. In the affine classification of real conics, this is a non-degenerate *empty affine conic*. When otherwise $\mathbb{K} = \mathbb{C}$, $Z_a(f) \neq \emptyset$ and its support is a non-degenerate *conic with a center* in the complex affine plane (or a non-degenerate *ellipse* when we consider the complexification of the affine plane, namely classification of conics in the complex affine plane up to affine transformations with coefficients from the subfield $\mathbb{R}$). Identifying $\mathbb{A}_{\mathbb{C}}^2$ with $\mathbb{A}_{\mathbb{R}}^4$, one can easily show that the locus $Z_a((f))$ (endowed with the topology induced by the euclidean topology of $\mathbb{A}_{\mathbb{R}}^4$) is homeomorphic to a 2-sphere in $\mathbb{A}_{\mathbb{R}}^3$ minus 2 points. This is called the *open Riemann surface* associated to $Z_a(f)$ (cf. Exercise 2.1 for full details).

Other examples of AAS's are the following.

Example 2.1.17 (Coordinate subspaces). Let $\mathbb{K}$ be any field, which we will assume infinite for simplicity, and let $0 < m \leqslant n$ be integers. For any sequence of integers $0 < i_1 < i_2 < i_3 < \cdots < i_m \leqslant n$, one has an injective map $\varphi := \varphi_{i_1, i_2, i_3, \cdots, i_m} : \mathbb{A}^m \hookrightarrow \mathbb{A}^n$ defined as follows:

$$(b_1, b_2, \ldots, b_m) \xrightarrow{\varphi} (0, \ldots, 0, \underset{i_1}{b_1}, 0, \ldots, 0, \underset{i_2}{b_2}, 0, \ldots, 0, \underset{i_m}{b_m}, 0, \ldots, 0).$$

φ is a homeomorphism between $(\mathbb{A}^m, \mathfrak{Zar}_a^m)$ and $\mathrm{Im}(\varphi)$ endowed with the topology induced by $\mathfrak{Zar}_a^n$. This image is the (Zariski) closed subset of $\mathbb{A}^n$ $Z_a(x_j \mid \forall j \neq i_1, \ldots, i_m)$, which is called a *m-dimensional coordinate affine subspace* of $\mathbb{A}^n$. The case $m = 1$ gives rise to the *ith-coordinate axis* of $\mathbb{A}^n$, for $1 \leqslant i \leqslant n$.

Example 2.1.18 (Affine subspaces). Let $\mathbb{K}$ be any field, which we will assume infinite for simplicity, and let $\ell_1, \ldots, \ell_k \in A^{(n)}$ be linearly independent linear forms, i.e. $\ell_i = a_{i1}x_1 + \cdots + a_{in}x_n$, where $a_{ij} \in \mathbb{K}$, $1 \leqslant i \leqslant k$, $1 \leqslant j \leqslant |, n$, and $b_1, \ldots, b_k \in \mathbb{K}$. Consider $Y := Z_a(\ell_1 - b_1, \ldots, \ell_k - b_k)$, i.e. the set of solutions of the linear system $A\underline{x} = \underline{b}$, where A is the $k \times n$ matrix whose *ith*-row is filled-up by the coefficients of the linear form ℓ_i, $1 \leqslant i \leqslant k$, and where $\underline{x}$ and $\underline{b}$ are the $n \times 1$ and $k \times 1$ column vector of indeterminates and constant terms, respectively.

By assumptions on the linear forms, $\mathrm{rank}(A) = k$, so $A\underline{x} = \underline{b}$ is compatible. Compatibility condition is independent from the field $\mathbb{K}$ and Rouché–Capelli's theorem ensures that the set of solutions Y is a linear variety in $\mathbb{A}^n$ of *dimension* $n - k$, i.e. Y has a linear *parametric representation* of the form

$$x_1 = c_1 + \lambda_1(t_1, \ldots, t_{n-k}), \ldots, \quad x_n = c_n + \lambda_n(t_1, \ldots, t_{n-k}),$$

where $c_i \in \mathbb{K}$ are constant, λ_i are linear homogeneous polynomials in the parameters t_j and with coefficients from $\mathbb{K}$, $1 \leqslant i \leqslant n$, $1 \leqslant j \leqslant n - k$.
Y is called a $(n - k)$-*dimensional affine subspace of* $\mathbb{A}^n_{\mathbb{K}}$. If

$$I = (\ell_1 - b_1, \ldots, \ell_k - b_k) \subset A^{(n)}$$

is the ideal generated by the linear equations defining Y, then I is a radical ideal: consider indeed the $\mathbb{K}$-algebra homomorphism

$$\pi : \mathbb{K}[x_1, \ldots, x_n] \longrightarrow \mathbb{K}[t_1, \ldots, t_{n-k}],$$

$$x_i \xrightarrow{\pi} c_i + \lambda_i(t_1, \ldots, t_{n-k}), \quad 1 \leqslant i \leqslant n.$$

The map π is surjective, as it follows from Gauss–Jordan elimination theory, and its kernel is given by I, as its generators are representatives of the equivalence class of linear systems whose zero-set is exactly Y. Thus, I is prime, since $\mathbb{K}[t_1, \ldots, t_{n-k}]$ is an integral domain, so it is radical (cf. Lemma 1.1.2).

Example 2.1.19 (Affine hypersurfaces). Let $\mathbb{K}$ be any field and let $f \in A^{(n)}$ be a non-constant polynomial. The AAS $Y := Z_a(f) = Z_a((f))$ is called the *affine hypersurface of equation* $f = 0$ in $\mathbb{A}^n$. Since $A^{(n)}$ is a UFD then, up to units, f factorizes as

$$f = f_1^{r_1} f_2^{r_2} \cdots f_\ell^{r_\ell}, \tag{2.5}$$

where $f_1, \ldots, f_\ell \in A^{(n)}$ are non-proportional, irreducible polynomials and $r_1, \ldots, r_\ell$ are positive integers. From Proposition 2.1.11, it is clear that

$$Z_a(f) = Z_a(f_1 \cdots f_\ell) = \bigcup_{i=1}^{\ell} Z_a(f_i); \tag{2.6}$$

indeed, by the primary decomposition of ideals in a UFD, one has $\sqrt{(f)} = (f_1 \cdots f_\ell)$ (cf. Atiyah and McDonald, 1969, Chapter 4, for more details). Y is called the *affine hypersurface* determined by $f \in A^{(n)}$ and the polynomial

$$f_{\text{red}} = f_1 f_2 \cdots f_\ell \tag{2.7}$$

is called the *reduced equation of* Y, which is indeed another equation for Y. The integer $\deg(f_{red})$ is called the *degree of* Y. When $n = 1$, $d = 1$ gives a point, $d = 2$ gives either the empty-set or two distinct points (it depends on the field $\mathbb{K}$), etcetera. If $n = 2$, $d = 1$ gives a line, $d = 2$ gives a conic, $d = 3$ gives a plane cubic curve, etcetera. For $n = 3$, $d = 1$ is a plane, $d = 2$ is a quadric, $d = 3$ is a cubic surface, and so on.

Definition 2.1.20. Let $\mathbb{K}$ be an infinite field and let $f \in A^{(n)}$ be a non-constant polynomial. The open set $U_a(f) := Z_a(f)^c = \mathbb{A}^n \setminus Z_a(f)$ is called a *principal open (affine) set* of $\mathfrak{Zar}_a^n$.

Lemma 2.1.21. *Principal open (affine) sets form a basis for the topology $\mathfrak{Zar}_a^n$.*

Proof. Any open set of $\mathfrak{Zar}_a^n$ is of the form $U = Z_a(I)^c$, for some ideal $I \subset A^{(n)}$. Since $A^{(n)}$ is Noetherian, $I = (f_1, \dots, f_m)$, for some $f_i \in A^{(n)}$, $1 \leqslant i \leqslant m$, and $Z_a(I) = \cap_{i=1}^m Z_a(f_i)$. This shows that $U = \cup_{i=1}^m U_a(f_i)$. $\qquad\square$

Example 2.1.22 (Products of AAS's. Cylinders).

(i) Let $Z_1 \subseteq \mathbb{A}^r$ and $Z_2 \subseteq \mathbb{A}^s$ be (Zariski) closed sets. Denote by $A_{\underline{x}}^{(r)} := \mathbb{K}[x_1, \dots, x_r]$ and $A_{\underline{y}}^{(s)} := \mathbb{K}[y_1, \dots, y_s]$ the ring of polynomials which are (evaluating) functions operating on the affine spaces $\mathbb{A}^r$ and $\mathbb{A}^s$, respectively. As a set, it is clear that $\mathbb{A}^r \times \mathbb{A}^s = \mathbb{A}^{r+s}$, and the ring of functions operating on this affine space will be denoted by $A_{\underline{x},\underline{y}}^{(r+s)} := \mathbb{K}[x_1, \dots, x_r, y_1, \dots, y_s]$. Now, $Z_1 = Z_a(I_1)$, for some ideal $I_1 = (f_{1,1}, \dots, f_{1,n}) \subseteq A_{\underline{x}}^{(r)}$, and $Z_2 = Z_a(I_2)$, for some ideal $I_2 = (f_{2,1}, \dots, f_{2,m}) \subseteq A_{\underline{y}}^{(s)}$, as it follows from Noetherianity. Since we have natural ring inclusions

$$A_{\underline{x}}^{(r)} \hookrightarrow A_{\underline{x},\underline{y}}^{(r+s)} \hookleftarrow A_{\underline{y}}^{(s)},$$

the polynomials generating the ideals I_1 and I_2 can be regarded as elements in $A_{\underline{x},\underline{y}}^{(r+s)}$. One poses

$$Z_1 \times Z_2 := Z_a((f_{1,1}, \dots, f_{1,n}, f_{2,1}, \dots, f_{2,m})) \subset \mathbb{A}^{r+s}, \qquad (2.8)$$

where the ideal $(f_{1,1}, \dots, f_{1,n}, f_{2,1}, \dots, f_{2,m})$ is intended as an ideal in $A_{\underline{x},\underline{y}}^{(r+s)}$. $Z_1 \times Z_2$ is called the *product of the two AAS's*; it is a closed subset of $\mathbb{A}^{r+s}$. For example, if $p_1 \in \mathbb{A}_{x_1}^1$ and $p_2 \in \mathbb{A}_{y_1}^1$, their product is given by the point $P = Z_a((x_1 - p_1, y_1 - p_2)) = (p_1, p_2) \in \mathbb{A}^2$.

(ii) Particular cases of products are given by *cylinders*. If $Z_1 \subset \mathbb{A}^r$ is a (Zariski) closed proper subset as above, then $Z_1 \times \mathbb{A}^1 \subset \mathbb{A}^{r+1}$ is called the *cylinder* over Z_1 in $\mathbb{A}^{r+1}$ whereas Z_1 is called a *directrix* of the cylinder. Since $\mathbb{A}^1$ in itself is determined by the ideal (0) (moreover this is the only possibility, when $\mathbb{K}$ is algebraically closed), the cylinder is simply given by

$$Z \times \mathbb{A}^1 := Z_a((f_{1,1}, \ldots, f_{1,n}, 0)) = Z_a((f_{1,1}, \ldots, f_{1,n})) \subset \mathbb{A}^{r+1};$$

$$(2.9)$$

in other words, equations defining Z_1 in $\mathbb{A}^r$ coincide with equations defining the cylinder over Z_1 in $\mathbb{A}^{r+1}$. For example, if $p_1 \in \mathbb{A}^1_x$, the cylinder in $\mathbb{A}^2$ over p_1 is $\{p_1\} \times \mathbb{A}^1_y$, whose defining equation is $Z_a(x_1 - p_1)$; this is nothing but the line $x_1 = p_1$ in the plane. The closed subsets $Z_a(x_1^2 + x_2^2 - 1)$, $Z_a(x_1^2 - x_2^2 - 1)$ and $Z_a(x_1^2 - x_2)$ regarded in $\mathbb{A}^3$ are the elliptic, hyperbolic and parabolic *quadric cylinders* in the real affine space $\mathbb{A}^3$ (see Figure 2.4).

(iii) When $\mathbb{K}$ is infinite, note that the topology $\mathfrak{Zar}_a^{r+s}$ is finer than the topology $\mathfrak{Z}_a^r \times \mathfrak{Z}_a^s$. Consider as a set $\mathbb{A}^2 = \mathbb{A}^1 \times \mathbb{A}^1$; from Remark 2.1.14-(ii), closed subsets in $\mathfrak{Zar}_a^1$ are only $\emptyset$, finite number of points and $\mathbb{A}^1$. Therefore, if we endow $\mathbb{A}^2$ with the topology $\mathfrak{Zar}_a^1 \times \mathfrak{Zar}_a^1$, the only closed subsets in this topology are $\emptyset$, $\mathbb{A}^2$, finite unions of points, finite unions of parallel lines to the x_1-axis and finite unions of parallel lines to the x_2-axis. On the contrary, since $\mathbb{K}$ is infinite, the set $Z_a(x_1 - x_2) \subset \mathbb{A}^2$ is a closed subset for $\mathfrak{Zar}_2^a$, which is not closed in $\mathfrak{Zar}_a^1 \times \mathfrak{Zar}_a^1$. However, given $Z_1 \times Z_2 \subset \mathbb{A}^{r+s}$, endowed with the (natural) Zariski topology $\mathfrak{Zar}_{a,Z_1 \times Z_2}^{r+s}$ (recall notation as in

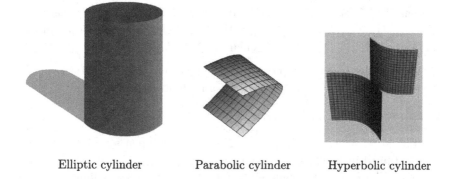

Elliptic cylinder   Parabolic cylinder   Hyperbolic cylinder

Fig. 2.4   Elliptic, hyperbolic and parabolic cylinders in the real affine 3-space.

Remark 2.1.14-(i)), the topology induced by $3\mathfrak{ar}_{a,Z_1 \times Z_2}^{r+s}$ on $\{p_1\} \times Z_2$ ($Z_1 \times \{p_2\}$, respectively), with $p_1 \in Z_1$ ($p_2 \in Z_2$, respectively), coincides with $3\mathfrak{ar}_{a,Z_2}^{s}$ ($3\mathfrak{ar}_{a,Z_1}^{r}$, respectively).

Up to now, the "operation" $Z_a(-)$ allows one to bridge from ideals in $A^{(n)}$ to (Zariski closed) subsets of $\mathbb{A}^n$. We now want to "drive" from subsets of $\mathbb{A}^n$ to ideals in $A^{(n)}$. To do this, we introduce the following.

Definition 2.1.23. Let $\mathbb{K}$ be any field. For any subset $Y \subseteq \mathbb{A}^n$, one denotes by $I_a(Y) := \{f \in A^{(n)} \mid f(P) = 0, \ \forall P \in Y\}$. It is an ideal, which is called the *ideal of Y in $A^{(n)}$*.

Remark 2.1.24. Note that for any $f \in I_a(Y)$, one has $Y \subseteq Z_a((f))$; geometrically speaking, if $Y \neq \emptyset$, $\mathbb{A}^n$, elements in $I_a(Y)$ give rise to hypersurfaces in $\mathbb{A}^n$ containing Y.

From Definition 2.1.23 it follows that

$$Y_1 \subseteq Y_2 \Rightarrow I_a(Y_1) \supseteq I_a(Y_2), \tag{2.10}$$

for any subsets Y_1 and Y_2 of $\mathbb{A}^n$.

Proposition 2.1.25. *Let $\mathbb{K}$ be any field and let Y be any subset of $\mathbb{A}^n$. Then $Z_a(I_a(Y)) = \overline{Y}$, where $\overline{Y}$ is the* closure *(in the topology $3\mathfrak{ar}_a^n$) of Y in $\mathbb{A}^n$. In particular, for any subset Y of $\mathbb{A}^n$ one has $Y \subseteq Z_a(I_a(Y))$, where equality holds if and only if Y is (Zariski) closed in $\mathbb{A}^n$.*

Proof. By Definition 2.1.23, $Z_a(I_a(Y))$ is a closed subset in $\mathbb{A}^n$ such that $Y \subseteq Z_a(I_a(Y))$; then $\overline{Y} \subseteq Z_a(I_a(Y))$. Conversely, let $W := Z_a(J)$ be any closed subset containing Y, J some ideal in $A^{(n)}$. Since $Y \subseteq W$, from (2.10) one has $J \subseteq I_a(W) \subseteq I_a(Y)$. By Proposition 2.1.11 (i), one has therefore $W = Z_a(J) = Z_a(I_a(W)) \supseteq Z_a(I_a(Y))$, i.e. any closed subset containing Y contains also $Z_a(I_a(Y))$. In particular, $\overline{Y} \supseteq Z_a(I_a(Y))$. $\qquad\square$

We consider some examples where the inclusion $Y \subset \overline{Y}$ is strict. In all the examples below, we will consider $\mathbb{K}$ an infinite field.

Example 2.1.26.

(i) Let $Y := \mathbb{A}^1 \setminus \{0\}$. Then $Y = Z_a(x_1)^c$ is an open set of $\mathbb{A}^1$ strictly contained in it. On the other hand $I_a(Y) = I_a(\mathbb{A}^1) = (0)$, since non-zero, non-constant polynomials $f \in A^{(1)}$ have at most finitely many roots in $\mathbb{K}$. In particular $\overline{Y} = \mathbb{A}^1$. The same conclusion holds for e.g. any principal open set $U_a(f) \subset \mathbb{A}^1$, for any $f \in A^{(1)} \setminus \mathbb{K}$.

(ii) Let $T \subset \mathbb{K}$ be an infinite subset of $\mathbb{K}$ and, for any integer $n \geq 2$, let T^n denote the n-tuple product $T \times \cdots \times T \subset \mathbb{A}^n$. If $U \subsetneq \mathbb{A}^n$ is a subset containing T^n, one has $I_a(U) = (0)$, as it follows from Theorem 1.3.13, and so $\overline{U} = \mathbb{A}^n$ as in (i). The same conclusion holds when, e.g. $\mathbb{K}$ is either $\mathbb{R}$ or $\mathbb{C}$, and U contains an open polydisk of the euclidean topology on $\mathbb{A}^n$.

By Proposition 2.1.11-(i) and (2.10), one is therefore left with *reversing–inclusion maps*:

$$\{\text{Subsets of } \mathbb{A}^n\} \xrightarrow{I_a(-)} \{\text{Ideals of } A^{(n)}\}$$
$$Y \;\; \to \;\; I_a(Y)$$

$$\tag{2.11}$$

and

$$\{\text{Subsets of } \mathbb{A}^n\} \xleftarrow{Z_a(-)} \{\text{Ideals of } A^{(n)}\}$$
$$Z_a(J) \;\; \leftarrow \;\; J.$$

$$\tag{2.12}$$

Remark 2.1.27.

(i) By definition of $Z_a(-)$, the map (2.12) is neither surjective nor injective. The non-surjectivity follows from the fact that $\text{Im}(Z_a(-)) = \mathcal{C}_a^n \subsetneq \{\text{subsets of } \mathbb{A}^n\}$. Similarly, for any field $\mathbb{K}$, the non-injectivity of the map (2.12) is a direct consequence of Proposition 2.1.11-(v), as it follows for any choice of ideal I which is not radical.

(ii) Note that, when $\mathbb{K}$ is not algebraically closed, injectivity of (2.12) fails also on radical ideals. Consider, e.g. $\mathbb{K} = \mathbb{R}$ and $\mathfrak{m} = (x^2 + 1) \subsetneq A^{(1)}$; $\mathfrak{m}$ is maximal so radical (cf. Lemma 1.1.2), on the other hand $\emptyset = Z_a(\mathfrak{m}) = Z_a((1))$. Similarly, for any prime $p \in \mathbb{Z}$, the ideal $I = (x^p - x) \subset \mathbb{Z}_p[x]$ is a proper, radical ideal (cf. Remark 2.1.16 (ii)) on the other hand $Z_a(I) = Z_a((0)) = \mathbb{A}^1_{\mathbb{Z}_p}$.

(iii) Examples 2.1.26 show that (2.11) is not injective; on the other hand, $\mathcal{C}_a^n$ surjects onto $\{\text{Ideals of } A^{(n)}\}$.

An important task is to understand how "domain" and "target" of the maps $I_a(-)$ and $Z_a(-)$ have to be modified in order to get bijections. A second task is to understand whether $I_a(-)$ is the inverse map of $Z_a(-)$. Results of the next sections will give answers to these questions, when $\mathbb{K}$ is algebraically closed (cf. Corollary 2.2.5).

2.2   Hilbert "Nullstellensatz"

In Proposition 2.1.11-(iv), we proved that, for any field $\mathbb{K}$ and for any $P \in \mathbb{A}^n$, the ideal $\mathfrak{m}_P = (x_1 - p_1, \ldots, x_n - p_n) \subset A^{(n)}$ is maximal. The next fundamental result will show that, when $\mathbb{K}$ is algebraically closed, all maximal ideals of $A^{(n)}$ are of this form.

Theorem 2.2.1 (Hilbert "Nullstellensatz"-weak form). *Let $\mathbb{K}$ be an algebraically closed field. Then*

$\mathfrak{m} \subset A^{(n)}$ *is a maximal ideal* $\Leftrightarrow$ $\mathfrak{m} = \mathfrak{m}_P$, *for some* $P = (p_1, \ldots, p_n) \in \mathbb{A}^n$.

Remark 2.2.2.

(i) The assumption $\mathbb{K}$ algebraically closed is essential for the $(\Rightarrow)$–part. Indeed, taking, e.g. $\mathfrak{m}_1 = (x^2 + 1) \subset A_{\mathbb{R}}^{(1)}$ and $\mathfrak{m}_2 = (x^2 + x + 1) \subset A_{\mathbb{Z}_2}^{(1)}$, one gets maximal ideals in both cases but none of them is of the form prescribed by Theorem 2.2.1.

(ii) Theorem 2.2.1 can be viewed as the analogue in $A^{(n)}$, for any integer $n \geqslant 1$ and for any algebraically closed field $\mathbb{K}$, of the *Fundamental theorem of Algebra* for $A_{\mathbb{C}}^{(1)}$ (recall Theorem 1.0.1). Indeed, if $I \subset A^{(n)}$ is a proper ideal then $I \subseteq \mathfrak{m}$, for some maximal ideal $\mathfrak{m} \subset A^{(n)}$. From Theorem 2.2.1, $\mathfrak{m} = \mathfrak{m}_P$ for some $P \in \mathbb{A}^n$, therefore $\emptyset \neq Z_a(I) \supseteq Z_a(\mathfrak{m}_P) = \{P\}$, as it follows from Proposition 2.1.11-(i) and (iv). In other words, if $\mathbb{K}$ is algebraically closed, $Z_a(I) = \emptyset \Leftrightarrow I = (1)$. Both examples in (i) are such that $\emptyset = Z_a(\mathfrak{m}_1) \subset \mathbb{A}_{\mathbb{R}}^1$ and $\emptyset = Z_a(\mathfrak{m}_2) \subset \mathbb{A}_{\mathbb{Z}_2}^1$ even if the ideals are maximal.

Proof of Hilbert "Nullstellensatz"-weak form. $(\Leftarrow)$ This is Proposition 2.1.11-(iv), which holds for any field $\mathbb{K}$.

$(\Rightarrow)$ Let $\mathfrak{m} \subset A^{(n)}$ be any maximal ideal; then $\mathbb{F} := A^{(n)}/\mathfrak{m}$ is a field. The composition of $\mathbb{K}$-algebra homomorphisms $\mathbb{K} \overset{\iota}{\hookrightarrow} A^{(n)} \overset{\pi}{\twoheadrightarrow} \mathbb{F}$, where ι the natural inclusion and π the canonical projection, gives rise to a $\mathbb{K}$-algebra homomorphism $\varphi : \mathbb{K} \to \mathbb{F}$ which is necessarily injective, since $\mathbb{K}$ is a field. Then $\mathbb{K} \overset{\varphi}{\hookrightarrow} \mathbb{F}$ is a field extension. Since $\mathbb{F}$ is, by construction, also a $\mathbb{K}$-algebra of finite type then, by Zariski's Lemma (cf. Lemma 1.7.2), the field extension $\mathbb{K} \subset \mathbb{F}$ is algebraic of finite degree. Thus, $\mathbb{F} = \mathbb{K}$, since $\mathbb{K}$ is algebraically closed. Posing $p_i := \pi(x_i) \in \mathbb{F} = \mathbb{K}$, for any $1 \leqslant i \leqslant n$, then

$$\mathfrak{m}_P := (x_1 - p_1, \ldots, x_n - p_n) \subseteq \mathrm{Ker}(\pi) = \mathfrak{m}.$$

On the other hand, $\mathfrak{m}_P$ is maximal in $A^{(n)}$ from Proposition 2.1.11-(iv), therefore equality must hold. $\square$

Theorem 2.2.3 (Hilbert "Nullstellensatz"-strong form). *Let* $\mathbb{K}$ *be an algebraically closed field and let* $n \geqslant 1$ *be any integer. For any ideal* $I \subseteq A^{(n)}$ *one has* $I_a(Z_a(I)) = \sqrt{I}$.

Remark 2.2.4. The assumption that $\mathbb{K}$ is algebraically closed is essential for the equality $I_a(Z_a(I)) = \sqrt{I}$. Indeed, taking, e.g. $\mathbb{A}^1_\mathbb{R}$ one has $\emptyset = Z_a((1)) = Z_a((x^2+1))$ even if (x^2+1) is radical since maximal in $A^{(1)}_\mathbb{R}$. Similarly, in $\mathbb{A}^1_{\mathbb{Z}_p}$, $p \in \mathbb{Z}$ a prime, $Z_a((0)) = Z_a((x^p - x))$ even if $(x^p - x)$ is radical in $A^{(1)}_{\mathbb{Z}_p}$ (recall Remark 2.1.16-(ii)).

Proof of Hilbert "Nullstellensatz"-strong form. For simplicity denote $Y := Z_a(I)$. We can assume that I is a proper ideal. Note that $\sqrt{I} \subseteq I_a(Y)$. Indeed, for any $g \in \sqrt{I}$ there exists a positive integer $s = s(g)$ such that $g^s \in I$ so, for any $P \in Y$, $0 = g^s(P) = (g(P))^s$ which implies $g \in I_a(Y)$, since $\mathbb{K}$ is a field.

We now prove the other inclusion $I_a(Y) \subseteq \sqrt{I}$. Let $g \in I_a(Y)$ be any element; by Noetherianity of $A^{(n)}$, let $I = (f_1, \ldots, f_m)$, where $f_i \in A^{(n)}$ suitable non-constant polynomials. Consider y another indeterminate over $\mathbb{K}$. In the polynomial ring $A^{(n+1)}_{\underline{x},y} := \mathbb{K}[x_1, \ldots, x_n, y] \cong A^{(n+1)}$ consider the ideal $J := (f_1, \ldots, f_m, yg - 1)$. Since $g \in I_a(Y)$, then $Z_a(J) = \emptyset \subset \mathbb{A}^{n+1}$. The fact that $\mathbb{K}$ is algebraically closed implies that $J = (1)$, as it follows from the *Hilbert "Nullstellensatz"-weak form*. Thus, there exist polynomials $q_1, \ldots, q_m, p \in A^{(n+1)}_{\underline{x},y}$ such that

$$1 = q_1 f_1 + \cdots + q_m f_m + p(yg - 1), \tag{2.13}$$

as a polynomial identity in $A^{(n+1)}_{\underline{x},y}$. If one poses $y = g^{-1}$ in (2.13) (this is the so called *Rabinowithch's trick*), one gets

$$1 = \tilde{q}_1 f_1 + \cdots + \tilde{q}_m f_m, \tag{2.14}$$

where $\tilde{q}_1, \ldots, \tilde{q}_m \in \mathbb{K}[x_1, \ldots, x_n, g^{-1}]$. Taking any sufficiently large positive integer N and multiplying both members of (2.14) by g^N yields

$$g^N = q_1^* f_1 + \cdots + q_m^* f_m,$$

with $q_i^* \in A^{(n)}$, for $1 \leqslant i \leqslant m$. This implies $g^N \in I$ so $g \in \sqrt{I}$, i.e.
$I_a(Y) \subseteq \sqrt{I}$. $\square$

Important consequences of the previous results are the following.

Corollary 2.2.5. *Let $\mathbb{K}$ be any algebraically closed field. Then:*

(i) *for any subset $Y \subseteq \mathbb{A}^n$, one has $I_a(Y) = I_a(\overline{Y})$;*
(ii) *the maps (2.11) and (2.12) induce bijections:*

$$\{(\text{Zariski}) \text{ closed subsets of } \mathbb{A}^n\} \overset{1-1}{\longleftrightarrow} \{\text{Radical ideals of } A^{(n)}\}$$
$$\cup \qquad\qquad\qquad\qquad\qquad \cup \qquad\qquad (2.15)$$
$$\{\text{Points of } \mathbb{A}^n\} \qquad \overset{1-1}{\longleftrightarrow} \{\text{Maximal ideals of } A^{(n)}\}.$$

Proof.

(i) By Proposition 2.1.25, for any subset $Y \subseteq \mathbb{A}^n$, $\overline{Y} = Z_a(I_a(Y))$. On the
other hand, one has also $\overline{Y} = Z_a(I_a(\overline{Y}))$, i.e. $Z_a(I_a(Y)) = Z_a(I_a(\overline{Y}))$.
Since $\mathbb{K}$ is algebraically closed, by the Hilbert "Nullstellensatz"-strong
form we have $\sqrt{I_a(Y)} = \sqrt{I_a(\overline{Y})}$ which implies $I_a(Y) = I_a(\overline{Y})$ since,
by definitions, they are both radical ideals.

(ii) The correspondence between points in $\mathbb{A}^n$ and maximal ideals is the
content of *Hilbert "Nullstellensatz"-weak form.* The bijective correspondence between radical ideals of $A^{(n)}$ and (Zariski) closed subsets
of $\mathbb{A}^n$ follows from the fact that, if I is a radical ideal, by the
Hilbert "Nullstellensatz"-strong form one has $I_a(Z_a(I)) = I$ whereas,
if Y is (Zariski) closed then $Z_a(I_a(Y)) = Y$, as it follows from
Proposition 2.1.25. $\square$

Remark 2.2.6.

(i) In Chapter 4, we will see that Corollary 2.2.5-(i) more generally holds
for any field $\mathbb{K}$. This will be a direct consequence of Theorem 1.3.13
and the fact that Y is *dense* in its closure $\overline{Y}$.

(ii) The algebraic counter-part of the Hilbert "Nullstellensatz"-strong form
is the following.

Claim 2.2.7. *If $\mathbb{K}$ is any algebraically closed field and $I \subseteq A^{(n)}$ is any
ideal, then $\sqrt{I} = \cap_{\substack{\mathfrak{m} \text{ maximal,} \\ \mathfrak{m} \supseteq I}} \mathfrak{m} = \cap_{P \in Z_a(I)} \mathfrak{m}_P.$*

Proof. The second equality follows from the Hilbert "Nullstellensatz"-weak form.

For the first equality, let $M_I := \{\mathfrak{m} \subset A^{(n)} \text{ maximal ideal} \,|\mathfrak{m} \supseteq I\}$. If $\mathfrak{m} \in M_I$ then $\sqrt{I} \subseteq \sqrt{\mathfrak{m}} = \mathfrak{m}$ (cf. Lemma 1.1.2), so $\sqrt{I} \subseteq \cap_{M_I} \mathfrak{m}$. To prove the other inclusion, observe that since $\mathbb{K}$ is algebraically closed any maximal ideal of $A^{(n)}$ is of the form $\mathfrak{m}_P$, for $P \in \mathbb{A}^n$, as the Hilbert "Nullstellensatz"-weak form states. Thus, $f \in \cap_{M_I} \mathfrak{m}$ if and only if $f(P) = 0$ for any $P \in Z_a(I)$, i.e. if and only if $f \in I_a(Z_a(I)) = \sqrt{I}$, the last equality following from Hilbert "Nullstellensatz"-strong form. $\qquad\square$

Proposition 2.2.8. *Let $\mathbb{K}$ be any algebraically closed field. Then* (2.15) *is a reversing-inclusion bijective correspondence for which the following further properties hold:*

(i) *for any closed subsets $Y_1, Y_2 \in \mathcal{C}_a^n$, one has $I_a(Y_1 \cup Y_2) = I_a(Y_1) \cap I_a(Y_2)$. The same holds for any finite union of closed subsets of $\mathbb{A}^n$;*
(ii) *for any collection $\{Y_\alpha\}_{\alpha \in A}$ of closed subsets, one has $I_a\left(\cap_{\alpha \in A} Y_\alpha\right) = \sqrt{\Sigma_{\alpha \in A} I_a\left(Y_\alpha\right)}$.*

Proof. Statement about reversing-inclusion directly follows from Proposition 2.1.11-(i) and formula (2.10).

(i) By definition $I_a(Y_1) \cap I_a(Y_2) := \left\{f \in A^{(n)} \,|\, Y_1, Y_2 \subseteq Z_a(f)\right\}$, which coincides with $I_a(Y_1 \cup Y_2)$.
(ii) Let $Y_\alpha := Z_a(J_\alpha)$, for $J_\alpha \subseteq A^{(n)}$ radical ideal for any $\alpha \in A$. By Proposition 2.1.11, $I_a(\cap_{\alpha \in A} Y_\alpha) = I_a(Z_a(\Sigma_{\alpha \in A} J_\alpha))$. Since $\mathbb{K}$ is algebraically closed, by the Hilbert "Nullstellensatz"-strong form, the latter equals $\sqrt{\Sigma_{\alpha \in A} J_\alpha}$, as decided. $\qquad\square$

Remark 2.2.9.

(a) Reversing-inclusion in Proposition 2.2.8 holds true for any field $\mathbb{K}$, for any pair of ideals $I \subseteq J$ in $A^{(n)}$ and for any pair of subsets $Y_1 \subseteq Y_2$ in $\mathbb{A}^n$ (cf. Proposition 2.1.11-(i) and formula (2.10)).
(b) Proposition 2.2.8-(i) more generally holds for any field $\mathbb{K}$ and for any pairs Y_1, Y_2 of subsets of $\mathbb{A}^n$, as the proof of Proposition 2.2.8 shows.
(c) The use of $\sqrt{\Sigma_{\alpha \in A} I_a\left(Y_\alpha\right)}$ in part (ii) is unavoidable, even if $I_a\left(Y_\alpha\right)$ is a radical ideal, for any $\alpha \in A$. For example, for any field $\mathbb{K}$, $I_1 = (x_1^2 - x_2)$ and $I_2 = (x_1^2 + x_2)$ are both radical ideals in $A^{(2)}$, since they are prime ideals; on the other hand $I_1 + I_2$ is not radical as $x_1^2 = \frac{1}{2}(x_1^2 - x_2) + \frac{1}{2}(x_1^2 + x_2) \in I_1 + I_2$ but $x_1 \notin I_1 + I_2$.

(d) From Example 2.1.19, when $\mathbb{K}$ is algebraically closed (2.15) implies that hypersurfaces in $\mathbb{A}^n$ are in $1-1$ correspondence with principle radical ideals in $A^{(n)}$ (which are not maximal when $n \geqslant 2$). Recalling the decomposition (2.7), the closed subsets $Z_a(f_1), \ldots, Z_a(f_\ell)$ are called the *irreducible components* of the hypersurface $Z_a(f) = Z_a(f_{\text{red}})$. These are the hypersurfaces corresponding to the irreducible factors of f_{red} in $A^{(n)}$. Note that, since $A^{(n)}$ is a UFD, each such irreducible factor is also a prime element in $A^{(n)}$ therefore it generates a prime ideal in $A^{(n)}$. The closed subset $Z_a(f_i) \subset \mathbb{A}^n$ is also called an *irreducible hypersurface* in $\mathbb{A}^n$, for any $1 \leqslant i \leqslant \ell$. Thus, (2.15) implies that irreducible hypersurfaces in $\mathbb{A}^n$ are in $1-1$ correspondence with principal prime ideals in $A^{(n)}$.

2.3   Some Consequences of Hilbert "Nullstellensatz" and of Elimination Theory

We will discuss some nice consequences of the "machinery" developed up to this point.

2.3.1   *Study's principle*

In this section, we will focus on the case of $\mathbb{K}$ an algebraically closed field.

Theorem 2.3.1 (Study's principle). *Let $\mathbb{K}$ be an algebraically closed field and let $n \geqslant 1$ be an integer. Let $f, g \in A^{(n)}$ be non-constant polynomials, with f irreducible. If $g(p_1, \ldots, p_n) = 0$, for any $P = (p_1, \ldots, p_n) \in Z_a(f)$, then f divides g in $A^{(n)}$.*

Proof. By the assumptions on f and g, one has $g \in I_a(Z_a(f))$. Since $\mathbb{K}$ is algebraically closed, from Theorem 2.2.3, $I_a(Z_a((f))) = \sqrt{(f)}$. On the other hand, since $f \in A^{(n)}$ is irreducible and $A^{(n)}$ is a UFD, then f is a prime element. This implies that (f) is a prime ideal and so a radical ideal (cf. Lemma 1.1.2) therefore $g \in (f)$, i.e. $f|g$ in $A^{(n)}$. $\qquad\qquad \square$

Remark 2.3.2. If $\mathbb{K}$ is not algebraically closed, Study's principle does not hold. Take, e.g. $f, g \in \mathbb{R}[x]$ any two distinct, non-constant, irreducible polynomials; then $Z_a(f) = Z_a(g) = \emptyset$ even if neither $f|g$ or $g|f$. Similar considerations can be done, e.g. in the ring $\mathbb{Z}_2[x]$.

2.3.2 *Intersections of affine plane curves*

Interesting application of Elimination Theory, is related to intersection of affine plane curves (i.e. hypersurfaces in $\mathbb{A}^2$) and classification of proper (Zarisky) closed subsets of $\mathbb{A}^2$. In what follows, the term *curve* is always meant to be an affine (possibly reducible) plane curve.

Theorem 2.3.3. *Let $\mathbb{K}$ be any infinite field. The intersection of two curves, if not empty, consists of either finitely many points, or of a curve, or of a union of a curve and of finitely many points.*

Corollary 2.3.4. *For any infinite field $\mathbb{K}$, non-empty, proper closed subsets in $3\mathfrak{ar}_a^2$ consist only of finitely many points, curves and unions of a curve and finitely many points.*

Proof of Theorem 2.3.3. Let $Z_a(f), Z_a(g)$ be two curves, with f, $g \in A^{(2)}$ non-constant polynomials. If f and g have a common, non-constant factor $h \in A^{(2)}$, the curve $Z_a(h)$ is contained in the intersection $Z_a(f,g)$ of the two originary curves. Therefore, from now on we will assume that f and g have no common, non-constant factor.

If both f and g are constant with respect to the same indeterminate, e.g. x_1, then $Z_a(f,g) = \emptyset$: if f and g have some powers of linear factors in their own factorizations, these factors represent lines which are parallel to the x_1 axis and, by the assumptions on f and g, one concludes. If otherwise f is constant with respect to x_1 whereas g is constant with respect to x_2, then $Z_a(f,g)$ consists of at most finitely many points ($Z_a(f,g) = \emptyset$ if no powers of linear factors appear in the factorization of either f or g), and we are done also in this case.

We can therefore assume that f and g are both non-constant with respect to the same indeterminate, e.g. x_2. By the assumption on f and g, the resultant $R := R_{x_2}(f,g) \in \mathbb{K}[x_1]$ is non-zero (cf. Theorem 1.3.16 and Section 1.3.4). Thus, if $(p_1, p_2) \in Z_a(f,g)$, one must have $R(p_1) = 0$ as it follows from Theorem 1.3.16 applied to $f'(x_2) := f(p_1, x_2)$, $g'(x_2) = g(p_1, x_2) \in \mathbb{K}[x_2]$, which must have $(x_2 - p_2) \in \mathbb{K}[x_2]$ as a common, non-constant factor. Since $R \in \mathbb{K}[x_1]$ has at most finitely many roots in $\mathbb{K}$, one can have at most finitely many choices for $p_1 \in \mathbb{K}$. If we make same considerations with respect to the other indeterminate, we arrive at the same conclusion for $p_2 \in \mathbb{K}$. Thus, $Z_a(f,g)$ consists of at most finitely many points. $\qquad\square$

A direct consequence of the previous results is the following.

Theorem 2.3.5 (Bezout's theorem (weak form)). *Let* $Z_1 = Z_a(f)$ *and* $Z_2 = Z_a(g)$ *be curves in* $\mathbb{A}^2$, *with* $f, g \in A^{(2)} \setminus \mathbb{K}$. *If* $\phi := \text{g.c.d.}(f, g)$ *in* $A^{(2)}$ *then* $Z_1 \cap Z_2 = Z_a(\phi) \cup Z_3$, *where* Z_3 *is either empty or a finite set of points.*

Exercises

Exercise 2.1. Let $f = x_1^2 + x_2^2 + 1 \in \mathbb{C}[x_1, x_2]$. Consider $Z_a(f) \subset \mathbb{A}_{\mathbb{C}}^2$ and identify $\mathbb{A}_{\mathbb{C}}^2$ with the Euclidean topological space $\mathbb{R}^4$, via the identifications

$$x_1 = a + ib, \quad x_2 = c + id, \quad a, b, c, d \in \mathbb{R}, \quad i^2 = -1.$$

Endowing $\mathbb{A}_{\mathbb{C}}^2$ with the Euclidean topology of $\mathbb{R}^4$, show that $Z_a(f)$ is homeomorphic to a two-sphere in $\mathbb{R}^3$ minus two points.

Exercise 2.2. Let $\mathbb{K}$ be any infinite field. Let $Y := \{(t, t^2) \in \mathbb{A}^2 \mid t \in \mathbb{K}\}$. Show that Y is an AAS. Find explicit generators of $I_a(Y)$ and show that $I_a(Y)$ is a prime ideal.

Exercise 2.3. In $A^{(3)} := \mathbb{R}[x_1, x_2, x_3]$ consider the ideal $I := (x_2 - x_1^2, x_3 - x_1 x_2)$.

 (i) Prove that the quotient ring $\frac{A^{(3)}}{I}$ is an integral domain.
 (ii) Show that $x_1 x_3 - x_2^2 \in I$.
 (iii) Let $J := (x_2 - x_1^2, x_1 x_3 - x_2^2)$. Show that $Z_a(I) \subsetneq Z_a(J)$ and that $\frac{A^{(3)}}{J}$ is not an integral domain.

Exercise 2.4. Let $\mathbb{K}$ be an algebraically closed field. Let $Y_k \subset \mathbb{A}^3$ be the union of the two lines

$$Y' := Z_a(x_2, x_3) \quad \text{and} \quad Y_k'' = Z_a(x_3 - k, x_1),$$

where $k \in \mathbb{K} \setminus \{0\}$ a non-zero element.

 (i) Determine the ideal $I_k := I_a(Y_k) \subset A^{(4)}$, for any $k \in \mathbb{K} \setminus \{0\}$.
 (ii) Consider $k \in \mathbb{K}$ as a movable parameter, so that one has a family of AAS's Y_k, consisting of "pairs" of affine lines, and a family of ideals $I_k = I_a(Y_k)$. Show that the family of AAS's Y_k, $k \in \mathbb{K}$, describes pairs of skew lines in $\mathbb{A}^3$ approaching each other when k approaches to 0. Show that, for any $k \neq 0$, the ideal I_k is radical but not prime.

(iii) Set I_0 the ideal where one poses $k = 0$ among the generators of the ideal I_k as in (i). Show that I_0 is neither prime nor radical. Find an element in $\sqrt{I_0} \setminus I_0$.

Exercise 2.5. In $A^{(2)} := \mathbb{R}[x_1, x_2]$ consider the two polynomials $f = (x_1 + 1)^2 + x_2^2 - 1$ and $g = (x_1 - 1)^2 + x_2^2 - 1$ and let $J := (f, g)$ be the ideal they generate.

(i) Determine $Z_a(J) \subseteq A^2_{\mathbb{R}}$.
(ii) Prove that the quotient ring $\frac{A^{(2)}}{J}$ is not a reduced ring.
(iii) Deduce that J cannot be a radical ideal and determine $\sqrt{J}$.

Chapter 3

Algebraic Projective Sets

From now on, unless otherwise stated, $\mathbb{K}$ will be always considered once and for all to be an algebraically closed field.

Let V be a $(n + 1)$-dimensional $\mathbb{K}$-vector space. One can define the following relation on $V \setminus \{\underline{0}\}$:

$$\underline{v} \sim \underline{w} \iff \exists\, t \in \mathbb{K}^* := \mathbb{K} \setminus \{0\} \text{ s.t. } \underline{w} = t\underline{v}.$$

It is easy to see that this is an equivalence relation, which is called *proportionality*. We will denote by $[\underline{v}]$ the equivalence class of $\underline{v} \in V \setminus \{\underline{0}\}$. The quotient set

$$\mathbb{P}(V) := \frac{(V \setminus \{\underline{0}\})}{\sim}$$

is called the *projective space* associated to V. We will denote by π_V (or simply π if no confusion arises) the canonical projection

$$\pi : V \setminus \{\underline{0}\} \twoheadrightarrow \mathbb{P}(V). \tag{3.1}$$

The integer n will be called the *dimension* of the projective space, which is denoted by $\dim(\mathbb{P}(V))$ (the empty-set is considered as the projective space of dimension -1, which is associated to the vector space $V = \{\underline{0}\}$). Elements of $\mathbb{P}(V)$ will be called *points*.

When in particular $V = \mathbb{K}^{n+1}$ is the $(n + 1)$-dimensional numerical (standard) $\mathbb{K}$-vector space, the associated projective space will be simply denoted by $\mathbb{P}^n$ and called the *n-dimensional numerical projective space* over $\mathbb{K}$. If $\underline{v} = (v_0, \ldots, v_n) \in \mathbb{K}^{n+1}$ is a non-zero numerical vector, its proportionality equivalence class $[\underline{v}]$ will be also denoted by $[v_0, \ldots, v_n]$

whereas the $(n+1)$-tuple $(v_0, \ldots, v_n)$ is called a *vector of homogeneous coordinates* for $[\underline{v}]$. Thus, homogeneous coordinates of a point $P \in \mathbb{P}^n$ are such that: (a) not all of them equal zero, and (b) they are defined up to proportionality by a factor $t \in \mathbb{K}^*$.

For any $0 \leqslant i \leqslant n$, we will denote by $P_i \in \mathbb{P}^n$ the point whose homogeneous coordinates are proportional to the components of the vector $\underline{e}_i$ of the canonical basis of $\mathbb{K}^{n+1}$. The points P_i are called the *vertices of the fundamental $(n+1)$-hedron (or pyramid)* of $\mathbb{P}^n$. The point $P_{n+1} = [1, \ldots, 1]$, is called the *unit point* of $\mathbb{P}^n$ whereas the points P_i, with $0 \leqslant i \leqslant n$, are called the *fundamental points* of $\mathbb{P}^n$.

3.1   Algebraic Projective Sets

Consider the projective space $\mathbb{P}(V)$ of dimension n and fix a homogeneous element $F \in S(V^*)_d$ (recall Example 1.10.20 and Section 1.10.1).

Remark 3.1.1. Once we fix a (ordered) basis on V, we have a natural identification of $\mathbb{K}^{n+1}$ with V. This induces a projectivity $\varphi : \mathbb{P}^n \to \mathbb{P}(V)$, which introduces homogeneous coordinates on $\mathbb{P}(V)$. Recalling notation (1.27), for any $d \geqslant 0$, in this case $S(V^*)$ naturally identifies with $\mathcal{S}^{(n)}$ by means of the isomorphism $\widetilde{\varphi} : \mathcal{S}^{(n)} \longrightarrow S(V^*)$ induced by the choice of the (ordered) basis in V. If $F \in S(V^*)_d$, then $\widetilde{\varphi}^{-1}(F)$ is a homogeneous polynomial of degree d. Therefore, F vanishes at $P \in \mathbb{P}(V)$ (where P the point corresponding to the equivalence class $[\underline{P}] = [p_0, \ldots, p_n]$ of the vector $\underline{P} \in V$ and the homogeneous coordinates $[p_0, \ldots, p_n]$ are w.r.t. the given basis of V) means that $\widetilde{\varphi}^{-1}(F)(p_0, \ldots, p_n) = 0$ which holds if and only if $\widetilde{\varphi}^{-1}(F)(tp_0, \ldots, tp_n) = 0$, for any $t \in \mathbb{K}^*$ (recall (1.29)). One says that $\widetilde{\varphi}^{-1}(F)(\underline{X}) = 0$ is a *homogeneous equation* of $Z(F)$ in $\mathbb{P}^n$, where $\underline{X} := (X_0, \ldots, X_n)$ indeterminates (cf. also Section 3.3.8).

If one introduces two different sets of homogeneous coordinates on $\mathbb{P}(V)$,

$$\varphi : \mathbb{P}^n \to \mathbb{P}(V), \quad \psi : \mathbb{P}^n \to \mathbb{P}(V),$$

$S(V^*)$ is identified to $\mathcal{S}^{(n)}$ in two different ways. In other words, there are two isomorphisms

$$\widetilde{\varphi} : \mathcal{S}^{(n)} \to S(V^*), \quad \widetilde{\psi} : \mathcal{S}^{(n)} \to S(V^*)$$

determined by the two distinct bases of V. If $\underline{X} = (X_0, \ldots, X_n)$ and $\underline{Y} := (Y_0, \ldots, Y_n)$ denote the row-vectors of induced indeterminates, respectively, and if A denotes the matrix associated to the base change in V, the map

$$\omega_A := \widetilde{\psi}^{-1} \circ \widetilde{\varphi} : \mathcal{S}^{(n)} \to \mathcal{S}^{(n)}$$

is an isomorphism of graded rings, which sends $\widetilde{\varphi}^{-1}(F)(\underline{X})$ to the polynomial

$$\omega_A\left(\widetilde{\varphi}^{-1}(F)(\underline{X})\right) := \widetilde{\psi}^{-1}(F)\left(\underline{Y}\, A^t\right),$$

where $\underline{X}^t = A\, \underline{Y}^t$. This isomorphism depends on the matrix A, so it is determined up to a non-zero proportionality factor. Any such isomorphism of $\mathcal{S}^{(n)}$ is said to be a *homogeneous linear substitution* of the indeterminates. In particular the homogeneous equation $\widetilde{\psi}^{-1}(F)(\underline{Y}) = 0$ can be deduced from $\widetilde{\varphi}^{-1}(F)(\underline{X}) = 0$ by means of a homogeneous linear substitution in the indeterminates, which preserves the degrees.

From (1.29) and what discussed above, differently from the affine case, an element $F \in S(V^*)_d$ does not determine a function on $\mathbb{P}(V)$. On the other hand, it makes sense to ask for $F \in S(V^*)_d$ to *vanish at P*: if $P = [\underline{p}] \in \mathbb{P}(V)$, one has $F(\underline{p}) = 0$ if and only if $F(t\,\underline{p}) = 0$, for any $t \in \mathbb{K} \setminus \{0\}$. In such a case, the point P is called a *zero* of the polynomial F. Given $F \in S(V^*)_d$, therefore it makes sense to consider $Z_p(F) \subseteq \mathbb{P}(V)$, which is called the *zero-set of F in* $\mathbb{P}(V)$.

Remark 3.1.2.

(i) Given any identification of $\mathbb{P}(V)$ with $\mathbb{P}^n$ as above, any polynomial $f \in \mathcal{S}^{(n)}$ of degree d can be uniquely decomposed into its homogeneous components $f = F_0 + \cdots + F_d$, where $F_i \in \mathcal{S}_i^{(n)}$, $0 \leqslant i \leqslant d$. Since $\mathbb{K}$ has infinitely many elements (Corollary 1.3.12), by Theorem 1.3.13 the polynomial f vanishes at $P \in \mathbb{P}^n$ (in the above sense) if and only if all of its homogeneous components F_i do, $0 \leqslant i \leqslant d$.

(ii) When $\mathbb{K}$ is finite, what stated in (i) does not hold in general. Take, e.g. $\mathbb{K} = \mathbb{Z}_p$, $x := x_0$ and $f = x^p - x \in \mathbb{Z}_p[x]$: f vanishes at all $\overline{a} \in \mathbb{Z}_p$ (cf. Example 2.1.10) but its homogeneous components vanish only at $\overline{0} \in \mathbb{Z}_p$.

Recalling notation (1.22), from Remark 3.1.2-(i) one poses:

Definition 3.1.3. For any subset $T \subseteq \mathcal{S}^{(n)}$ the subset of $\mathbb{P}^n$, $Z_p(T) := \cap_{F \in H(T)} Z_p(F)$, is called the *zero-set of T in* $\mathbb{P}^n$.

The subscript p in the above definition stands for the term *projective*, to make distinction with the affine case in Definition 2.1.1. Note that

$$Z_p(T) = Z_p((H(T))), \tag{3.2}$$

where $(H(T)) \subseteq S^{(n)}$ is a homogeneous ideal (cf. Proposition 1.10.7).

Since $S^{(n)}$ is Noetherian, there exist $F_1, \ldots, F_m \in H(T)$ s.t. $(H(T)) = (F_1, \ldots, F_m)$. Thus, as in Definition 2.1.4, the set $\{F_1, \ldots, F_m\}$ is called a *system of homogeneous equations defining* $Z_p(T)$. In particular, one has

Definition 3.1.4. A subset $Z \subseteq \mathbb{P}^n$ is called an *Algebraic Projective Set* (APS), if there exist polynomials $F_1, \ldots, F_m \in H(S^{(n)})$ s.t. $Z = Z_p(F_1, \ldots, F_m)$; equivalently, $Z = Z_p(\mathfrak{I})$ for a homogeneous ideal $\mathfrak{I} \subseteq S^{(n)}$. More generally, $Z \subseteq \mathbb{P}(V)$ is a *APS* if there exists a subset T of $S(V^*)$ such that $Z = Z_p(T)$.

Recalling Propositions 1.10.8 and 2.1.11, one has

Proposition 3.1.5.

(i) *Let $T_1 \subseteq T_2$ be any subsets of $S^{(n)}$; then one has $Z_p(T_2) \supseteq Z_p(T_1)$. In particular for any pair of homogeneous ideals $\mathfrak{I}_1 \subseteq \mathfrak{I}_2$ in $S^{(n)}$, then $Z_p(\mathfrak{I}_1) \supseteq Z_p(\mathfrak{I}_2)$.*

(ii) *For any homogeneous ideals $\mathfrak{I}_1, \mathfrak{I}_2 \subseteq S^{(n)}$, one has $Z_p(\mathfrak{I}_1 \cap \mathfrak{I}_2) = Z_p(\mathfrak{I}_1) \cup Z_p(\mathfrak{I}_2)$.*

(iii) *For any family $\{\mathfrak{I}_\alpha\}_{\alpha \in A}$ of homogeneous ideals in $S^{(n)}$, one has $Z_p\left(\sum_{\alpha \in A} \mathfrak{I}_\alpha\right) = \cap_{\alpha \in A} Z_p(\mathfrak{I}_\alpha)$.*

Proof. One applies verbatim strategies as in the proofs of Proposition 2.1.11-(i), (ii) and (iii). $\qquad\square$

From the previous proposition,

$$\mathcal{C}_p^n := \{\text{APS's in } \mathbb{P}^n\} \tag{3.3}$$

is the set of *closed subsets* of a topology on $\mathbb{P}^n$. This topology will be called the *Zariski topology on* $\mathbb{P}^n$, which will be denoted by $\mathfrak{Zar}_{\mathbb{P}^n}$ (or simply $\mathfrak{Zar}_p^n$). Proposition 3.1.5 can be stated in more generality for APS's in $\mathbb{P}(V)$, using $S(V^*)$ instead of $S^{(n)}$. Therefore, $\mathcal{C}_p^V$ will denote the *set of APS's of* $\mathbb{P}(V)$, which is the set of closed subsets of the *Zariski topology on* $\mathbb{P}(V)$.

This topology will be denoted by $\mathfrak{Zar}_{\mathbb{P}(V)}$. Open sets in the Zariski topology will be complementary sets of APS's. In particular, for any $F \in S(V^*)_d$,

$$U_p(F) = Z_p(F)^c \tag{3.4}$$

is called a *principal open set of* $\mathbb{P}(V)$. As in Lemma 2.1.21, principal open sets form a basis for $\mathfrak{Zar}_{\mathbb{P}(V)}$.

Any non-empty subset $Y \subseteq \mathbb{P}(V)$ will be endowed, from now on, with the topology induced on Y by $\mathfrak{Zar}_{\mathbb{P}(V)}$. This will be called the *Zariski topology of Y* and will be denoted by $\mathfrak{Zar}_{\mathbb{P}(V),Y}$ (respectively, $\mathfrak{Zar}^n_{p,Y}$ or simply $\mathfrak{Zar}_Y$, when the inclusion $Y \subseteq \mathbb{P}^n$ is clearly understood). From Remark 3.1.1, in what follows we will focus for simplicity on the case of $\mathbb{P}^n$; statements can be easily adapted to $\mathbb{P}(V)$, replacing $S^{(n)}$ with $S(V^*)$.

From Remark 3.1.2-(i) and the Noetherianity of $S^{(n)}$, one has:

Definition 3.1.6. For any subset $Y \subseteq \mathbb{P}^n$, $I_p(Y) := \{f \in S^{(n)} \mid f(P) = 0, \ \forall \ P \in Y\}$ is a homogeneous, finitely generated ideal, which is called the *homogeneous ideal of Y in $S^{(n)}$*.

Same strategies as in the affine case give the following.

Proposition 3.1.7.

(i) *For any subsets $Y_1 \subseteq Y_2 \subseteq \mathbb{P}^n$, $I_p(Y_1) \supseteq I_p(Y_2)$;*
(ii) *For any subsets $Y_1, Y_2 \subseteq \mathbb{P}^n$, $I_p(Y_1 \cup Y_2) = I_p(Y_1) \cap I_p(Y_2)$.*

3.2 Homogeneous "Hilbert Nullstellensatz"

From the previous section, as in the affine case one has natural reversing–inclusion maps

$$\{\text{Subsets of } \mathbb{P}^n\} \xrightarrow{I_p(-)} \{\text{Homogeneous ideals of } S^{(n)}\} \tag{3.5}$$
$$Y \ \rightarrow \ I_p(Y)$$

and

$$\{\text{Subsets of } \mathbb{P}^n\} \xleftarrow{Z_p(-)} \{\text{Homogeneous ideals of } S^{(n)}\} \tag{3.6}$$
$$Z_p(\mathfrak{J}) \ \leftarrow \ \mathfrak{J}.$$

To better understand their behaviour on (Zariski) closed subsets and to prove the homogeneous version of *Hilbert "Nullstellensatz"*, we need to first introduce some useful definitions. Use notation as in (3.1).

Definition 3.2.1. For any subset $Y \subseteq \mathbb{P}^n$, let

$$C_a(Y) := \pi^{-1}(Y) \cup \{\underline{0}\}$$

$$= \{(p_0, \ldots, p_n) \in \mathbb{K}^{n+1} \setminus \{\underline{0}\} \mid [p_0, \ldots, p_n] \in Y\} \cup \{\underline{0}\}.$$

This subset of $\mathbb{A}^{n+1}$ is called the *affine cone* over Y.

Remark 3.2.2.

(i) It is clear from the definition that

$$C_a(Y) = C_a(Y') \Leftrightarrow Y = Y'. \tag{3.7}$$

Moreover, for any subset $Y \subseteq \mathbb{P}^n$, by definition of affine cone one has

$$I_a(C_a(Y)) = I_p(Y). \tag{3.8}$$

(ii) By Proposition 2.1.25 and by (3.8), one has $\overline{C_a(Y)}^a = Z_a(I_a(C_a(Y))) = Z_a(I_p(Y))$, where $\overline{C_a(Y)}^a$ denotes the (affine) closure of $C_a(Y)$ in $\mathbb{A}^{n+1}$. Since $I_p(Y)$ is a homogeneous ideal, then

$$Z_a(I_p(Y)) = C_a(Z_p(I_p(Y))) = C_a(\overline{Y}^p), \tag{3.9}$$

where $\overline{Y}^p$ denotes the (projective) closure of $Y \subseteq \mathbb{P}^n$. In particular one has $\overline{C_a(Y)}^a = C_a(\overline{Y}^p)$. From this equality and from (3.7), it follows that $C_a(Y) \in \mathcal{C}_a^{n+1}$ if and only if $Y \in \mathcal{C}_p^n$ and that the map π in (3.1) is continuous.

Corollary 3.2.3. *For any homogeneous ideal* $\mathcal{J} \subseteq \mathcal{S}^{(n)}$, $Z_p(\mathcal{J}) = Z_p(\sqrt{\mathcal{J}})$.

Proof. It follows from Remark 3.2.2 and from Proposition 2.1.11-(v). $\quad\square$

Differently from the affine case (cf. (2.15)), even if $\mathbb{K}$ is algebraically closed and even if one restricts the maps (3.5) and (3.6) to the sets of homogeneous radical ideals and of (Zariski) closed subsets of $\mathbb{P}^n$, one still does not have bijective correspondences. Indeed, one always has $Z_p((1)) = \emptyset$; on the other hand the following result, which can be considered as the homogeneous analogue of the *Hilbert "Nullstellensatz"-weak form* Theorem 2.2.1, shows that this is not the only case, even on radical ideals.

Theorem 3.2.4 (Homogeneous Hilbert "Nullstellensatz" weak form). *Let* $\mathcal{J} \subset \mathcal{S}^{(n)}$ *be a proper, homogeneous ideal. Then*

$$Z_p(\mathcal{J}) = \emptyset \iff \sqrt{\mathcal{J}} = \mathcal{S}_+, \tag{3.10}$$

where $\mathcal{S}_+$ *is the irrelevant maximal (so radical) ideal of* $\mathcal{S}^{(n)}$ *(cf. Definition 1.10.4).*

Proof. First note that, by the maximality of $\mathcal{S}_+$ and by Lemma 1.1.2, condition $\sqrt{I} = \mathcal{S}_+$ is equivalent to the fact that there exists an integer $t \geqslant 1$ s.t. $\mathcal{S}_+^t \subseteq \mathcal{I}$. Thus, (3.10) reduces to showing:

$$Z_p(\mathcal{I}) = \emptyset \;\Leftrightarrow\; \exists\, t > 0 \text{ s.t. } \mathcal{S}_+^t \subseteq \mathcal{I}. \tag{3.11}$$

($\Leftarrow$) Suppose $\mathcal{S}_+^t \subseteq \mathcal{I}$, for some positive integer t. Since $(X_0^t, X_1^t, \ldots, X_n^t) \subseteq \mathcal{S}_+^t \subseteq \mathcal{I}$, from Proposition 3.1.5-(i), we have $\emptyset \supseteq Z_p(\mathcal{S}_+^t) \supseteq Z_p(\mathcal{I})$.

($\Rightarrow$) Assume $Z_p(\mathcal{I}) = \emptyset$, with $\mathcal{I} = (F_1, \ldots, F_m)$, any F_i a homogeneous polynomial. One can assume that not all polynomials are linear forms, otherwise the assumption $Z_p(\mathcal{I}) = \emptyset$ would give $\mathcal{I} = \mathcal{S}_+$ and we are done. Using notation as in Definition 1.10.16, consider $f_i := \delta_0(F_i)$, $1 \leqslant i \leqslant m$. Thus $Z_p(F_1, \ldots, F_m) = \emptyset$ implies $Z_a(f_1, \ldots, f_m) = \emptyset$.

Since $\mathbb{K}$ is algebraically closed, by Theorem 2.2.1, one has therefore $(f_1, \ldots, f_m) = (1) \subset A^{(n)}$ so there exists polynomials $g_1, \ldots, g_m \in A^{(n)}$ such that

$$1 = \sum_{i=1}^{m} f_i\, g_i. \tag{3.12}$$

Recall that $\mathfrak{h}_0(f_i) | F_i$, for any $1 \leqslant i \leqslant m$ (more precisely the two polynomials coincide if and only if X_0 does not divide F_i). Hence, there exists a nonnegative integer s_i s.t. $X_0^{s_i}\, \mathfrak{h}_0(f_i) = F_i$, for any $1 \leqslant i \leqslant m$. This implies there exists a positive integer N_0 such that (3.12) becomes

$$X_0^{N_0} = \sum_{i=1}^{m} F_i\, G_i \in \mathcal{I}, \tag{3.13}$$

where $G_i = \mathfrak{h}_0(g_i)$. Reasoning in the same way for all the other indexes $j = 1, \ldots, n$, one deduces that there also exist positive integers $N_1, \ldots, N_n$ such that also $X_1^{N_1}, X_2^{N_2}, \ldots, X_n^{N_n} \in \mathcal{I}$. Let $N := \mathrm{Max}\{N_0, N_1, \ldots, N_n\}$. By the assumptions on the degrees of the F_i's, one has $N \geqslant 2$. Take $t := (n+1)(N-1)+1 > n+1$. Thus, any monomial of the form $X_0^{a_0} X_1^{a_1} \cdots X_n^{a_n}$, with $\Sigma_{i=0}^{n} a_i \geqslant t$, is such that there exists at least one a_j for which $a_j \geqslant N \geqslant N_j$, for some $j \in \{0, \ldots, n\}$, i.e. $X_0^{a_0} X_1^{a_1} \cdots X_n^{a_n} \in \mathcal{I}$. Since $\mathcal{S}_+^t$ is generated by monomials $X_0^{a_0} X_1^{a_1} \cdots X_n^{a_n}$, where $\Sigma_{i=0}^{n} a_i = t$, one can conclude. $\qquad\square$

To sum-up the ideals (1) and $\mathcal{S}_+$ both map to the empty-set via

$$\{\text{Homogeneous ideals of } \mathcal{S}^{(n)}\} \xrightarrow{Z_p(-)} C_p^n,$$

whereas S_+ is not in the image of the map

$$\mathcal{C}_p^n \xrightarrow{I_p(-)} \{\text{Homogeneous ideals of } S^{(n)}\},$$

since $I_p(\emptyset) = (1)$. On the other hand, one has

Theorem 3.2.5 (Homogeneous Hilbert "Nullstellensatz"-strong form). *Let* $\mathfrak{I} \subset S^{(n)}$ *be a homogeneous ideal such that* $Z_p(\mathfrak{I}) \neq \emptyset$. *Then* $I_p(Z_p(\mathfrak{I})) = \sqrt{\mathfrak{I}}$.

Proof. From Remark 3.2.2, we have $I_p(Z_p(\mathfrak{I})) = I_a(C_a(Z_p(\mathfrak{I}))) = I_a(Z_a(\mathfrak{I}))$. Since $\mathbb{K}$ is algebraically closed, from Theorem 2.2.3, the latter equals $\sqrt{\mathfrak{I}}$, as desired. $\qquad\square$

As for Corollary 2.2.5-(ii), one has therefore

Corollary 3.2.6. *The maps* (3.5) *and* (3.6) *induce bijections*:

$$\mathcal{C}_p^n \xleftrightarrow{1-1} \{\text{Radical ideals of } S^{(n)}\} \setminus \{S_+\}.$$

3.3   Fundamental Examples and Remarks

3.3.1   *Points*

When $\mathbb{K}$ is algebraically closed, (2.15) gave bijective correspondence between points in $\mathbb{A}^n$ and maximal ideals $\mathfrak{m}_P \subset A^{(n)}$; the situation is different for points in $\mathbb{P}^n$. As in the affine case, points are still closed sets in $\mathfrak{Zar}_p^n$; indeed, if $P = [p_0, \ldots, p_n] \in \mathbb{P}^n$ with e.g. $p_i \neq 0$, then $\{P\} = Z_p(p_i X_0 - p_0 X_i, \ldots, p_i X_n - p_n X_i)$. On the other hand, the homogeneous ideal $\mathfrak{I} := (p_i X_0 - p_0 X_i, \ldots, p_i X_n - p_n X_i) \subset S^{(n)}$ is prime but not maximal, as $\frac{S^{(n)}}{\mathfrak{I}} \cong \mathbb{K}[X_i]$ is an integral domain but not a field. Using Theorem 3.2.5 and Example 2.1.18, one can easily show the following more precise result.

Proposition 3.3.1. *Let* $\mathfrak{I} \subset S^{(n)}$ *be a radical ideal. Then* $Z_p(\mathfrak{I}) = \{P\}$, *for some point* $P \in \mathbb{P}^n$, *if and only if* $\mathfrak{I}$ *is generated by linearly independent linear forms* $L_0, \ldots, L_{n-1} \in S_1^{(n)}$.

In particular, $\mathfrak{Zar}_p^n$ is T_1 and $\mathcal{C}_p^n$ contains all finite subsets of $\mathbb{P}^n$. Moreover, from Proposition 1.10.19, when $n = 1$ these (together with the empty-set) are the only proper closed subsets of $\mathbb{P}^1$. As in the affine case, $\mathfrak{Zar}_p^1$ is not T_2. This will more generally hold for any $\mathfrak{Zar}_p^n$, $n \geqslant 1$ (cf. Remark 4.1.1).

3.3.2 Coordinate linear subspaces

Let $0 \leqslant m \leqslant n$ be integers. For any sequence of integers $0 < i_0 < i_1 < \cdots < i_m \leqslant n$, one has an injective map

$$\varphi := \varphi_{i_1, i_2, i_3, \cdots, i_m} : \mathbb{P}^m \longrightarrow \mathbb{P}^n$$

defined as follows:

$$[p_0, p_1, \ldots, p_m] \overset{\varphi}{\longrightarrow} [0, \ldots, 0, \underset{i_0}{p_0}, 0, \ldots, 0, \underset{i_1}{p_2}, 0, \ldots, 0, \underset{i_m}{p_m}, 0, \ldots, 0).$$

As in Example 2.1.17, φ is a homeomorphism between $(\mathbb{P}^m, 3\mathfrak{ar}_p^m)$ and $\text{Im}(\varphi) \subset \mathbb{P}^n$ endowed with the induced topology by $3\mathfrak{ar}_p^n$. This image is the closed subset $Z_p(X_j \mid \forall j \neq i_0, \ldots, i_m)$, which is called a *m-dimensional coordinate linear subspace* of $\mathbb{P}^n$. For $m = 0$, we get the points P_i which are the vertices of the fundamental pyramid of $\mathbb{P}^n$; for $m = 1$ we get *coordinate axes* of $\mathbb{P}^n$. For $m = n - 1$, we get *fundamental hyperplanes of* $\mathbb{P}^n$. The hyperplane given by $Z_p(X_j)$ will be also denoted by H_j, $0 \leqslant j \leqslant n$.

3.3.3 Hyperplanes and the dual projective space

Let V be a $(n + 1)$-dimensional $\mathbb{K}$-vector space. If we consider $V^* := \text{Hom}(V, \mathbb{K})$, then $\mathbb{P}(V^*)$ is called the *dual projective space* of $\mathbb{P}(V)$. Recalling Example 1.10.20 and Remark 3.1.1, the choice of an (ordered) basis $(\underline{e}_0, \ldots, \underline{e}_n)$ for V allows one to identify $\mathbb{P}(V)$ with $\mathbb{P}^n$ and V^* with $\mathcal{S}_1^{(n)} = \mathbb{K}[X_0, \ldots, X_n]_1$, where $(X_0, \ldots, X_n)$ the dual basis of $(\underline{e}_0, \ldots, \underline{e}_n)$. For simplicity, $\mathbb{P}(\mathcal{S}_1^{(n)})$ is denoted by $(\mathbb{P}^n)^*$. Since any hyperplane of $\mathbb{P}^n$ is of the form

$$H_{\underline{a}} := Z_p(a_0 X_0 + \ldots + a_n X_n), \text{ for some } \underline{a} \in \mathbb{K}^{n+1} \setminus \{\underline{0}\}$$

and since $H_{\underline{a}} = H_{t\underline{a}}$, for any $t \in \mathbb{K}^*$, it is clear that $(\mathbb{P}^n)^*$ is identified with the set of hyperplanes in $\mathbb{P}^n$, the correspondence given by $[\underline{a}] \leftrightarrow H_{\underline{a}}$. In this correspondence, the fundamental hyperplane $H_i \subset \mathbb{P}^n$ corresponds to the vertex P_i^* of the fundamental pyramid of $(\mathbb{P}^n)^*$, $0 \leqslant i \leqslant n$.

3.3.4 Fundamental affine open sets (or affine charts) of $\mathbb{P}^n$

Consider the principal open set $U_p(X_i) = H_i^c = \mathbb{P}^n \setminus H_i$, $0 \leqslant i \leqslant n$. We denote it by

$$U_i := \{[p_0, \ldots, p_n] \in \mathbb{P}^n \mid p_i \neq 0\}. \tag{3.14}$$

Since $\mathbb{P}^n = \cup_{i=0}^n U_i$, then $\{U_0, \ldots, U_n\}$ is a (finite) open covering of $\mathbb{P}^n$, where each open set is principal (cf. (3.4)). Similarly, $\mathbb{P}^n$ can be also written as a disjoint union $\mathbb{P}^n = U_i \cup H_i$, $0 \leqslant i \leqslant n$, where each H_i is homeomorphic to $\mathbb{P}^{n-1}$. For any $i \in \{0, \ldots, n\}$, consider now the map

$$P = [p_0, \ldots, p_n] \in U_i \xrightarrow{\phi_i} \left(\frac{p_0}{p_i}, \ldots, \frac{p_{i-1}}{p_i}, \frac{p_{i+1}}{p_i}, \ldots, \frac{p_n}{p_i} \right) \in \mathbb{A}^n, \quad (3.15)$$

which is well-defined.

Proposition 3.3.2. *For any $i \in \{0, \ldots, n\}$, the map ϕ_i is a homeomorphism.*

Proof. We refer to the case $i = 0$, since the case $i > 0$ can be proved similarly. To ease notation, set $\phi := \phi_0$ and $U := U_0$. It is clear that ϕ is bijective, whose inverse is given by $\phi^{-1}(c_1, \ldots, c_n) = [1, c_1, \ldots, c_n]$. It suffices to prove that ϕ and ϕ^{-1} are closed maps.

For the map ϕ, note that for any closed subset $Y \subseteq \mathbb{P}^n$, $Y \cap U$ is closed in U. Therefore, to show that ϕ is closed, it suffices to prove that $\phi(Z_p(F) \cap U)$ is closed in $\mathbb{A}^n$, for any $F \in H(\mathcal{S}^{(n)})$; this is obvious since, by definition of ϕ, $\phi(Z_p(F) \cap U) = Z_a(\delta_0(F))$.

Similarly, to show that ϕ^{-1} is closed it suffices to showing that, for any non-constant polynomial $g \in A^{(n)}$, the set $\phi^{-1}(Z_a(g))$ is closed in U. By the definition of ϕ one observes that $\phi^{-1}(Z_a(g)) = Z_p(\mathfrak{h}_o(g)) \cap U$ which is closed in U, so ϕ^{-1} is closed. □

Remark 3.3.3.

(i) Since any U_i is homeomorphic to $\mathbb{A}^n$, the principal open sets U_i are also called *principal open affine sets*, or even *affine charts*, of $\mathbb{P}^n$. Thus, $\mathbb{P}^n$ has a finite open covering by affine charts.

(ii) The projective space $\mathbb{P}^n$ can be therefore viewed as an extension of $\mathbb{A}^n$, by identifying $\mathbb{A}^n$ with any of the affine charts U_i, i.e.

$$\mathbb{A}^n \quad \hookrightarrow \quad \mathbb{P}^n$$
$$(c_1, c_2, \ldots, c_n) \longrightarrow [c_1, c_2, \ldots c_i, 1, c_{i+1}, \ldots c_n]$$

(in the sequel, for simplicity, we will usually identify $\mathbb{A}^n$ with the affine chart U_0). In this identification, the fundamental hyperplane $H_i \subset \mathbb{P}^n$ will be called the *hyperplane at infinity* of U_i and each point $P \in H_i$ is called a *point at infinity* (or *improper point*) of U_i, $0 \leqslant i \leqslant n$.

(iii) More generally, for any subset $Y \subset \mathbb{P}^n$, $Y_i := Y \cap U_i$ is an open subset of Y, $0 \leqslant i \leqslant n$, and $\{Y_0, \ldots, Y_n\}$ is an open covering of Y. If in particular Y is moreover closed in $\mathbb{P}^n$, say $Y = Z_p(\mathfrak{I})$ for some homogeneous ideal $\mathfrak{J}$, then Y_i is closed in U_i, $0 \leqslant i \leqslant n$. Indeed, by Noetherianity $\mathfrak{J} = (F_1, \dots, F_m)$, for some $F_j \in H(\mathcal{S}^{(n)})$. From the proof of Proposition 3.3.2, one has

$$Y_i = Z_a(\delta_i(F_1), \dots, \delta_i(F_m)). \tag{3.16}$$

In particular any Y_i is a *fundamental affine open set* of Y (since it is open in Y and homeomorphic to an affine closed set) and contemporarly it is *locally closed* in $\mathbb{P}^n$, i.e. it is the intersection of a closed set and of an open set of $\mathbb{P}^n$.

3.3.5 *Projective closure of affine sets*

As already mentioned above, we will usually identify $\mathbb{A}^n$ with U_0 via the map ϕ_0. In this case, for any subset $Y \subset \mathbb{A}^n$, its closure in $\mathbb{P}^n$, denoted by $\overline{Y}^p$ will be called the *projective closure* of Y. Moreover, one poses

$$Y_\infty := \overline{Y}^p \cap H_0 \tag{3.17}$$

and call it *the set of points at infinity (or improper points)* of $Y \subset \mathbb{A}^n$.

Suppose that $Y \in \mathcal{C}_a^n$; let therefore $I = (f_1, \dots, f_m) \subset A^{(n)}$ be such that $Y = Z_a(I)$. Differently from (3.16), in general the m polynomials obtained by homogenizing the generators of I are not sufficient to determine $\overline{Y}^p$, as the following easy example shows.

Example 3.3.4. Identify $\mathbb{A}^2$ with the affine chart $U_0 \subset \mathbb{P}^2$ and consider $Y = Z_a(I) \subset \mathbb{A}^2$, where $I = (x_1, x_2 + x_1^2)$. One has $\mathfrak{h}_0(x_1) = X_1$, $\mathfrak{h}_0(x_2 + x_1^2) = X_0 X_2 + X_1^2$ and let $\mathfrak{J} = (\mathfrak{h}_0(x_1), \mathfrak{h}_0(x_2 + x_1^2)) = (X_1, X_0 X_2 + X_1^2)$, which is a homogeneous ideal. Note that Y is the origin $O \in \mathbb{A}^2$, indeed $I = \mathfrak{m}_O$, as $x_2 = (x_2 + x_1^2) - x_1(x_1) \in I$. The point O is also closed in $\mathbb{P}^2$ (corresponding to the fundamental vertex $P_0 = [1, 0, 0]$ of $\mathbb{P}^2$). Thus, $\overline{O}^p = O = P_0$ whereas $Z_p(\mathfrak{J}) = \{[0, 0, 1], [1, 0, 0]\}$. In particular, $\overline{Z_a(I)}^p \subsetneq Z_p(\mathfrak{J})$.

Proposition 3.3.5. *Let $Y = Z_a(J) \subseteq \mathbb{A}^n$ be a closed subset, for some ideal $J \subset A^{(n)}$. Identify $\mathbb{A}^n$ with the affine chart $U_0 \subset \mathbb{P}^n$ and define the homogeneous ideal*

$$\mathfrak{J}^* := (\mathfrak{h}_0(f), \ \forall f \in J) \subseteq \mathcal{S}^{(n)}. \tag{3.18}$$

Then $\overline{Y}^p = Z_p(\mathfrak{J}^*)$.

Proof. By definition of $\mathfrak{J}^*$, $Z_p(\mathfrak{J}^*)$ is a closed set of $\mathbb{P}^n$ containing Y. Then $\overline{Y}^p \subseteq Z_p(\mathfrak{J}^*)$.

To prove the opposite inclusion, let $W \subseteq \mathbb{P}^n$ be any closed subset containing Y. For any $F \in H(I_p(W))$, the polynomial $\delta_0(F)$ vanishes along Y, i.e. $Y \subseteq Z_a(\delta_0(F))$. By Theorem 2.2.3, since $\mathbb{K}$ is algebraically closed, one has $\delta_0(F) \in \sqrt{J}$, i.e. there exists an integer $r \geqslant 1$ such that $\delta_0(F)^r \in J$. Assume $\deg(F) = d$ and let $\deg(\delta_0(F)) := d - s$, for some integer $0 \leqslant s < d$ (recall that $s = 0$ if and only if X_0 does not divide F). Then $F = X_0^s \, \mathfrak{h}_0(\delta_0(F))$ and so $F^r = X_0^{sr} \, \mathfrak{h}_0(\delta_0(F))^r$. From Lemma 1.10.17-(ii), $\mathfrak{h}_0$ is multiplicative so $\mathfrak{h}_0(\delta_0(F))^r = \mathfrak{h}_0(\delta_0(F)^r)$. By definition of $\mathfrak{J}^*$, this implies that $F^r \in \mathfrak{J}^*$. Thus, $I_p(W) \subseteq \sqrt{\mathfrak{J}^*}$; from Proposition 3.1.5-(i) and Corollary 3.2.3, it follows that $W = Z_p(I_p(W)) \supseteq Z_p(\sqrt{\mathfrak{J}^*}) = Z_p(\mathfrak{J}^*)$. In other words, any projective closed subset containing Y contains also $Z_p(\mathfrak{J}^*)$. Since this must hold also for $\overline{Y}^p$, one has proved the other inclusion. $\qquad\square$

Note the following nice consequences of the previous result.

Remark 3.3.6.

(i) With same notation and assumptions as in Proposition 3.3.5, take $J := (f_1, \ldots, f_m) \subseteq A^{(n)}$ and let $F_i := \mathfrak{h}_0(f_i) \in H(\mathbb{S}^{(n)})$, $1 \leqslant i \leqslant m$. Then, even if in general $\overline{Y}^p \subseteq Z_p(F_1, \ldots, F_m)$, one has

$$Y = \overline{Y}^p \cap U_0 = Z_p(F_1, \ldots, F_m) \cap U_0, \qquad (3.19)$$

as it follows from the proof of Proposition 3.3.2 (therein we showed that $\phi_0^{-1}(Y) = Z_p(F_1, \ldots, F_m) \cap U_0$).

(ii) *If $Y \subseteq \mathbb{A}^n$ is closed, then*

$$\overline{Y}^p = Y \cup Y_\infty. \qquad (3.20)$$

Proof. By the induced topology $\mathfrak{Zar}^n_{p,\mathbb{A}^n}$, the closure of Y in $U_0 = \mathbb{A}^n$ is given by $\overline{Y}^p \cap U_0$. On the other hand, since Y is already closed in $U_0 = \mathbb{A}^n$, one has $Y = \overline{Y}^p \cap U_0$. $\qquad\square$

Note that the previous proof is another way to see that any closed subset of $\mathbb{A}^n$ is locally closed in $\mathbb{P}^n$ (cf. Remark 3.3.3-(iii))

(iii) *For any subset $Y \subseteq \mathbb{P}^n$, one has*

$$I_p(Y) = I_p(\overline{Y}^p). \qquad (3.21)$$

Proof. Let indeed $F \in H(\mathcal{S}^{(n)})$ such that $Y \subseteq Z_p(F)$; then $\overline{Y}^p \subseteq Z_p(F)$ so $\overline{Y}^p = Z_p(I_p(Y))$. On the other hand, one has also $\overline{Y}^p = Z_p(I_p(\overline{Y}^p))$. By Theorem 3.2.5, one has $\sqrt{I_p(Y)} = \sqrt{I_p(\overline{Y}^p)}$ and so $I_p(Y) = I_p(\overline{Y}^p)$, since by definition both ideals are radical ideals. $\qquad\square$

A different proof follows from Remark 3.2.2 and Corollary 2.2.5-(i); indeed one has $I_p(Y) = I_a(C_a(Y)) = I_a(\overline{C_a(Y)}^a) = I_a(C_a(\overline{Y}^p)) = I_p(\overline{Y}^p)$.

(iv) If $Y \subset U_0$ is a finite set of points, then clearly $Y_\infty = \emptyset$. When $\mathbb{K}$ is not algebraically closed, it may happen that $Y_\infty = \emptyset$ even if Y is not reduced to a finite sets of points. Consider, e.g. the real ellipse $C = Z_a(x_1^2 + x_2^2 - 1) \subset \mathbb{A}_\mathbb{R}^2$. Its projective closure is the non-degenerate conic $Z_p(X_1^2 + X_2^2 - X_0^2) \subset \mathbb{P}_\mathbb{R}^2$ so that $C_\infty = Z_p(X_0, X_1^2 + X_2^2) = \emptyset$. The same conic, considered instead in $\mathbb{A}_\mathbb{C}^2$, is such that $C_\infty = \{[0, 1, i], [0, 1, -i]\}$, with $i^2 = -1$ (cf. Corollary 6.2.6 for more general results).

(v) *Identifying $\mathbb{A}^n$ with U_0, let $\overline{\mathbb{A}^n}^p$ denote the projective closure of $\mathbb{A}^n$ in $\mathbb{P}^n$. Then $\overline{\mathbb{A}^n}^p = \mathbb{P}^n$, i.e. $\mathbb{A}^n$ is dense in $\mathbb{P}^n$.*

Proof. For any integer $d \geqslant 1$, let $F \in \mathcal{S}_d^{(n)}$ be any homogeneous polynomial such that $\mathbb{A}^n \subseteq Z_p(F)$. Up to a permutation of the indeterminates, we can assume that F contains a monomial proportional to X_0^d. Therefore, we can write F as $F = X_0^d F_0 + X_0^{d-1} F_1 + \ldots + F_d$, where $F_i \in \mathcal{S}_i^{(n-1)} = \mathbb{K}[X_1, \ldots, X_{n-1}]_i$, $0 \leqslant i \leqslant d$. We get

$$\delta_0(F) = F_0 + F_1 + \cdots F_d \in A^{(n)}, \tag{3.22}$$

which is the decomposition of $\delta_0(F)$ in its homogeneous components. By the assumptions on F, one has $\mathbb{A}^n = U_0 \subseteq Z_a(\delta_0(F))$. Since $\mathbb{K}$ is infinite (because algebraically closed), by Theorem 1.3.13, we get that $\delta_0(F) = 0$. Thus, $F_i = 0$, for any $0 \leqslant i \leqslant d$, and so also $F = 0$. Therefore $Z_p(F) = Z_p((0)) = \mathbb{P}^n$, as desired. $\qquad\square$

3.3.6 *Projective subspaces and their ideals*

Let V be a $(n + 1)$-dimensional $\mathbb{K}$-vector space and let $W \subset V$ be a subvector space of dimension $m + 1 > 0$, with $m < n$. Then $\mathbb{P}(W)$ is called *(projective) subspace* of $\mathbb{P}(V)$ of *dimension* m and *codimension* $c := n - m$. The empty-set is the only subspace of dimension -1; points in $\mathbb{P}(V)$ are subspaces of dimension 0; codimension-1 subspaces are hyperplanes (cf. Section 3.3.3).

The following statements are easy consequences of standard Linear Algebra:

(i) $Z \subseteq \mathbb{P}(V)$ is a subspace if and only if $Z = Z_p(L_0, \ldots, L_h)$, where $L_0, \ldots, L_h \in V^* = S(V^*)_1$;

(ii) If $L, L_0, \ldots, L_h \in V^*$, then $Z_p(L_0, \ldots, L_h) \subseteq Z_p(L)$ if and only if L is linearly dependent from $L_0, \ldots, L_h$ in V^*;

(iii) $Z_p(L_0, \ldots, L_h) = Z_p(G_0, \ldots, G_k)$ if and only if $L_0, \ldots, L_h$ and $G_0, \ldots, G_k$ span the same sub-vector space of V^*;

(iv) $Z_p(L_0, \ldots, L_h) = Z_p(F_{i_1}, \ldots, F_{i_c})$, where $F_{i_1}, \ldots, F_{i_c}$ is a basis of $W^* := \mathrm{Span}\{L_0, \ldots, L_h\} \subseteq V^*$ and where c is the codimension of W^* in V^*;

(v) if $\Lambda_1, \Lambda_2 \subseteq \mathbb{P}(V)$ are subspaces s.t. $\Lambda_i := \mathbb{P}(W_i)$ for $i = 1, 2$, then $\Lambda_1 \cap \Lambda_2 \subset \mathbb{P}(V)$ is a subspace, which is called the *intersection subspace* of the two subspaces; more precisely $\Lambda_1 \cap \Lambda_2 = \mathbb{P}(W_1 \cap W_2)$.

(vi) Let $S \subset \mathbb{P}(V)$ be any subset. Since the family $\mathcal{L}_S$ of projective subspaces containing S is not empty, one can consider $\langle S \rangle := \cap_{\Lambda \in \mathcal{L}_S} \Lambda$. By (v), this is a subspace of $\mathbb{P}(V)$; more precisely, it is the smallest subspace of $\mathbb{P}(V)$ containing S. This is called the *subspace generated by S* (or even the *linear envelope of S*) in $\mathbb{P}(V)$. S is said to be *non-degenerate* in $\mathbb{P}(V)$ if $\langle S \rangle = \mathbb{P}(V)$, i.e. if S is not contained in any proper projective subspace. If $\Lambda_1, \ldots, \Lambda_h$ are subspaces, one uses the symbol $\Lambda_1 \vee \cdots \vee \Lambda_h$ instead of $\langle \Lambda_1 \cup \cdots \cup \Lambda_h \rangle$.

(vii) *(projective) Grassmann formula: for any subspaces Λ_1, Λ_2 in $\mathbb{P}(V)$ one has*

$$\dim(\Lambda_1) + \dim(\Lambda_2) = \dim(\Lambda_1 \vee \Lambda_2) + \dim(\Lambda_1 \cap \Lambda_2). \qquad (3.23)$$

The proof is a straightforward application of linear Grassmann formula for subspaces in V.

(viii) *Let $\dim(V) = n + 1$. If Λ_1, Λ_2 are projective subspaces in $\mathbb{P}(V)$ such that $\dim(\Lambda_1) + \dim(\Lambda_2) \geqslant n$ then $\Lambda_1 \cap \Lambda_2 \neq \emptyset$.*

The previous result is a direct consequence of (vii) and it is a natural generalization of the well-known fact that two lines in the projective plane always intersect. It is moreover clear that, by the existence of parallelism, no Grassmann formula can exist for affine subspaces in $\mathbb{A}^n$ defined by non-homogeneous linear systems (cf. Example 2.1.18).

(ix) Given $(m+1)$-points $P_0, P_1, \ldots, P_m \in \mathbb{P}(V)$, by Grassmann formula, one has $\dim(P_0 \vee P_1 \vee \cdots P_m) \leqslant m$. The points are said to be *linearly independent* (or even *in general linear position*) if equality holds.

It is then clear that, for any m-dimensional subspace Λ in $\mathbb{P}(V)$, with $m \leqslant n$, there exist $(m + 1)$ points $P_0, P_1, \ldots, P_m \in \Lambda$ which are linearly independent, i.e. such that $\Lambda = P_0 \vee P_1 \vee \cdots P_m$. Fixing an (ordered) basis of V induces a choice of homogeneous coordinates on $\mathbb{P}(V)$, which identifies $\mathbb{P}(V)$ with $\mathbb{P}^n$ and $\Lambda = P_0 \vee P_1 \vee \cdots P_m$ with an APS of the form $Z_p(L_0, \ldots L_{c-1}) \subset \mathbb{P}^n$, where c is the codimension of Λ in $\mathbb{P}^n$ and where $L_0, \ldots, L_{c-1} \in \mathcal{S}_1^{(n)}$. It is easy to see that the natural map

$$\mathbb{P}^m \xrightarrow{\phi} \Lambda \subset \mathbb{P}^n, \quad \text{defined by} \quad [\lambda_0, \ldots, \lambda_m] \xrightarrow{\phi} \sum_{i=0}^{m} \lambda_i P_i, \quad (3.24)$$

is a homeomorphism between $(\mathbb{P}^m, \mathfrak{Zar}_p^m)$ and $(\Lambda, \mathfrak{Zar}_{p,\Lambda}^n)$ (this extends what discuss in Section 3.3.2). The map ϕ is called a *parametric representation* of Λ.

(x) If $\Lambda \subseteq \mathbb{P}(V)$ is a subspace, one defines

$$\Lambda^\perp := \{[L] \in \mathbb{P}(V)^* \mid \Lambda \subseteq Z_p(L)\}.$$

This is a subspace of $\mathbb{P}(V)^*$, which is called the *orthogonal to* Λ. One has that $\dim(\Lambda^\perp) = \operatorname{codim}(\Lambda) = c$.

(xi) One has the following properties:

$$(\Lambda^\perp)^\perp = \Lambda, \quad (\Lambda_1 \vee \Lambda_2)^\perp = \Lambda_1^\perp \cap \Lambda_2^\perp, \quad (\Lambda_1 \cap \Lambda_2)^\perp = \Lambda_1^\perp \vee \Lambda_2^\perp.$$

As for homogeneous ideals defining subspaces, let Λ be a subspace of codimension c in the projective space $\mathbb{P}^n$. By (i) and (iv) above, $\Lambda = Z_p(L_0, \ldots, L_{c-1})$, where $L_0, \ldots, L_{c-1}$ linearly independent in $\mathcal{S}_1^{(n)}$.

Claim 3.3.7. *The ideal* $\mathfrak{I} := (L_0, \ldots, L_{c-1})$ *is prime.*

Assuming for a moment the content of the claim, one gets that $\mathfrak{I}$ is radical (cf. Lemma 1.1.2) thus it coincides with $I_p(\Lambda)$. In other words, subspaces $\Lambda \subset \mathbb{P}^n$, of codimension c, are APS's of $\mathbb{P}^n$ whose homogeneous ideal $I_p(\Lambda)$ is generated by c linearly independent linear forms (cf. Example 2.1.18 for the affine case).

Proof. Since $L_0, \ldots, L_{c-1}$ are linearly independent linear forms, there exist $L_c, \ldots, L_n \in \mathcal{S}_1^{(n)}$ such that $\{L_0, \ldots, L_{c-1}, L_c, \ldots, L_n\}$ is a basis for $\mathcal{S}_1^{(n)}$ (cf. (1.26)). The map $\varphi : \mathcal{S}_1^{(n)} \to \mathcal{S}_1^{(n)}$, defined by $X_i \xrightarrow{\varphi} L_i$, is an automorphism of the vector space $\mathcal{S}_1^{(n)}$ which extends to an automorphism of the $\mathbb{K}$-algebra $S^{(n)}$, always denoted by φ, by the rule: $\varphi(f(X_0, \ldots, X_n)) = f(L_0(\underline{X}), \ldots, L_n(\underline{X}))$, where we posed for brevity $\underline{X} = (X_0, \ldots, X_n)$ (note that φ preserves the graduation of $S^{(n)}$). In particular, φ bijectively maps the maximal ideal S_+ to the ideal $(L_0(\underline{X}), \ldots, L_n(\underline{X}))$ and, consequently, the ideal $(X_0, \ldots, X_{c-1})$ to $\mathfrak{I}$. Since $\frac{S^{(n)}}{(X_0, \ldots, X_{c-1})} \cong S^{(n-c)}$, it follows that $\mathfrak{I}$ is prime. $\square$

By using Example 1.10.20 and Remark 3.1.1, one can more generally state previous results for linear subspaces in $\mathbb{P}(V)$, replacing $S^{(n)}$ with $S(V^*)$.

3.3.7   *Projective and affine subspaces*

As usual, we will identify $\mathbb{A}^n$ with the affine chart U_0 of $\mathbb{P}^n$ and we will denote by H_0 the hyperplane at infinity for U_0.

Let $\Lambda \subset \mathbb{P}^n$ be any non-empty subspace of codimension $c > 0$ and denote by $\Lambda_0 := \Lambda \cap U_0$. If $\Lambda \subseteq H_0$, it is clear that $\Lambda_0 = \emptyset$; therefore, we may assume that Λ is not contained in H_0. In such a case, from Section 3.3.6, Λ is defined by c linearly independent, homogeneous, linear equations of the form:

$$\begin{cases} a_{10}X_0 + a_{11}X_1 + \cdots + a_{1n}X_n = 0 \\ \ldots\ldots \quad \ldots\ldots \quad \ldots\ldots \\ a_{c0}X_0 + a_{c1}X_1 + \cdots + a_{cn}X_n = 0. \end{cases} \tag{3.25}$$

The previous linear system can be written as $A^*\underline{X}^t = \underline{0}$, where A^* is a $c \times (n+1)$ matrix with entries in $\mathbb{K}$, of maximal rank c, whereas $\underline{X} := (X_0, \ldots, X_n)$ and $\underline{0}$ the $(c \times 1)$-matrix with zero entries. From (3.16), Λ_0 is defined by

$$\begin{cases} a_{11}x_1 + \cdots + a_{1n}x_n + a_{10} = 0 \\ \ldots\ldots \quad \ldots\ldots \quad \ldots\ldots \\ a_{c1}x_1 + \cdots + a_{cn}x_n + a_{c0} = 0, \end{cases} \tag{3.26}$$

where as customary $x_i := \frac{X_i}{X_0}$, $1 \leqslant i \leqslant n$. The non-homogeneous linear system (3.26) can be written as

$$A \cdot \underline{x}^t + \underline{a} = \underline{0},$$

where A is the $(c \times n)$-matrix obtained by selecting the last n columns and all the rows of A^*, whereas $\underline{a}$ is the $(c \times 1)$-matrix defined by the first column of A^*, whereas $\underline{x} := (x_1, \ldots, x_n)$.

From the assumptions on A^*, the rank of A is c; in particular, (3.26) is compatible and $\Lambda_0 \neq \emptyset$ is an affine subspace in the sense of Example 2.1.18. The homogeneous linear system obtained by adding the equation $X_0 = 0$ to (3.25) defines the projective subspace $\Lambda \cap H_0$ of dimension $m - 1$, where $m = n - c$. This subspace is called the *direction* of Λ_0.

Let $\underline{\xi}_0$ be a solution of (3.26), i.e. a point $P_0 \in \Lambda_0$. Let moreover $\underline{\xi}_1, \ldots, \underline{\xi}_m$ be independent solutions of the homogeneous linear system associated to (3.26). The bijective map

$$\phi_0 : (\lambda_1, \ldots, \lambda_m) \in \mathbb{A}^m \to \underline{\xi}_0 + \lambda_1 \underline{\xi}_1 + \cdots + \lambda_m \underline{\xi}_m \in \Lambda_0$$

is the restriction to $\mathbb{A}^m = U_0 \subset \mathbb{P}^m$ of the map $\phi : \mathbb{P}^m \to \Lambda$ as in (3.24), when in Λ we choose the $m + 1$ points $P_0, P_1, \ldots, P_m$ with relative homogeneous coordinates

$$[1, \underline{\xi}_0], \ [0, \underline{\xi}_i], \quad 1 \le i \le m,$$

respectively. Thus, ϕ_0 is a homeomorphism between $(\mathbb{A}^m, \mathfrak{Zar}^m_a)$ and $(\Lambda_0, \mathfrak{Zar}^n_{p, \Lambda_0})$ and it is a *parametric representation* of Λ_0 (cf. Example 2.1.18). The map ϕ_0 can be interpreted as a way to introduce a *system of coordinates* in Λ_0, in such a way that the point P_0 coincides with its origin.

Conversely, from Section 3.3.5, for any affine subspace Λ_0 as in (3.26) it follows that Λ as in (3.25) is its projective closure. As for Section 3.3.6 (vi), one has therefore a natural notion of *non-degenerate* subset S of an affine space $\mathbb{A}^n$.

3.3.8 *Homographies, projectivities and affinities*

Let V and W be two $\mathbb{K}$-vector spaces. A map $\varphi : \mathbb{P}(V) \to \mathbb{P}(W)$ is called a *homography* if there exists an injective linear map $f : V \to W$ s.t.

$$\pi_W \circ f = \varphi \circ \pi_V,$$

where π_V and π_W as in (3.1); i.e. $\varphi([\underline{v}]) = [f(\underline{v})]$ for any $\underline{v} \in V \setminus \{\underline{0}\}$. Note that the existence of a homography $\varphi : \mathbb{P}(V) \to \mathbb{P}(W)$ implies $\dim(\mathbb{P}(V)) \le \dim(\mathbb{P}(W))$; moreover, it is straightforward to verify that the composition of homographies is still a homography.

As a matter of notation, we will put $\varphi = \varphi_f$ when we want to stress that the homography φ depends on the linear map f. Let $\text{Hom}_{\mathbb{K}}(V, W)$ be the vector space of $\mathbb{K}$-linear maps from V to W. It is easy to see that

$$\varphi_f = \varphi_g \Longleftrightarrow [f] = [g] \quad \text{in } \mathbb{P}(\text{Hom}_{\mathbb{K}}(V,W)). \tag{3.27}$$

Thus, the set of homographies from $\mathbb{P}(V)$ to $\mathbb{P}(W)$, denoted by

$$\mathbb{O}(\mathbb{P}(V),\mathbb{P}(W)),$$

identifies with the subset of $\mathbb{P}(\text{Hom}_{\mathbb{K}}(V,W))$ whose elements represent equivalence classes of injective linear maps from V to W.

A homography $\varphi = \varphi_f$ is called *projectivity* if $f : V \to W$ is an isomorphism; in such a case, $\mathbb{P}(V)$ and $\mathbb{P}(W)$ are said to be *projectively isomorphic*. Note that homographies in $\mathbb{O}(\mathbb{P}(V),\mathbb{P}(V))$ are all projectivities of $\mathbb{P}(V)$ onto itself. In particular, $\mathbb{O}(\mathbb{P}(V),\mathbb{P}(V))$ forms a group with respect to the composition, which is denoted by $\text{PGL}(V)$. This group is the image of the linear group $\text{GL}(V)$ under the canonical quotient map $\pi_{\text{End}(V)}$ as in (3.1). If in particular $\mathbb{P}(V)$ and $\mathbb{P}(W)$ are projectively isomorphic, then $\text{GL}(V) \cong \text{GL}(W)$ and $\text{PGL}(V) \cong \text{PGL}(W)$.

If $V = \mathbb{K}^{n+1}$, then $\text{GL}(V)$ and $\text{PGL}(V)$ will be denoted by $\text{GL}(n+1,\mathbb{K})$ and $\text{PGL}(n+1,\mathbb{K})$, respectively; the first group identifies with the group of non-degenerate, $(n+1) \times (n+1)$ matrices with entries in $\mathbb{K}$ whereas the second one with the quotient of $\text{GL}(n+1,\mathbb{K})$ modulo its center, i.e. modulo the group of *scalar matrices* of the form tI_{n+1}, where $t \in \mathbb{K}^*$ and I_{n+1} the identity matrix of order $n+1$.

If V has dimension $n+1$, a projectivity $\varphi : \mathbb{P}^n \to \mathbb{P}(V)$ assigns to any point $P \in \mathbb{P}(V)$ a proportionality class $[p_0,\dots,p_n]$ of a numerical vector in $\mathbb{K}^{n+1}$; in other words, $\varphi : \mathbb{P}^n \to \mathbb{P}(V)$ can be viewed as introducing a *system of homogeneous coordinates* on $\mathbb{P}(V)$. In this correspondence, *fundamental points* in $\mathbb{P}(V)$ are the images via φ of the fundamental points in $\mathbb{P}^n$. In the system of homogeneous coordinates on $\mathbb{P}(V)$ induced by φ, the fact that P has homogeneous coordinates $[p_0,\dots,p_n]$ will be denoted by $P =_\varphi [p_0,\dots,p_n]$, or simply by $P = [p_0,\dots,p_n]$, if no confusion arises.

With the choice of two different systems of homogeneous coordinates $\varphi : \mathbb{P}^n \to \mathbb{P}(V)$ and $\psi : \mathbb{P}^n \to \mathbb{P}(V)$ on $\mathbb{P}(V)$, there exists a non-degenerate, $(n+1) \times (n+1)$ matrix A with entries in $\mathbb{K}$ such that, for any $P \in \mathbb{P}(V)$ with

$$P =_\varphi [\underline{x}] = [x_0,\dots,x_n] \quad \text{and} \quad P =_\psi [\underline{y}] = [y_0,\dots,y_n],$$

then $\underline{y}^t = A \cdot \underline{x}^t$, where A denotes any representative of the proportionality class $[A]$ of matrices determined by A. Note that $[A]$ determines the projectivity $\psi^{-1} \circ \varphi \in \text{PGL}(n+1,k)$.

Theorem 3.3.8 (Fundamental theorem of projectivities). *Let* $\mathbb{P}_1$
and $\mathbb{P}_2$ *be two projective spaces, both of dimension* n. *Consider* $(n+2)$-*tuples*
of points $(P_0, \ldots, P_{n+1})$ *and* $(Q_0, \ldots, Q_{n+1})$ *in* $\mathbb{P}_1$ *and* $\mathbb{P}_2$, *respectively,*
which are in general *position* (*i.e. any* $(n + 1)$-*tuple of such points*
are linearly independent points). *Then there exists a unique projectivity*
$\varphi : \mathbb{P}_1 \to \mathbb{P}_2$ *such that* $\varphi(P_i) = Q_i$, $0 \leqslant i \leqslant n + 2$.

Proof. Left to the reader as Exercise 3.1. $\qquad\square$

Corollary 3.3.9 (Fundamental theorem on systems of homogeneous coordinates). *Let* $\mathbb{P}(V)$ *be a projective space of dimension* n *and*
let $P_0, \ldots, P_{n+1} \in \mathbb{P}(V)$ *be points in general position. Then, there exists a*
unique system of homogeneous coordinates on $\mathbb{P}(V)$ *in such a way that the*
ordered $(n+2)$-*tuple* $(P_0, \ldots, P_{n+1})$ *corresponds to the ordered* $(n+2)$-*tuple*
formed by fundamental points and the unit point of $\mathbb{P}^n$.

Let $\mathbb{P}(V)$ be a projective space of dimension n. If $Z := \mathbb{P}(W)$ is a m-dimensional linear subspace of $\mathbb{P}(V)$, to explicitly construct a homography
$\varphi : \mathbb{P}^m \to Z$ one can proceed as follows. Consider any basis $\underline{w}_0, \ldots, \underline{w}_m$ of
W and then the linearly independent points $P_i = [\underline{w}_i] \in \mathbb{P}(V)$, $0 \leq i \leq m$.
One has $Z = P_0 \vee \ldots \vee P_m$ and one can construct the map

$$\varphi : [\lambda_0, \ldots, \lambda_m] \in \mathbb{P}^m \to [\lambda_0 \underline{w}_0 + \cdots + \lambda_m \underline{w}_m] \in Z$$

which is a projectivity sending the (natural) fundamental points of $\mathbb{P}^m$
to $P_0, \ldots, P_m$, respectively, and the unit point to $[\underline{w}_0 + \cdots + \underline{w}_m]$; the
homography $\varphi \in \mathbb{O}(\mathbb{P}^m, \mathbb{P}(V))$ as above is said to be a *homogeneous parametric representation* of Z. In particular, if Z_1, Z_2 are two m-dimensional
linear subspaces of $\mathbb{P}(V)$, the previous discussion shows there exists a
projectivity of $\mathbb{P}(V)$ into itself which sends Z_1 onto Z_2. This also implies
that, given any m-dimensional linear subspace Z of $\mathbb{P}(V)$, there always
exists a system of homogeneous coordinates on $\mathbb{P}(V)$ in such a way that
$Z = Z_p(X_{m+1}, \ldots, X_n)$.

Proposition 3.3.10. *Any homography* $\varphi : \mathbb{P}(V) \to \mathbb{P}(W)$ *is a continuous*
map in the Zariski topologies of the projective spaces $\mathbb{P}(V)$ *and* $\mathbb{P}(W)$,
respectively. If in particular φ *is a projectivity, then it is a homeomorphism.*

Proof. Let $f : V \to W$ be the injective linear map inducing φ; its
transpose linear map $f^t : W^* \to V^*$ extends to a degree-zero, surjective,

graded algebra homomorphism $f^t : S(W^*) \to S(V^*)$. For any $G \in H(S(W^*))$, one has $\varphi^{-1}(Z_p(G)) = Z_p(f^t(G))$, which implies that φ is continuous. $\square$

Proposition 3.3.11. *A homography $\varphi : \mathbb{P}(V) \to \mathbb{P}(W)$ is a homeomorphism of $\mathbb{P}(V)$ onto its image, which is a linear subspace of $\mathbb{P}(W)$ of the same dimension of $\mathbb{P}(V)$.*

Proof. The image of $\varphi = \varphi_f$ is the linear subspace $Z := \mathbb{P}(f(V)) \subseteq \mathbb{P}(W)$. From Proposition 3.3.10, it therefore suffices to show that φ is closed. Consider the surjective algebra homomorphism $f^t : S(W^*) \to S(V^*)$ introduced in the proof of Proposition 3.3.10. For any $F \in H(S(V^*))$, there exists $G \in H(S(W^*))$ such that $f^t(G) = F$. Since $\varphi^{-1}(Z_p(G)) = Z_p(f^t(G)) = Z_p(F)$ (cf. the proof of Proposition 3.3.10), one has $\varphi(Z_p(F)) = Z_p(G) \cap Z$, which implies that φ is closed. $\square$

Definition 3.3.12. Given APS's $X, Y \subseteq \mathbb{P}^n$, they are said to be *projectively equivalent* if X and Y are transformed into each other by a linear change of coordinates in the projective space $\mathbb{P}^n$. Namely, there exists a projectivity $\varphi \in \mathrm{PGL}(n+1, \mathbb{K})$, associated to a class $[A]$ of an invertible square-matrix of order $n+1$, which establishes an isomorphism between X and Y.

Remark 3.3.13. *Projective equivalence* obviously induces an equivalence relation among APS's. We will see in Chapter 6 that projective equivalence is a stronger relation than that induced by *isomorphism* of *APS's*. Indeed, in Section 4.1.1 we will introduce the notion of *homogeneous coordinate ring* for an APS and we will see that homogeneous coordinate rings depend on the embedding into the ambient projective space; therefore isomorphic APS's might have non-isomorphic homogeneous coordinate rings (cf. Example 6.5.11). This cannot occur for projectively equivalent APS's, i.e. projectively equivalent APS's have isomorphic homogeneous coordinate rings (cf. Remark 6.5.10).

A map $\psi : \mathbb{A}^n \to \mathbb{A}^m$, with $n \leqslant m$, is called an *affinity* if there exists a homography $\Psi : \mathbb{P}^n \to \mathbb{P}^m$ such that, identifying $\mathbb{A}^n$ with the affine chart U_0, the restriction of Ψ to U_0 coincides with ψ. In particular, any affinity is a homeomorphism onto its image and it sends affine subspaces of $\mathbb{A}^n$ to affine subspaces of $\mathbb{A}^m$ of the same dimension. From Section 3.3.7 and what discussed above, $\psi : \mathbb{A}^n \to \mathbb{A}^m$ is an affinity if and only if there exist a $(n \times m)$-matrix A of maximal rank n and a $(m \times 1)$-matrix $\underline{a}$ such that

$$\psi(\underline{x}^t) = A \cdot \underline{x}^t + \underline{a}, \tag{3.28}$$

where $\underline{x} = (x_1, \ldots, x_n)$. In particular, ψ is a homomorphism of $\mathbb{K}$-vector space if and only if $\psi(\underline{0}^t) = \underline{0}^t$.

Affinities of $\mathbb{A}^n$ onto itself form a group under composition, which is called the *affine group* of $\mathbb{A}^n$ and which is denoted by $\text{Aff}(\mathbb{A}^n)$. It is then clear that elements of $\text{Aff}(\mathbb{A}^n)$ transform affine subspaces to affine subspaces of the same dimension. In particular, if Z is any affine subspace of dimension m of $\mathbb{A}^n$, there always exists an affinity in $\text{Aff}(\mathbb{A}^n)$ which maps Z to the affine subspace $Z_a(x_{m+1}, \ldots, x_n) \subset \mathbb{A}^n$.

3.3.9 *Projective cones*

Let $Z \subset \mathbb{P}^n$ be a (Zariski) closed subset and let $C_a(Z) \subset \mathbb{A}^{n+1}$ be its affine cone. We can always identify $\mathbb{A}^{n+1}$ as an affine open subset of $\mathbb{P}^{n+1}$ in such a way that the given $\mathbb{P}^n$ coincides with the hyperplane at infinity of $\mathbb{A}^{n+1}$. Thus, we can consider the projective closure of $C_a(Z)$ in $\mathbb{P}^{n+1}$, which we will denote by $C_p(Z)$ and call *the projective cone over Z with vertex O*, where $O = [1, 0, \ldots, 0] \in U_0 = \mathbb{A}^{n+1} \subset \mathbb{P}^{n+1}$. From (3.17), we have that $C_p(Z) \cap \mathbb{P}^n = C_a(Z)_\infty = Z \subset \mathbb{P}^n$. Therefore,

$$C_p(Z) = C_a(Z) \cup C_a(Z)_\infty = C_a(Z) \cup Z,$$

so $C_p(Z)$ is the union of (projective) lines passing through O and through a point of Z. Finally one has

$$I_p(C_p(Z)) = I_p(Z), \tag{3.29}$$

where the latter ideal is considered as a homogeneous ideal in $\mathbb{S}^{(n+1)}$.

3.3.10 *Projective hypersurfaces and projective closure of affine hypersurfaces*

Similarly to Example 2.1.19, for any non-constant polynomial $F \in H(\mathbb{S}^{(n)})$ we can consider its decomposition $F = F_1^{r_1} F_2^{r_2} \cdots F_\ell^{r_\ell}$ as in (2.5), where $F_1, \ldots, F_\ell \in \mathbb{S}^{(n)}$ are all its non-proportional, irreducible factors (all homogeneous, from Proposition 1.10.15-(iii)) and where $r_1, \ldots, r_\ell$ are positive integers. Then $Y := Z_p(F)$ is called the *projective hypersurface* determined by F and the polynomial

$$F_{\text{red}} = F_1 F_2 \cdots F_\ell \tag{3.30}$$

is the *reduced equation of Y*. As in the affine case, $\deg(F_{\text{red}})$ is the *degree of Y* and $I_p(Y) = (F_{\text{red}})$ is its radical ideal.

One can consider *principal open (projective) sets $U_p(F)$* and show that they form a basis for $\mathfrak{3ar}_p^n$ (the proof is identical to that of Lemma 2.1.21). Moreover, using the *Homogeneous Hilbert "Nullstellensatz"-strong form* (cf. Theorem 3.2.5), one can easily adapt the proof of Theorem 2.3.1 to prove:

Theorem 3.3.14 (Homogeneous Study's principle). *Let $F, G \in H(\mathcal{S}^{(n)})$ be non-constant polynomials and let $Y = Z_p(F)$ be the projective hypersurface defined by F. If $Z_p(G) \supseteq Y$, then G is divided by F_{red} in $\mathcal{S}^{(n)}$.*

The previous result allows us to consider projective closures of affine hypersurfaces. Indeed one has:

Proposition 3.3.15. *Let $Y = Z_a(f) \subset \mathbb{A}^n$ be an affine hypersurface, for some non-constant polynomial $f \in A^{(n)}$. Identifying $\mathbb{A}^n$ with the affine chart U_0 of $\mathbb{P}^n$, one has $I_p(\overline{Y}^p) = (\mathfrak{h}_0(f_{\text{red}}))$, where $\overline{Y}^p$ the projective closure of Y and f_{red} the reduced equation of Y.*

Proof. Since $I_a(Y) = (f_{\text{red}})$ (cf. Remark 2.2.9-(d)), we can directly assume that $f = f_{\text{red}}$. From Proposition (3.3.5), $\overline{Y}^p = Z_p(\mathfrak{J}^*)$, where $\mathfrak{J}^* := (\mathfrak{h}_0(g) \mid g \in (f))$. Since any $g \in (f)$ is of the form $g = fh$, for some $h \in A^{(n)}$, and since $\mathfrak{h}_0$ is multiplicative (cf. Lemma 1.10.17-(ii)), then $\mathfrak{J}^* = (\mathfrak{h}_0(f))$, i.e. $(\mathfrak{h}_0(f)) \subseteq I_p(\overline{Y}^p)$. Note that if f factors as in (2.7), the fact that $\mathfrak{h}_0$ is multiplicative implies that $\mathfrak{h}_0(f) = \mathfrak{h}_0(f_1)\mathfrak{h}_0(f_2)\cdots\mathfrak{h}_0(f_\ell)$ where all the factors are square-free homogeneous polynomials, i.e. $\mathfrak{h}_0(f)$ is a *reduced* homogeneous polynomial.

On the other hand, for any homogeneous polynomial $G \in I_p(\overline{Y}_p)$, one has $Z_p(G) \supseteq \overline{Y}^p$. Thus, from the homogeneous Study's principle, we have $G \in (\mathfrak{h}_0(f))$, i.e. $I_p(\overline{Y}^p) \subseteq (\mathfrak{h}_0(f))$ and we are done. □

3.3.11　*Proper closed subsets of $\mathbb{P}^2$*

As in Corollary 2.3.4, we can classify all proper closed subsets of the projective plane. In what follows, the term *curve* is always meant to be any projective hypersurface in $\mathbb{P}^2$.

Proposition 3.3.16. *Proper closed subsets in $\mathfrak{3ar}_p^2$ consist only of finitely many points, curves and unions of a curve and finitely many points.*

Proof. If $Z = \emptyset$, there is nothing to prove. Therefore, let $Z \subset \mathbb{P}^2$ be any proper, non-empty closed subset. We can proceed as follows. One has $Z = Z_1 \cup Z_2$, where $Z_1 := Z \cap U_0$ is a AAS in $U_0 \cong \mathbb{A}^2$ whereas $Z_2 := Z \cap H_0 = (Z_1)_\infty$, with $H_0 = Z_p(X_0)$ the line at infinity of U_0. Since $Z_2 \subseteq H_0$ and since H_0 is homeomorphic to $\mathbb{P}^1$, Z_2 is either empty, or a finite number of points or it is the whole line H_0. As for Z_1, since it is closed in $U_0 \cong \mathbb{A}^2$, we can apply Corollary 2.3.4. Then $Z_1 = Z_1' \cup Z_2'$, where Z_1' either consists of finitely many points or it is empty and Z_2' either consists of finitely many affine curves or it is empty. Then one concludes by using Sections 3.3.1, 3.3.7 and Proposition 3.3.15. $\square$

3.3.12 *Affine and projective twisted cubics*

Let $n \geqslant 2$ be an integer, t an indeterminate over $\mathbb{K}$ and $h_2(t), \ldots, h_n(t) \in \mathbb{K}[t]$, not all of them constant polynomials. Consider

$$C := \{(t, h_2(t), \ldots, h_n(t)) \in \mathbb{A}^n \mid t \in \mathbb{K}\} \subset \mathbb{A}^n.$$

It is clear that C is the AAS defined by

$$C = Z_a(x_2 - h_2(x_1), \ldots, x_n - h_n(x_1)). \tag{3.31}$$

Lemma 3.3.17. *One has $I_a(C) = \big(x_2 - h_2(x_1), \ldots, x_n - h_n(x_1)\big)$, which is a prime ideal in $A^{(n)}$.*

Proof. Consider the $\mathbb{K}$-algebra homomorphism

$$\rho_n : A^{(n)} \to \mathbb{K}[t] \quad \text{defined by} \quad \rho_n(x_1) = t, \ \rho_n(x_i) = h_i(t), \ 2 \leqslant i \leqslant n,$$

which is obviously surjective, so $\mathrm{Ker}(\rho_n)$ is a prime ideal. Now, $g \in \mathrm{Ker}(\rho_n)$ if and only if

$$0 = \rho_n(g(x_1, \ldots, x_n)) = g(t, h_2(t), \ldots, h_n(t)), \quad \forall t \in \mathbb{K},$$

i.e. if and only if $g \in I_a(C)$, which means $I_a(C) = \mathrm{Ker}(\rho_n)$. The ideal $J := \big(x_2 - h_2(x_1), \ldots, x_n - h_n(x_1)\big)$ is contained in $\mathrm{Ker}(\rho_n)$ and is such that $A^{(n)}/J \cong \mathbb{K}[x_1] \cong \mathbb{K}[t]$. This implies that $J = \mathrm{Ker}(\rho_n)$, as desired. $\square$

Any curve C as above is called an *affine rational curve with polynomial parametrization*. In what follows, we will study a particular case of the previous set-up.

Consider the injective map

$$\phi : t \in \mathbb{A}^1 \to (t, t^2, t^3) \in \mathbb{A}^3. \tag{3.32}$$

From (3.31), $C := \mathrm{Im}(\phi)$ is a rational curve with polynomial parametrization, which is called *(standard) affine twisted cubic*. The image of C under any affinity of $\mathbb{A}^3$ will be called *affine twisted cubic*. From Lemma 3.3.17, the ideal $I_a(C) = (f_1, g_1)$, where

$$f_1 := x_1^2 - x_2, \quad g_1 := x_1^3 - x_3,$$

is prime so radical (cf. Lemma 1.1.2). By straightforward computations, if one takes $f_2 := g_1 - x_1 f_1$, one has

$$I_a(C) = (f_1, f_2) := (x_2 - x_1^2, x_3 - x_1 x_2). \tag{3.33}$$

Consider now the map

$$\psi : [\lambda, \mu] \in \mathbb{P}^1 \to [\lambda^3, \lambda^2 \mu, \lambda \mu^2, \mu^3] \in \mathbb{P}^3 \tag{3.34}$$

which is well-defined, injective and whose image Z is called the *(standard) projective twisted cubic*. Note that ϕ is the restriction of ψ to $\mathbb{A}^1 \cong U_0 \subset \mathbb{P}^1$, thus $C \subset Z$. More precisely

$$Z = C \cup \{P\}$$

where $P = [0, 0, 0, 1] = \psi([0, 1])$. The set Z is an APS in $\mathbb{P}^3$; indeed, $Z = Z_p(F_1, F_2, F_3)$, where

$$F_1 := X_0 X_2 - X_1^2, \quad F_2 := X_0 X_3 - X_1 X_2, \quad F_3 := X_1 X_3 - X_2^2, \tag{3.35}$$

are three irreducible, homogeneous polynomials which are linearly independent in $\mathcal{S}_2^{(3)}$. These polynomials are determined by the maximal minors of the matrix of linear forms

$$A := \begin{pmatrix} X_0 & X_1 & X_2 \\ X_1 & X_2 & X_3 \end{pmatrix}. \tag{3.36}$$

For this reason Z is said to be *determinantal* and the condition $\mathrm{rank}(A) = 1$ is said to be a *matrix equation* of Z (cf. also Harris, 1995, Example 1.10, p. 9).

Claim 3.3.18. *One has* $\overline{C}^p = Z$, *where* $\overline{C}^p$ *the projective closure of* C *in* $\mathbb{P}^3$.

Even if this is a consequence of a more general fact (cf. Remark 9.2.2-(iii)), we give here a constructive proof.

Proof. The inclusion $\overline{C}^p \subseteq Z$ is trivial. On the other hand, if we consider any homogeneous $G \in I_p(\overline{C}^p)$ and if we identify $\mathbb{A}^3$ with $U_0 \subset \mathbb{P}^3$, one has $\delta_0(G) \in I_a(C)$, i.e. $G(1, t, t^2, t^3) = 0$ for any $t \in \mathbb{K}$. Thus, $G(\lambda^3, \lambda^2\mu, \lambda\mu^2, \mu^3) = 0$, for any $\mu \in \mathbb{K}$ and for any $\lambda \in \mathbb{K}^*$, which implies therefore that $G(\lambda^3, \lambda^2\mu, \lambda\mu^2, \mu^3)$ is identically zero, i.e. $G \in I_p(Z)$. □

From (3.33) and (3.35), we note that

$$\delta_0(F_1) = f_1, \ \delta_0(F_2) = f_2, \ \delta_0(F_3) = f_3 := x_1 f_2 - x_2 f_1 \in I_a(C).$$

and viceversa

$$\mathfrak{h}_0(f_1) = F_1, \ \mathfrak{h}_0(f_2) = F_2, \ \mathfrak{h}_0(f_3) = F_3.$$

Even if $I_a(C)$ is generated by f_1 and f_2, the polynomials F_1 and F_2 do not generate $I_p(Z)$. More precisely, they do not even define Z as an APS, since

$$Z_p(F_1, F_2) = Z_p(\mathfrak{h}_0(f_1), \mathfrak{h}_0(f_2)) = Z \cup Z_p(X_0, X_1), \tag{3.37}$$

(cf. also Eisenbud, 1995, § 21.10, pp. 539–540).

Claim 3.3.19. $I_p(\overline{C}^p) = (F_1, F_2, F_3)$.

Proof. To prove the claim, we use the following notation:

$$B_1 := F_1, \ B_2 := -F_2, \ B_3 := F_3,$$

i.e. we consider the maximal minors of the matrix A in (3.36) with a suitable sign. Let us consider the free, graded $\mathcal{S}^{(3)}$-module

$$\mathcal{M} := \left(\mathcal{S}^{(3)}(-2)\right)^{\oplus 3},$$

whose homogeneous degree-d part $\mathcal{M}_d$ is the $\mathbb{K}$-vector space

$$\left(\mathcal{S}^{(3)}(-2)^{\oplus 3}\right)_d = (\mathcal{S}^{(3)}_{d-2})^{\oplus 3}$$

(cf. Section 1.10.2). One is therefore reduced to proving that the homogeneous, degree-0 homomorphism

$$(G_1, G_2, G_3) \in \mathcal{M} \xrightarrow{\phi} G_1 B_1 + G_2 B_2 + G_3 B_3 \in I_p(\overline{C}^p)$$

is surjective. This is equivalent to showing that the induced $\mathbb{K}$-vector space homomorphism

$$(G_1, G_2, G_3) \in \mathcal{M}_d \xrightarrow{\phi_d} G_1 B_1 + G_2 B_2 + G_3 B_3 \in I_p(\overline{C}^p)_d$$

is surjective for any integer d. Let $\mathcal{K} := \mathrm{Ker}(\phi)$, which is still a graded $\mathcal{S}^{(3)}$-module. Its homogeneous, degree-d piece $\mathcal{K}_d$ consists of triples $(G_1, G_2, G_3) \in (\mathcal{S}^{(3)}_{d-2})^{\oplus 3}$ such that $\Sigma_{i=1}^3 G_i B_i = 0$, i.e. $\mathcal{K}_d = \mathrm{Ker}(\phi_d)$. These triples are called *syzygies* of (F_1, F_2, F_3) and $\mathcal{K}$ is said to be the *syzygy module*.

One trivially has $\mathcal{K}_d = \{0\}$ if $d < 2$; moreover $\mathcal{K}_2 = \{0\}$ since, as observed above, the polynomials F_1, F_2 and F_3 are linearly independent in $\mathcal{S}^{(3)}_2$. On the other hand,

$$\underline{a}_1 := (X_0, X_1, X_2), \quad \underline{a}_2 := (X_1, X_2, X_3)$$

are two linearly independent elements in $\mathcal{K}_3$, as it clearly follows from the matrix A in (3.36). Consider the free, graded $\mathcal{S}^{(3)}$-module $\mathcal{N} := \left(\mathcal{S}^{(3)}(-3)\right)^{\oplus 2}$ and the degree-0, homogeneous homomorphism

$$(D_1, D_2) \in \mathcal{N} \xrightarrow{\psi} D_1 \underline{a}_1 + D_2 \underline{a}_2 \in \mathcal{K}.$$

We first show that ψ is an isomorphism of $\mathcal{S}^{(3)}$- graded modules. To prove this it suffices to show that, for any positive integer d, the induced $\mathbb{K}$-vector space homomorphism

$$(D_1, D_2) \in (\mathcal{S}^{(3)}_{d-3})^{\oplus 2} \xrightarrow{\phi_d} D_1 \underline{a}_1 + D_2 \underline{a}_2 \in \mathcal{K}_d \subseteq (\mathcal{S}^{(3)}_{d-2})^{\oplus 3}$$

is an isomorphism. To prove this, consider any $(G_1, G_2, G_3) \in \mathcal{K}_d$. By Laplace's rule and the definition of B_i, $1 \leqslant i \leqslant 3$, this is equivalent to

$$\det \begin{pmatrix} G_3 & G_2 & G_1 \\ X_0 & X_1 & X_2 \\ X_1 & X_2 & X_3 \end{pmatrix} = 0, \tag{3.38}$$

which implies that the three rows in (3.38) are linearly dependent over the field $\mathbb{K}(X_0, X_1, X_2, X_3)$, whereas the rows of A are linearly independent, since the polynomials $F_1 = B_1$, $F_2 = -B_2$ and $F_3 = B_3$ are non-zero. There exist therefore rational functions

$$\frac{a_1}{a_0}, \quad \frac{b_1}{b_0} \in \mathbb{K}(X_0, X_1, X_2, X_3),$$

with a_0, a_1 (respectively, b_0, b_1) relatively prime elements, such that

$$\begin{aligned} G_1 &= \frac{a_1}{a_0} X_2 + \frac{b_1}{b_0} X_3 = \frac{a_1 b_0 X_2 + a_0 b_1 X_3}{a_0 b_0}, \\ G_2 &= \frac{a_1}{a_0} X_1 + \frac{b_1}{b_0} X_2 = \frac{a_1 b_0 X_1 + a_0 b_1 X_2}{a_0 b_0}, \\ G_3 &= \frac{a_1}{a_0} X_0 + \frac{b_1}{b_0} X_1 = \frac{a_1 b_0 X_0 + a_0 b_1 X_1}{a_0 b_0}. \end{aligned} \tag{3.39}$$

Since the G_i's are polynomials in $\mathbb{K}[X_0, X_1, X_2, X_3]$, if p is any prime factor of a_0, then it must divide $b_0 X_0, b_0 X_1, b_0 X_2$, i.e. p must divide b_0. Recursively applying the same argument and changing the roles between a_0 and b_0, one can assume $a_0 = b_0$, so that (3.39) become

$$G_1 = \frac{a_1 X_2 + b_1 X_3}{b_0}, \quad G_2 = \frac{a_1 X_1 + b_1 X_2}{b_0}, \quad G_3 = \frac{a_1 X_0 + b_1 X_1}{b_0}.$$

Thus, b_0 divides the polynomials

$$\alpha_1 := a_1 X_2 + b_1 X_3, \quad \alpha_2 := a_1 X_1 + b_1 X_2, \quad \alpha_3 := a_1 X_0 + b_1 X_1,$$

which implies that b_0 divides therefore the polynomials

$$X_1 \alpha_3 - X_0 \alpha_2 = -b_1 B_1, \quad X_2 \alpha_3 - X_0 \alpha_1 = b_1 B_2, \quad X_2 \alpha_2 - X_1 \alpha_1 = -b_1 B_3.$$

Since B_1, B_2, B_3 are all irreducible and distinct, this implies that b_0 divides b_1. Since, by assumption, b_0 and b_1 are relatively prime elements, one concludes that $b_0 \in \mathbb{K}^*$. Therefore, we can assume $a_0 = b_0 = 1$.

For simplicity of notation, let $a := a_1$, $b := b_1$. We want to show that $(a, b) \in (S_{d-3}^{(3)})^{\oplus 2}$. Indeed, if $j \neq d - 3$ and if A_j, B_j are the homogeneous, degree-j components of a, b, from (3.39) one gets

$$A_j X_2 + B_j X_3 = 0, \; A_j X_1 + B_j X_2 = 0, \; A_j X_0 + B_j X_1 = 0$$

so, reasoning as above, one deduces $A_j = B_j = 0$, for any integer $j \neq d - 3$. This proves that ψ is bijective.

In particular, if $\mathfrak{J} := (F_1, F_2, F_3)$, we proved the existence of an *exact sequence* of $S^{(3)}$-graded modules

$$0 \to \mathcal{N} \xrightarrow{\psi} \mathcal{M} \xrightarrow{\phi} \mathfrak{J} \to 0,$$

where $\mathcal{M}$ and $\mathcal{N}$ are free $S^{(3)}$-graded modules. Namely the maps ψ and ϕ are degree-0, homogeneous homomorphisms such that ψ is injective, ϕ is surjective and $\mathrm{Im}(\psi) = \mathrm{Ker}(\phi)$. Such an exact sequence is called a *homogeneous free resolution* of the homogeneous ideal $\mathfrak{J}$.

From Lemma 1.10.12 and from the exactness of the sequence, for any integer d one has

$$\dim(\mathfrak{J}_d) = 3\binom{d+1}{3} - 2\binom{d}{2} = \binom{d+3}{3} - (3d+1).$$

We now compute the dimension of $I_p(\overline{C}^p)_d$, for any integer $d \geqslant 0$ (for $d < 0$ this dimension is obviously 0). Consider the homogeneous substitution of indeterminates of degree 3

$$\nu : f(X_0, X_1, X_2, X_3) \in \mathcal{S}^{(3)} \rightarrow f(\lambda^3, \lambda^2\mu, \lambda\mu^2, \mu^3) \in \mathcal{S}^{(1)}$$

which induces surjective $\mathbb{K}$-vector space homomorphisms

$$\nu_d : \mathcal{S}_d^{(3)} \rightarrow \mathcal{S}_{3d}^{(1)},$$

for any non-negative integer d. One has $\mathrm{Ker}(\nu_d) = I_p(\overline{C}^p)_d$, for any $d \geqslant 0$, so

$$\dim \left(I_p(\overline{C}^p)_d \right) = \binom{d+3}{3} - (3d+1) = \dim(\mathfrak{I}_d).$$

This proves that $\mathfrak{I} = I_p(\overline{C}^p)$ as desired. $\qquad\qquad\square$

Finally, one more generally sets: a *projective twisted cubic* is any closed subset $Z \subset \mathbb{P}^3$ for which there exists a projectivity Φ of $\mathbb{P}^3$ such that $\Phi(Z)$ coincides with the image of (3.34). Thus, Z is a projective twisted cubic if and only if there exists a basis $F_i(\lambda, \mu)$ of $\mathcal{S}_3^{(1)}$, $0 \leqslant i \leqslant 3$, such that Z is the image of the following map:

$$\psi : [\lambda, \mu] \in \mathbb{P}^1 \rightarrow [F_0(\lambda, \mu), F_1(\lambda, \mu), F_2(\lambda, \mu), F_3(\lambda, \mu)] \in \mathbb{P}^3$$

which is called a *parametric representation* of the projective twisted cubic Z.

Exercises

Exercise 3.1. Prove Theorem 3.3.8.

Exercise 3.2. Prove Proposition 3.3.16 using resultants and Elimination Theory of homogeneous polynomials.

Exercise 3.3. Let $m < n$ be positive integers. Let V be a $\mathbb{K}$-vector space of dimension $n + 1$ and let $W \subset V$ be a vector subspace of V of dimension $m + 1$. Let $V^* = \mathrm{Hom}_{\mathbb{K}}(V, \mathbb{K})$ be the dual vector space of V. Consider

$$\mathrm{Ann}(W) := \{f \in V^* \mid f(\underline{w}) = 0, \ \forall \ \underline{w} \in W\} \subset V^*.$$

(i) Prove that $\mathrm{Ann}(W)$ is a vector subspace of V^* and compute its dimension.

(ii) Let $\Lambda := \mathbb{P}(W)$ be the linear subspace of $\mathbb{P}(V)$ associated to W. Write $\Lambda^\perp$ in terms of $\mathrm{Ann}(W)$.

Exercise 3.4.

(i) Let $Z := Z_p(X_1^2 + X_2^2 - X_0 X_1) \subset \mathbb{P}_{\mathbb{R}}^2$ be a projective conic. Give affine classification of the affine conics which are the traces cut-out by Z in the three different affine charts U_0, U_1 and U_2 of $\mathbb{P}_{\mathbb{R}}^2$.

(ii) Determine the affine (respectively, projective) cone $C_a(Z) \subset \mathbb{A}_{\mathbb{R}}^3$ (respectively, $C_p(Z) \subset \mathbb{P}_{\mathbb{R}}^3$) and $C_a(Z)_\infty$ when $\mathbb{A}_{\mathbb{R}}^3$ is identified with the affine chart U_3 of $\mathbb{P}_{\mathbb{R}}^3$.

Exercise 3.5. Let $X = Z_p(F_1, \ldots, F_m) \subset \mathbb{P}^n$ an APS, where $F_j \in \mathcal{S}_{d_j}^{(n)}$, d_j positive integers for any $1 \leqslant j \leqslant m$. Prove that Z can be written as a zero-locus of a set of homogeneous polynomials of the same degree $d > 0$.

Chapter 4

Topological Properties and Algebraic Varieties

4.1 Irreducible Topological Spaces

Let $(Y, \mathcal{T}_Y)$ be any topological space, where $\mathcal{T}_Y$ denotes a given topology on Y. Then Y is said to be *irreducible* if Y cannot be expressed as $Y = Y_1 \cup Y_2$, where Y_1, Y_2 are proper closed subsets of Y. Equivalently, Y is irreducible if and only if any two non-empty, open subsets $U_1, U_2 \in \mathcal{T}_Y$ are such that $U_1 \cap U_2 \neq \emptyset$. When Y is not irreducible, it is called *reducible*.

Remark 4.1.1.

(i) If Y contains more than one point and it is irreducible, then it cannot be Hausdorff (i.e. $\mathcal{T}_Y$ is not T_2).

(ii) If $\mathbb{K} = \mathbb{R}$ or $\mathbb{C}$, $\mathbb{A}_{\mathbb{K}}^n$ endowed with the euclidean topology is reducible for any $n \geqslant 1$.

(iii) Irreducibility is a *topological property*, i.e. it is invariant under homeomorphisms of topological spaces.

(iv) If Y is irreducible, then it is connected. The converse is not true in general (cf. Example 4.1.5-(iii)).

In what follows, any subset $W \subseteq Y$ will be considered as endowed with the induced topology by that of Y.

Proposition 4.1.2. *Let Y be a topological space and let $W \subseteq Y$ be any subset.*

(i) W *is irreducible if and only if for any pair of distinct points* $P_1, P_2 \in W$ *there exists an irreducible subset* $Z \subseteq W$ *such that* $P_1, P_2 \in Z$;

(ii) W *is irreducible if and only if any non-empty, open subset* $U \subseteq W$ *is dense, i.e.* $\overline{U} = W$ *where* $\overline{U}$ *denotes the closure of* U *in* W;

(iii) W *is irreducible if and only if* $\overline{W}$ *is irreducible, where* $\overline{W}$ *denotes the closure of* W *in* Y;

(iv) W *is irreducible if and only if any non-empty, open subset* $U \subseteq W$ *is irreducible.*

Proof.

(i) The implication ($\Rightarrow$) is trivial. For ($\Leftarrow$), suppose by contradiction that W is reducible and let $W = W_1 \cup W_2$, where W_i proper closed subsets of W, $1 \leqslant i \leqslant 2$. Let $P_i \in W_i \setminus W_{3-i}$, $1 \leqslant i \leqslant 2$, and let Z be any irreducible subset $Z \subseteq W$ s.t. $P_1, P_2 \in Z$. Then $Z = Z_1 \cup Z_2$, where $Z_i = W_i \cap Z$, $1 \leqslant i \leqslant 2$, and Z_1, Z_2 irreducible, proper closed subsets of Z, since $P_i \notin Z_{3-i}$. This contradicts the irreducibility of Z.

(ii) Let us prove the non-trivial implication of (ii). Let W be irreducible and let U be any non-empty open subset of W. If U were not dense in W, then one would have $W = (\overline{U} \cap W) \cup (W \setminus U)$, where $\overline{U} \cap W$ and $W \setminus U$ are both proper closed subsets of W; this is against the irreducibility assumption on W.

(iii) Let Z be any closed subset of Y and denote by $Z' := Z \cap W$ and $Z'' := Z \cap \overline{W}$, where $\overline{W}$ denotes the closure of W in Y. One has

$$Z' = W \Leftrightarrow W \subseteq Z \Leftrightarrow \overline{W} \subseteq Z \Leftrightarrow Z'' = \overline{W}.$$

In other words, Z' is a proper closed subset of W if and only if Z'' is a proper closed subset of $\overline{W}$. From (ii) we know that W is irreducible if and only if any of its proper closed subsets have empty interior set. Thus, if the interior of Z'' is empty, the same occurs for the interior of Z', so that if $\overline{W}$ is irreducible then W is. Viceversa, if W is irreducible, let $\overline{W} = W_1 \cup W_2$, where W_i closed subsets of $\overline{W}$, $1 \leqslant i \leqslant 2$. The irreducibility of W implies that $W \subseteq W_i$, for either $i = 1$ or $i = 2$. Consequently, $\overline{W} \subseteq W_i$, i.e. $\overline{W} = W_i$. Thus, $\overline{W}$ is irreducible.

(iv) The implication ($\Rightarrow$) is a direct consequence of (ii) and (iii), whereas ($\Leftarrow$) is trivial. $\square$

Corollary 4.1.3. *For any integer* $n \geqslant 1$:

(i) $(\mathbb{P}^n, \mathfrak{Zar}_p^n)$,

(ii) *any non-empty, open subset* $U \subseteq \mathbb{P}^n$,

(iii) $(\mathbb{A}^n, \mathfrak{Zar}_a^n)$,

(iv) *any non-empty, open subset* $U \subseteq \mathbb{A}^n$,

are irreducible.

Proof. From Proposition 4.1.2-(iv), it suffices to prove (i), which is a direct consequence of Proposition 4.1.2-(i). Indeed, for any pair of distinct points $P_1, P_2 \in \mathbb{P}^n$, the line $P_1 \vee P_2$ is irreducible, since homeomorphic to $\mathbb{P}^1$ (recall (3.24) and Remark 4.1.1-(iii)). □

Remark 4.1.4. From Remark 4.1.1-(i) and Corollary 4.1.3, it follows that for any integer $n \geqslant 1$ topologies $\mathfrak{Zar}_a^n$ and $\mathfrak{Zar}_p^n$ are not T_2, i.e. Hausdorff (cf. Remark 2.1.14-(i) and Section 3.3.1 for $n = 1$). Note further that irreducibility in Corollary 4.1.3 holds true since $\mathbb{K}$ is an infinite field; if otherwise $\mathbb{K} = \mathbb{Z}_p$, for $p \in \mathbb{Z}$ a prime, then all the topological spaces listed in Corollary 4.1.3 are reducible and both $\mathfrak{Zar}_a^n, \mathfrak{Zar}_p^n$ are the discrete topology.

Example 4.1.5.

(i) Any affine subspace of $\mathbb{A}^n$ is irreducible, since homeomorphic to $\mathbb{A}^m$ for some $m \leqslant n$. Same conclusion holds for any projective subspace of $\mathbb{P}^n$.

(ii) Any affine twisted cubic as in Section 3.3.12 is irreducible, since it is homeomorphic to $\mathbb{A}^1$. Therefore, any projective twisted cubic is irreducible too.

(iii) The affine hypersurface $Y = Z_a(x_1 x_2) \subset \mathbb{A}^2$ is reducible. Indeed, $I_a(Y) = (x_1 \, x_2) = (x_1) \cap (x_2)$ so $Y = Z_a(x_1) \cup Z_a(x_2) = Y_1 \cup Y_2$, where $Y_1, Y_2 \subset Y$ are proper closed subsets of Y. In particular, Y is connected but not irreducible. Recalling Proposition 4.1.2, note moreover that: (a) there exist pairs of distinct points in Y not contained in any irreducible subset of Y; (b) $U := Z_a(x_1)^c \cap Y$ is an open subset of Y whose closure in Y is $\overline{U} = Z_a(x_2) = Y_2$, so not dense in Y; (c) $U_1 := Z_a(x_1, x_2)^c \cap Y$ is a non-connected open subset of Y which is reducible.

Corollary 4.1.6. *Let* Y, W *be topological spaces such that* Y *is irreducible. Assume there exists a continuous map* $f : Y \to W$. *Then* $\mathrm{Im}(f) \subseteq W$ *is irreducible.*

Proof. Since $\mathrm{Im}(f) \subseteq W$ is endowed with the induced topology of W, it suffices to showing that, when f is assumed to be also surjective, then W is irreducible. To prove this, if $W = W_1 \cup W_2$, where W_1, W_2 closed subsets of W, then $Y = f^{-1}(W_1) \cup f^{-1}(W_2)$. Since Y is irreducible and f is continuous, one must have $Y = f^{-1}(W_i)$, for some $i \in \{1,2\}$. Consequently, by the surjectivity of f, $W = f(Y) = f(f^{-1}(W_i)) = W_i$. $\square$

4.1.1   Coordinate rings, ideals and irreducibility

Here, we focus on the case of Zariski topology.

Definition 4.1.7. For any subset $Y \subseteq \mathbb{A}^n$, consider the ideal $I_a(Y) \subseteq A^{(n)}$. The $\mathbb{K}$-algebra of finite type (cf. Definition 1.5.3)

$$A(Y) := \frac{A^{(n)}}{I_a(Y)} \tag{4.1}$$

is called the (*affine*) *coordinate ring of* Y. Similarly, for any subset $Y \subseteq \mathbb{P}^n$, with homogeneous ideal $I_p(Y) \subseteq \mathcal{S}^{(n)}$, the graded $\mathbb{K}$-algebra of finite type (recall Proposition 1.10.9-(ii))

$$\mathcal{S}(Y) := \frac{\mathcal{S}^{(n)}}{I_p(Y)} \tag{4.2}$$

is called the *homogeneous coordinate ring of* Y.

Proposition 4.1.8. *Let $Y \subseteq \mathbb{P}^n$ be any irreducible subset and let $U \subseteq Y$ be any non-empty open subset of Y, then $I_p(Y) = I_p(U)$. The same holds replacing above $\mathbb{P}^n$ with $\mathbb{A}^n$ and $I_p(-)$ with $I_a(-)$.*

Proof. We prove the first part, the proof in the affine case being identical. Let $\overline{Y}^p$ be the projective closure of Y in $\mathbb{P}^n$; then (3.21) gives $I_p(Y) = I_p(\overline{Y}^p)$. Since Y is irreducible, from Proposition 4.1.2-(ii), we have $\widetilde{U} = Y$, where $\widetilde{U}$ denotes the closure of U in Y. Thus $\overline{U}^p = \widetilde{\overline{U}}^p = \overline{Y}^p$ and, from (3.21), we get $I_p(U) = I_p(\overline{Y}_p) = I_p(Y)$. $\square$

Corollary 4.1.9. *For any subset $Y \subseteq \mathbb{A}^n$, let $\overline{Y}^a$ denote its closure in $\mathbb{A}^n$. Then $A(Y) = A(\overline{Y}^a)$. Similarly, for any subset $Y \subseteq \mathbb{P}^n$, let $\overline{Y}^p$ denote its projective closure. Then $\mathcal{S}(Y) = \mathcal{S}(\overline{Y}^p)$.*

When it is clear from the context and no confusion arises, for brevity we will sometimes use the symbol $I(Y)$ instead of $I_a(Y)$ or $I_p(Y)$. The rings $A(Y)$ and $\mathcal{S}(Y)$ are always reduced rings (cf. Definition 1.1.1) since $I(Y)$ is always radical. On the other hand, in some cases these rings are not integral domains. For example, if $I_p(Y) = (X_0 \, X_1) \subset \mathcal{S}^{(2)}$, then $\mathcal{S}(Y)$ contains X_0 and X_1 as zero-divisors. Indeed, $Z_p(I_p(Y)) = Y = Y_0 \cup Y_1$, where $Y_i = Z_p(X_i)$, $0 \leqslant i \leqslant 1$; namely Y is a *reducible conic* which is the union of two fundamental lines of $\mathbb{P}^2$.

The following result gives an algebraic criterion for irreducibility.

Proposition 4.1.10. *A subset* $Y \subseteq \mathbb{A}^n$ *(respectively,* $Y \subseteq \mathbb{P}^n$*) is irreducible if and only if* $I(Y)$ *is a prime ideal, i.e. if and only if* $A(Y)$ *(respectively,* $\mathcal{S}(Y)$*) is an integral domain.*

Proof. The second part of the statement is obvious by definition of $A(Y)$ (respectively, $\mathcal{S}(Y)$); therefore we need to prove the first equivalence. From Corollary 4.1.9, it suffices to consider the case with Y closed in the Zariski topology. Assume therefore Y to be irreducible. If $f g \in I(Y)$ then $(fg) \subseteq I(Y)$ so $Y = Z(I(Y)) \subseteq Z(fg) = Z(f) \cup Z(g)$ (cf. Propositions 2.1.11 and 3.1.5). Thus, either $Y \subseteq Z(f)$ or $Y \subseteq Z(g)$. This means that either $f \in I(Y)$ or $g \in I(Y)$, i.e. $I(Y)$ is prime.

Conversely let $I(Y)$ be a prime ideal and let $Y = Y_1 \cup Y_2$, where Y_1, Y_2 closed subsets of Y. Then $I(Y) = I(Y_1) \cap I(Y_2)$ (cf. Propositions 2.2.8 and 3.1.7). $\qquad\square$

Claim 4.1.11. *In the above assumptions,* $I(Y)$ *is either* $I(Y_1)$ *or* $I(Y_2)$.

Proof. If it were $I(Y) \subsetneq I(Y_i)$ for both $i = 1, 2$, there would exist $f_i \in I(Y_i) \setminus I(Y)$, $1 \leqslant i \leqslant 2$, such that

$$f_1 f_2 \in I(Y_1) \cdot I(Y_2) \subseteq I(Y_1) \cap I(Y_2) = I(Y),$$

contradicting that $I(Y)$ is a prime ideal (cf. Exercise 1.1).

From Claim 4.1.11, we have either $Y = Y_1$ or $Y = Y_2$ so Y is irreducible. $\qquad\square$

Recalling bijective correspondences in Corollaries 2.2.5 and 3.2.6, we have:

Corollary 4.1.12. *The maps* (2.11) *and* (2.12) *induce bijections:*

$$\{\textit{irreducible closed subsets of } \mathbb{A}^n\} \overset{1-1}{\longleftrightarrow} \{\textit{prime ideals of } A^{(n)}\}.$$

Similarly, the maps (3.5) *and* (3.6) *induce bijections:*

$$\{\textit{irreducible closed subsets of } \mathbb{P}^n\}$$
$$\overset{1-1}{\longleftrightarrow} \{\textit{homogeneous prime ideals of } \mathcal{S}^{(n)}\} \setminus \{\mathcal{S}_+\}.$$

Example 4.1.13.

(i) We find in this way an alternative proof of Corollary 4.1.3, via purely algebraic approach; indeed one has $A(\mathbb{A}^n) = A^{(n)}$ and $\mathcal{S}(\mathbb{P}^n) = \mathcal{S}^{(n)}$

which are integral domains, so one concludes by Proposition 4.1.10. For the open subsets, one uses Corollary 4.1.9.

(ii) For any subset $Z \subset \mathbb{P}^n$ we have that

$$Z \text{ irreducible } \Leftrightarrow C_a(Z) \text{ irreducible } \Leftrightarrow C_p(Z) \text{ irreducible.} \qquad (4.3)$$

Proof. This follows from (3.8), (3.29) and what discussed above. $\square$

(iii) If $Z \subset \mathbb{P}^n$ is any hypersurface, then Z is irreducible if and only if its reduced equation $F \in \mathcal{S}^{(n)}$ is an irreducible polynomial, as it follows from (3.30) and Claim 4.1.11. Same occurs in the affine case (cf. Example 2.1.19).

(iv) Any affine rational curve with polynomial representation as in (3.31) is irreducible, as it follows from Lemma 3.3.17.

(v) Recalling (3.37), the closed subset $K := Z_p(F_1, F_2)$ is reducible consisting of the (standard) projective twisted cubic together with the line at infinity of the affine plane $Z_a(x_1) \subset \mathbb{A}^3$. From Proposition 4.1.10, $\mathcal{I} = (F_1, F_2)$ cannot be a prime ideal so $\mathcal{S}(K)$ is not an integral domain; indeed, if we denote by $\overline{F_3}$, $\overline{X_0} \in \mathcal{S}(K)$ the images of F_3, $X_0 \in \mathcal{S}^{(3)}$ in $\mathcal{S}(K)$, respectively, since $I_p(K) = (F_1, F_2)$, then $\overline{F_3}$, $\overline{X_0} \in \mathcal{S}(K) \setminus \{0\}$ (F_3 is linearly independent from F_1, F_2 in $\mathcal{S}_2^{(3)}$) on the other hand $\overline{X_0} \cdot \overline{F_3} = 0$ in $\mathcal{S}(K)$, since $X_0 \cdot F_3 = X_1 F_2 - X_2 F_1 \in I_p(K)$.

(vi) Corollary 2.3.4 and Proposition 3.3.16 imply that any non-empty, irreducible, proper closed subset of $\mathbb{A}^2$ and of $\mathbb{P}^2$ is either one point or an irreducible curve.

4.1.2 Algebraic varieties

In the sequel, we will always endow topological spaces with Zariski topology and use the following terminology.

Definition 4.1.14. Y is said to be a *projective variety* (respectively, *affine variety*) if Y is irreducible and closed in a projective (respectively, affine) space. Y is said to be a *quasi-projective variety* (respectively, *quasi-affine variety*) if Y is irreducible and *locally closed* (i.e. it is the intersection of a closed set and of an open set) in a projective (respectively, affine) space.

Remark 4.1.15. From the previous definition, we note that Y is a quasi-projective (respectively, quasi-affine) variety if and only if it is an open subset of a projective (respectively, affine) variety, namely its closure in $\mathbb{P}^n$ (respectively, in $\mathbb{A}^n$).

Proposition 4.1.16. *Let $Z \in \mathcal{C}_p^n$. Then Z is a projective variety if and only if for any affine chart $U_i \subset \mathbb{P}^n$, $0 \leqslant i \leqslant n$, $Z_i := Z \cap U_i \subset U_i \cong \mathbb{A}^n$ is an affine variety.*

Proof. ($\Rightarrow$) This is obvious.

($\Leftarrow$) In this case one is reduced to showing that $I_p(Z)$ is a prime ideal. Since $I_p(Z)$ is a homogeneous ideal, the primality of $I_p(Z)$ can be proved by using Proposition 1.10.8-(ii) and the fact that, by assumptions, $I_a(Z_i)$ is prime, for any $0 \leqslant i \leqslant n$. $\qquad\square$

From Definition 4.1.14, any projective (respectively, affine, quasi-affine) variety is also a quasi-projective variety, in other words the notion of *quasi-projective variety* is the most general one.

Definition 4.1.17. The term *algebraic variety* will be used from now on to indicate any quasi-projective variety. Similarly, with the term *closed algebraic set* we will indicate any AAS or any APS, whereas with *locally-closed algebraic set* any locally-closed subset of either an affine or a projective space.

4.2 Noetherian Spaces: Irreducible Components

Definition 4.2.1. A non-empty topological space $(Y, \mathcal{T}_Y)$ is said to be *Noetherian* if it satisfies the following condition:

(*) for any sequence $\{Y_n\}_{n \in \mathbb{N}}$ of closed subsets of Y such that, for any $n \in \mathbb{N}$ one has $Y_n \supseteq Y_{n+1}$, there exists $m \in \mathbb{N}$ such that, for any $n \geqslant m$, one has $Y_n = Y_m$.

Condition (*) is called the *descending chain conditions* (denoted by *d.c.c.* for short) on closed subsets of Y.

Remark 4.2.2. By passing to complementary sets, one clearly has that Y is Noetherian if and only if it satisfies the *ascending chain conditions* (*a.c.c.*) on its open sets, that is:

(**) for any sequence $\{U_n\}_{n \in \mathbb{N}}$ of open subsets of Y such that, for any $n \in \mathbb{N}$ one has $U_n \subseteq U_{n+1}$, there exists $m \in \mathbb{N}$ s.t., for any $n \geqslant m$, one has $U_n = U_m$.

In what follows, any subset $W \subseteq Y$ of a topological space Y will be endowed with the induced topology, i.e. with $\mathcal{T}_{Y,W}$.

Proposition 4.2.3. *Let Y be a Noetherian topological space.*

(i) *Any non-empty subset $W \subseteq Y$ is Noetherian;*

(ii) *Y, as well as any non-empty subset $W \subseteq Y$, is compact.*

Proof. (i) is an easy consequence of the following observation: if $W \subseteq Y$ and if $W_1 \supseteq W_2$ are closed subsets of W, there exist Y_i closed subsets of Y such that $W_i = Y_i \cap W$, $1 \leqslant i \leqslant 2$, and $Y_1 \supseteq Y_2$; indeed it suffices to replace Y_2 with $Y_1 \cap Y_2$ if necessary.

As for (ii), by contradiction let $\{U_i\}_{i \in I}$ be an open covering of Y from which one cannot extract a finite sub-covering. Thus, there exist sequences $\{i_n\}_{n \in \mathbb{N}}$ of indexes in I and of distinct points $\{P_n\}_{n \in \mathbb{N}}$ of Y such that, for $U_n := \cup_{h=1}^{n} U_{i_h}$, one has $P_{n-1} \in U_n$ but $P_n \notin U_n$ for any $n \in \mathbb{N}$. This would contradict the a.c.c. on open subsets of Y. $\square$

Example 4.2.4. An easy example of Noetherian topological space is the following: consider Y an infinite set, $\mathcal{P}(Y)$ the power-set of Y and let $\mathcal{C}_Y \subset \mathcal{P}(Y)$ be the subset consisting of $\emptyset$, Y and all finite subsets of Y. Then $\mathcal{C}_Y$ can be taken as a family of closed subsets for a topology on Y, which turns out to be Noetherian.

The next result shows that also $\mathfrak{Zar}$ is Noetherian.

Corollary 4.2.5.

(i) *For any integer $n \geqslant 1$, $(\mathbb{A}^n, \mathfrak{Zar}_a^n)$ and $(\mathbb{P}^n, \mathfrak{Zar}_p^n)$ are Noetherian.*

(ii) *Any locally-closed algebraic set is Noetherian.*

Proof. (i) directly follows from the Noetherianity of the rings $A^{(n)}$ and $S^{(n)}$ (cf. Proposition 1.4.4) and from reversing inclusions (cf. (2.10) and Proposition 3.1.7-(i)).

To prove (ii) observe that, from Proposition 4.2.3-(i), statement (i) implies (ii). $\square$

The following result bridges Noetherianity with irreducibility.

Theorem 4.2.6 (Irredundant decomposition). *Let Y be a Noetherian topological space and let $W \subseteq Y$ be a non-empty, closed subset. Then:*

(i) *W can be expressed as a finite union*

$$W := W_1 \cup \cdots \cup W_n, \qquad (4.4)$$

where each W_i is a closed, irreducible subset of W;

(ii) *decomposition* (4.4) *is uniquely determined* (*up to reordering the* W_i's) *if it is* irredundant, *i.e. if* $W_i \nsubseteq W_j$ *for any* $i \neq j \in \{1, \ldots, n\}$.

Proof.

(i) If W is irreducible, we are done. Otherwise let $W = W_1 \cup W'$, where W_1, W' proper closed subsets of W. If both of them are irreducible, we are done (observe moreover that neither $W_1 \subseteq W'$ nor $W' \subseteq W_1$, otherwise W would be irreducible against the assumption). Otherwise, at least one of W_1, W' has to be reducible. Assume that W' is. Repeating the same argument, we can construct a sequence $\{W_n\}_{n \in \mathbb{N}}$ of closed subsets of W s.t., for any $n \in \mathbb{N}$, $W_{n+1} \subsetneq W_n$. By Noetherianity of Y and by Proposition 4.2.3, we conclude.

(ii) Let $W = W_1' \cup \cdots \cup W_m'$ be another decomposition, satisfying the same assumptions of that in (4.4); assume moreover that both of them are irredundant. One has

$$W_i = (W_i \cap W_1') \cup \cdots \cup (W_i \cap W_m'), \quad \text{for any} \quad i \in \{1, \ldots, n\}.$$

Since W_i is irreducible, there exists $j \in \{1, \ldots, m\}$ s.t. $W_i = W_i \cap W_j'$, i.e. $W_i \subseteq W_j'$. Similarly, there exists $h \in \{1, \ldots, n\}$ s.t. $W_j' \subseteq W_h$, i.e. $W_i \subseteq W_h$. Since (4.4) is irredundant, then $i = h$ and $W_i = W_j'$. Repeating the same arguments, one can conclude. $\square$

Definition 4.2.7. When (4.4) is an irredundant decomposition, $W_1, \ldots, W_n$ are called the *irreducible components* of W and (4.4) is said to be the *decomposition of* W *into its irreducible components*.

Corollary 4.2.8. *If* $(Y, \mathcal{T}_Y)$ *is a Noetherian and Hausdorff topological space, then* Y *is finite and* $\mathcal{T}_Y$ *is the discrete topology.*

Proof. By Theorem 4.2.6, Y can be decomposed as a finite union of its irreducible components. Therefore, to prove the statement, we can assume Y to be irreducible and conclude by using Remark 4.1.1-(i). $\square$

In the case of Zariski topology we have:

Corollary 4.2.9.

(i) *Any algebraic locally-closed subset can be decomposed into finitely many algebraic varieties.*

(ii) *Any radical ideal $I \subset A^{(n)}$ (respectively, homogeneous radical ideal $I \subset S^{(n)}$) can be uniquely written as $I = \cap_{i=1}^{n} I_j$, where I_j prime ideals in $A^{(n)}$ (respectively, homogeneous prime ideals in $S^{(n)}$).*

Proof. (i) is a consequence of Theorem 4.2.6, whereas (ii) follows from the bijective correspondences proved in Corollary 2.2.5, in the non-homogeneous case, and in Corollary 3.2.6, in the homogeneous one. In this latter situation, the only left case deals with the ideal S_+, which obviously is $S_+ = (X_0) \cap (X_1) \cap \cdots \cap (X_n)$. □

Remark 4.2.10. In general, the intersection of algebraic varieties is not necessarily an algebraic variety. Take, e.g. F_1, $F_2 \in S^{(3)}$ homogeneous polynomials as in (3.35). These polynomials are irreducible in $S^{(3)}$ so the quadric (hyper)surfaces $Z_p(F_1)$, $Z_p(F_2) \subset \mathbb{P}^3$ are projective varieties (cf. Example 4.1.13-(iii)). On the other hand, $Z_p(F_1) \cap Z_p(F_2) = Z_p(F_1, F_2)$ is reducible, consisting of the union of a projective twisted cubic and a line (cf. Example 4.1.13-(v)).

4.3   Combinatorial Dimension

Let $(Y, \mathcal{T}_Y)$ be a topological space.

Definition 4.3.1. The *combinatorial (or topological) dimension* of the topological space $(Y, \mathcal{T}_Y)$, denoted by $\dim_c(Y)$ (or even $\dim_t(Y)$) is defined to be:

$$\dim_c(Y) := \mathrm{Sup}_{n \in \mathbb{Z}_{\geqslant 0}} \{ Z_0 \subset Z_1 \subset \cdots \subset Z_n \},$$

where each Z_i is a non-empty, irreducible, closed subset of Y, $0 \leqslant i \leqslant n$, and where all the inclusions are strict.

Example 4.3.2.

(i) One obviously has $\dim_c(\mathbb{A}^0) = \dim_c(\mathbb{P}^0) = 0$.
(ii) Let $\mathbb{K}$ be an infinite field. Considering as customary $\mathbb{A}^1 := \mathbb{A}^1_{\mathbb{K}}$ endowed with the Zariski topology, one has $\dim_c(\mathbb{A}^1) = 1$, since irreducible, non-empty closed subsets are simply points and $\mathbb{A}^1$, cf. Remark 2.1.14-(ii). Therefore, for any $P \in \mathbb{A}^1$, one has $Z_0 = \{P\} \subset Z_1 = \mathbb{A}^1$. The same holds for $\mathbb{P}^1$ (cf. Section 3.3.1).
(iii) Similarly as in (ii), $\dim_c(\mathbb{A}^2) = 2$, respectively, $\dim_c(\mathbb{P}^2) = 2$ (cf. Corollary 2.3.4, respectively, Proposition 3.3.16), since in this case irreducible curves enter into the game. In Corollary 4.3.8, we will

show that more generally $\dim_c(\mathbb{A}^n) = \dim_c(\mathbb{P}^n) = n$ holds, for any non-negative integer n.

(iv) If $(Y, \mathcal{T}_Y)$ is an irreducible, Noetherian and T_1 (i.e. points in Y are closed) topological space, then $\dim_c(Y) = 0$ if and only if Y is a singleton, whereas $\dim_c(Y) = 1$ if and only if proper closed subsets of Y are finite subsets of Y.

(v) The following is an example of a Noetherian topological space having infinite combinatorial dimension. Let $Y = [0, 1]$ be the real interval and let $\mathcal{T}_Y$ be the topology on Y whose set of closed subsets is given by $\mathcal{C}_Y := \{Y\} \cup \{\emptyset\} \cup \{[\frac{1}{n}, 1], \, n \in \mathbb{N}\}$. Thus, Y is Noetherian and all closed subsets of Y are irreducible. For any $n \in \mathbb{N}$, the following:

$$\{1\} = [1, 1] \subset \left[\frac{1}{2}, 1\right] \subset \left[\frac{1}{3}, 1\right] \subset \cdots \subset \left[\frac{1}{n}, 1\right]$$

is a chain of non-empty, irreducible closed subsets, where each inclusion is strict. The supremum on n is therefore infinite.

From the very definition of combinatorial dimension, if Y is Noetherian one has that

$$\dim_c(Y) := \max_i \left\{ \dim_c(Y_i) \mid Y_i \text{ irreducible component of } Y \right\}. \tag{4.5}$$

Definition 4.3.3. Let $(Y, \mathcal{T}_Y)$ be a topological space. Y is said to have *pure dimension* (or even that Y is *pure*) if each of the irreducible components of Y have the same combinatorial dimension equal to $\dim_c(Y)$.

Example 4.3.4.

(i) Any irreducible topological space is obviously pure. There are also reducible topological spaces which are pure, as, e.g. the affine reducible conic $Y = Z_a(x_1^2 - x_2^2) \subset \mathbb{A}^2$. Nonetheless there are also reducible topological spaces which are not pure. Consider, e.g. $Y = Z_a((x_1 x_2 - x_1, x_2^2 - x_2)) \subset \mathbb{A}^2$. Since $I_a(Y) = (x_1, \; x_2) \cap (x_2 - 1)$, from Proposition 2.1.11-(ii), Y turns out the be the union of the origin $\{O\} = Z_a((x_1, x_2)) := Y_1$ and of the line $Y_2 = Z_a((x_2 - 1))$ not passing through O. Thus, Y_1 and Y_2 are the irreducible components of Y, $\dim_c(Y) = \dim_c(Y_2) = 1 > 0 = \dim_c(Y_0)$ and Y is not pure.

(ii) If Y is pure and $\dim_c(Y) = 1$, then Y is called *curve*; affine (or projective) plane curves are closed subsets in the Zariski topology which are pure of combinatorial dimension 1. If Y is pure and $\dim_c(Y) = 2$, then Y is called *surface*.

Proposition 4.3.5. *Let $(Y, \mathcal{T}_Y)$ be a topological space. Then:*

(i) *if $Z \subseteq Y$ is a subset, then $\dim_c(Z) \leqslant \dim_c(Y)$. If $\dim_c(Y) < +\infty$ then all subsets of Y have finite combinatorial dimension and the quantity*

$$\operatorname{codim}_{c,Y}(Z) := \dim_c(Y) - \dim_c(Z) \geqslant 0 \qquad (4.6)$$

is called the combinatorial (*or even* topological) codimension *of Z in Y;*

(ii) *if $\{U_j\}_{j \in J}$ is an open covering of Y, then $\dim_c(Y) = \operatorname{Sup}_{j \in J}\{\dim_c(U_j)\}$;*

(iii) *if Y is irreducible, with $\dim_c(Y) < +\infty$, and if $Z \subseteq Y$ is closed with $\dim_c(Z) = \dim_c(Y)$, then $Y = Z$ (and so Z is irreducible).*

Proof.

(i) If $\dim_c(Y) = +\infty$, there is nothing else to prove. Assume therefore $\dim_c(Y) < +\infty$; in this case Y is Noetherian and so $Z \subseteq Y$ is also Noetherian. With no loss of generality, we may assume Z to be irreducible (otherwise Z has finitely many irreducible components). If $Z_2 \subset Z_1 \subset Z$ is any collection of proper closed subsets of Z, there exist closed subsets $Y_1, Y_2 \subset Y$ such that $Z_i = Y_i \cap Z$, $1 \leqslant i \leqslant 2$, so in particular $Y_2 \subset Y_1$. With no loss of generality, we may assume Z_1 and Z_2 to be irreducible. In such a case, also Y_1 and Y_2 can be assumed to be irreducible; indeed, if by contradiction, e.g. Y_1 were reducible, say $Y_1 = Y_1' \cup Y_1''$ with $Z_1 \cap Y_1' \neq \emptyset \neq Z_1 \cap Y_1''$, then $Z_1 = Z_1 \cap Y_1 = (Z_1 \cap Y_1') \cup (Z_1 \cap Y_1'')$, where $Z_1 \cap Y_1'$ and $Z_1 \cap Y_1''$ are closed subsets of Z_1 for the induced topology, then Z_1 would be reducible. To complete the proof of (i), one simply applies the previous reasoning to any chain of closed, non-empty, proper subsets of Z and Y.

(ii) Assume to have a chain of proper inclusions of non-empty, closed irreducible subsets of Y

$$(*) \quad Z_0 \subset Z_1 \subset \cdots \subset Z_n$$

and let $P_0 \in Z_0$ be a point. There exists $j_0 \in J$ such that $P_0 \in U_{j_0}$, since $\{U_j\}_{j \in J}$ is an open covering of Y. Thus, $U_{j_0} \cap Z_i \neq \emptyset$, for any $0 \leqslant i \leqslant n$. Since $U_{j_0} \cap Z_i$ is an open set of Z_i which is irreducible, then $U_{j_0} \cap Z_i$ is dense in Z_i, for any $0 \leqslant i \leqslant n$ (cf. Proposition 4.1.2-(ii)). Therefore, $U_{j_0} \cap Z_i \subset U_{j_0} \cap Z_{i+1}$ where each inclusion is strict; indeed, if it were $U_{j_0} \cap Z_i = U_{j_0} \cap Z_{i+1}$ for some index i, then by density and irreducibility we would have $Z_i = Z_{i+1}$, contradicting $(*)$. Thus, from $(*)$ we have

$$U_{j_0} \cap Z_0 \subset U_{j_0} \cap Z_1 \subset \cdots \subset U_{j_0} \cap Z_n,$$

which is a chain of proper inclusions of non-empty, closed subsets of U_{j_0}, which can be assumed to be irreducible with no loss of generality. Considering the supremum, one finds $\dim_c(U_{j_0})$. Considering the supremum on $j \in J$, one concludes.

(iii) Let $Z \subseteq Y$ be closed and set $\dim_c(Y) := n < +\infty$. Since by assumption we have $\dim_c(Z) = n$, there exists a maximal chain of irreducible closed subsets of Z, say

$$(*) \quad Z_0 \subset Z_1 \subset \cdots \subset Z_n,$$

where all inclusions are strict. Since Z is closed in Y and each Z_j is closed in Z, then each Z_j is also closed and irreducible in Y, $0 \leqslant j \leqslant n$. Then $(*)$ is also a maximal chain of irreducible closed subsets of Y, therefore $Z_n = Y$. Thus, $Y \subseteq Z$, since Z_n is closed in Z, which implies $Z = Y$ and that Z is irreducible. $\qquad\square$

Remark 4.3.6.

(i) Note that the combinatorial codimension of affine (respectively, projective) linear subspaces of the affine (respectively, projective) space coincides with the usual notion of codimension.

(ii) Since $\mathbb{A}^n \cong U_0 \subset \mathbb{P}^n$ (cf. Remark 3.3.3-(ii)), from Proposition 4.3.5-(ii) we get $\dim_c(\mathbb{A}^n) = \dim_c(\mathbb{P}^n)$.

Recalling the definition of Krull-dimension and of affine coordinate ring of an affine variety (cf. Definitions 1.12.1 and 4.1.7), one has the following result.

Proposition 4.3.7. *Let $\mathbb{K}$ be an algebraically closed field and $n \geqslant 1$ be an integer. Let $Y \subseteq \mathbb{A}^n_{\mathbb{K}}$ be any affine variety and let $A(Y) = \frac{A^{(n)}}{I_a(Y)}$ be its affine coordinate ring. Then $\dim_c(Y) = K \dim(A(Y))$.*

Proof. Since $Y \subseteq \mathbb{A}^n_{\mathbb{K}}$ is an affine variety, then Y is irreducible, $I_a(Y) \subset A^{(n)}$ is a prime ideal and $A(Y)$ is an integral $\mathbb{K}$-algebra of finite type (cf. Proposition 4.1.10). Moreover, from Corollary 4.2.5, Y is Noetherian, therefore $\dim_c(Y) < +\infty$. Let

$$(*) \quad Y_0 \subset Y_1 \subset \cdots \subset Y_m = Y$$

be any chain of non-empty, irreducible, closed subsets of Y. Since Y is closed in $\mathbb{A}^n$ and each Y_j is closed in Y_{j+1}, for any $0 \leqslant j \leqslant m-1$, each Y_j is closed and irreducible in $\mathbb{A}^n$. Therefore, any Y_j is an affine variety of $\mathbb{A}^n$ so each ideal $I_a(Y_j)$ is a prime ideal in $A^{(n)}$. From the bijective correspondence between irreducible, closed subsets of $\mathbb{A}^n$ and prime ideals in $A^{(n)}$, together with the reversing inclusion among ideals, respectively closed subsets (cf. Corollary 4.1.12 and Proposition 2.2.8), $(*)$ gives rise to the chain of prime ideals in $A^{(n)}$

$$(**) \quad I_a(Y_0) \supset I_a(Y_1) \supset \cdots \supset I_a(Y_m) = I_a(Y),$$

each of which contains the prime ideal $I_a(Y)$. Let $\pi : A^{(n)} \twoheadrightarrow A(Y) = \frac{A^{(n)}}{I_a(Y)}$ be the canonical epimorphism, which maps $I_a(Y)$ to the prime ideal (0) in $A(Y)$ whereas, for any $0 \leqslant j \leqslant m-1$, $I_a(Y_j)^e = \pi(I_a(Y_j)) = \frac{I_a(Y_j)}{I_a(Y)} := \mathfrak{p}_j$.

Any $\mathfrak{p}_j$ is a prime ideal in $A(Y)$, as it directly follows from the fact that $I_a(Y_j)$ is a prime ideal in $A^{(n)}$ and from the epimorphism π. Taking $(*)$ and $(**)$ as maximal chains, one deduces that $\dim_c(Y) = K\dim(A(Y))$. $\square$

Corollary 4.3.8. *For any integer $n \geqslant 0$, $\dim_c(\mathbb{A}^n) = \dim_c(\mathbb{P}^n) = n$.*

Proof. The fact that $\dim_c(\mathbb{A}^n) = n$ follows from Proposition 4.3.7 and from Corollary 1.12.4-(i), whereas $\dim_c(\mathbb{P}^n) = n$ follows from Remark 4.3.6-(ii). $\square$

More generally, for projective varieties one has the following.

Proposition 4.3.9. *Let $\mathbb{K}$ be an algebraically closed field and $n \geqslant 1$ be an integer. Let $Y \subseteq \mathbb{P}^n_{\mathbb{K}}$ be any projective variety and let $\mathcal{S}(Y) = \frac{S^{(n)}}{I_p(Y)}$ be its homogeneous coordinate ring. Let $C_a(Y) \subset \mathbb{A}^{n+1}$ the affine cone over Y. Then $K\dim(\mathcal{S}(Y)) = \dim_c(C_a(Y))$.*

Proof. Since $I_p(Y) = I_a(C_a(Y))$, we have an isomorphism of integral $\mathbb{K}$–algebras of finite type $\mathcal{S}(Y) = \frac{S^{(n)}}{I_p(Y)} \cong \frac{A^{(n+1)}}{I_a(C_a(Y))} = A(C_a(Y))$ so $C_a(Y)$ is irreducible since Y is. Thus, $C_a(Y) \subset \mathbb{A}^{n+1}$ is an affine variety so, from Proposition 4.3.7, $\dim_c(C_a(Y)) = K\dim(A(C_a(Y))) = K\dim(\mathcal{S}(Y))$. $\square$

As a consequence of some other definitions of *dimension* (cf. Corollary 10.2.2), in the above set-up we will more precisely show that

$$K \dim(\mathcal{S}(Y)) = \dim_c(C_a(Y)) = \dim_c(Y) + 1. \qquad (4.7)$$

Exercises

Exercise 4.1. Find an example of an irreducible polynomial $f \in A_{\mathbb{R}}^{(2)}$ whose zero set $Z_a(f)$ is not irreducible in $\mathbb{A}_{\mathbb{R}}^2$ (cf. Hartshorne, 1977, Exercise 1.12, p. 18).

Exercise 4.2. Let $\mathbb{K}$ be an algebraically closed field. Consider in $A_{\mathbb{K}}^{(3)} = \mathbb{K}[x_1, x_2, x_3]$ the ideal $J := (x_1 x_3 - x_2^2, x_1^3 - x_2 x_3)$. Show that J is not a prime ideal and find irreducible components of $Z_a(J)$ in $\mathbb{A}_{\mathbb{K}}^3$ (cf. Reid, 1988, p. 57).

Exercise 4.3. Let $\mathbb{K}$ be an algebraically closed field. Show that if $X \subset \mathbb{P}_{\mathbb{K}}^n$ is a quasi-projective variety and if $Y \subset X$ is irreducible and locally closed in X, then $Y \subset \mathbb{P}_{\mathbb{K}}^n$ is a quasi-projective variety.

Exercise 4.4. Let $\mathbb{K}$ be an algebraically closed field and let $I = (x_1^2 - x_2^3, x_2^2 - x_3^3) \subset A_{\mathbb{K}}^{(3)}$ be an ideal. Consider the $\mathbb{K}$-algebra homomorphism $\alpha : A_{\mathbb{K}}^{(3)} \to \mathbb{K}[t]$, defined by $\alpha(x_1) = t^9$, $\alpha(x_2) = t^6$, $\alpha(x_3) = t^4$. Show that $\mathrm{Ker}(\alpha) = I$ and deduce that $Z_a(I)$ is irreducible, with $I_a(Z_a(I)) = I$ (cf. Fulton, 2008, Exercise 1.40, p. 12).

Exercise 4.5. Let $\mathbb{K}$ be an algebraically closed field and let $\mathcal{C}$ be the set parametrizing conics in the projective plane $\mathbb{P}_{\mathbb{K}}^2$.

(i) Show that $\mathcal{C}$ can be identified with the projective space $\mathbb{P}_{\mathbb{K}}^5$.
(ii) Let $\mathcal{D} \subset \mathcal{C}$ be the subset parametrizing degenerate conics in $\mathbb{P}_{\mathbb{K}}^2$. Show that $\mathcal{D}$ can be identified with an cubic hypersurface Σ_3 in $\mathbb{P}_{\mathbb{K}}^5$.
(iii) Show that Σ_3 is an irreducible hypersurface.

Chapter 5

Regular and Rational Functions on Algebraic Varieties

Here we study functions which are defined on algebraic varieties. To do this, we need first to introduce some basic facts concerning pre-sheaves and sheaves on a topological space.

5.1 Basics on Sheaves

Let $(Y, \mathcal{T}_Y)$ be a topological space.

Definition 5.1.1. A *pre-sheaf* $\mathcal{F}$ *(of sets) on* Y is the datum of:

(i) a non-empty set $\mathcal{F}(U)$, for any open set $U \subseteq Y$, and
(ii) a map $\rho_V^U : \mathcal{F}(U) \to \mathcal{F}(V)$, for any $V \subseteq U \subseteq Y$ open sets,

such that the following hold:

(F1) $\rho_U^U = \mathrm{Id}_{\mathcal{F}(U)}$, for any open set $U \subseteq Y$;
(F2) for any open sets $W \subseteq U \subseteq V$, one has $\rho_W^U \circ \rho_U^V = \rho_W^V$.

The maps ρ_V^U are called *restrictions* and, for any open set $U \subseteq Y$, the elements of $\mathcal{F}(U)$ are called *sections of* $\mathcal{F}$ *over* U. In particular, elements of $\mathcal{F}(Y)$ are called *global sections* of $\mathcal{F}$.

In the above definition, to any open set U of a topological space it is associated a non-empty set $\mathcal{F}(U)$; in particular, for any pre-sheaf $\mathcal{F}$, we have

$\mathcal{F}(\emptyset) \neq \emptyset$. In the sequel, we shall always consider pre-sheaves $\mathcal{F}$ such that $\mathcal{F}(\emptyset) = \{\cdot\}$ is a fixed singleton. This convention will be taken for granted and not mentioned anymore.

Definition 5.1.2. A *sheaf (of sets) on* Y is the datum of a pre-sheaf $\mathcal{F}$ on Y as above, satisfying moreover the following condition:

(F3) for any family $\{U_i\}_{i \in I}$ of open sets of Y, where I a set of indexes, and for any collection of sections $(f_i) \in \Pi_{i \in I} \mathcal{F}(U_i)$ s.t.

$$\rho^{U_i}_{U_i \cap U_j}(f_i) = \rho^{U_j}_{U_i \cap U_j}(f_j), \ \forall\, i, j \in I,$$

there exists a unique $f \in \mathcal{F}(U)$, where $U := \cup_{i \in I} U_i$, such that

$$\rho^U_{U_i}(f) = f_i, \ \forall\, i \in I.$$

Remark 5.1.3. Not all pre-sheaves are sheaves. Consider, e.g. $(Y, \mathcal{T}_Y)$ a topological space and let S be any non-empty set. For any non-empty open subset $U \subseteq Y$ set $\mathcal{C}_S(U) := \{S\}$ whereas $\mathcal{C}_S(\emptyset) = \{\cdot\}$, where $\{\cdot\} \neq S$. For any $\emptyset \neq V \subseteq U$ open set inclusion, set moreover $\rho^U_V = \mathrm{Id}_S$, whereas if $V = \emptyset$ set ρ^U_V to be the constant map to $\{\cdot\}$. Then $\mathcal{C}_S$ defined in this way is called *constant pre-sheaf* associated to S on Y. It is easy to observe that if Y contains two open sets U and U' such that $U \cap U' = \emptyset$ (e.g. when Y is Hausdorff), then $\mathcal{C}_S$ is a pre-sheaf but not a sheaf. If otherwise Y is an irreducible topological space, any constant pre-sheaf is a sheaf.

When the $\mathcal{F}(U)$'s are more precisely, e.g. abelian groups, rings, $\mathbb{K}$-algebras, R-modules for a given ring R, etc., one requires restrictions ρ^U_V to be homeomorphims of abelian groups, rings, $\mathbb{K}$-algebras, R-modules, etcetera, respectively. In this case, $\mathcal{F}$ is called a pre-sheaf (respectively a sheaf) of abelian groups, rings, $\mathbb{K}$-algebras, R-modules, etcetera, on Y.

Remark 5.1.4. Let $\mathcal{F}$ be a sheaf on the topological space Y and let U be any non-empty, open subset of Y; one can define the pre-sheaf $\mathcal{F}|_U$ by $\mathcal{F}|_U(V) := \mathcal{F}(V)$, for any non-empty, open set V of U. Since $\mathcal{F}$ is a sheaf, the $\mathcal{F}|_U$ is a sheaf too. This is called the *sheaf obtained by restriction of $\mathcal{F}$ to U*.

For further details on sheaves, see, e.g. Hartshorne (1977, § II.1, pp. 60–69; Mumford, 1988, § I.4, pp. 24–30).

5.2 Regular Functions

If we take a polynomial $f \in A^{(n)}$, we can consider the associated function $f : \mathbb{A}^n \to \mathbb{K}$ (recall (2.1)). Accordingly, if $g_1, g_2 \in A^{(n)}$ are polynomials, with $g_2 \neq 0$, the *rational function* $\frac{g_1}{g_2}$ can be considered as a function defined on the principal open subset $U_a(g_2) := \mathbb{A}^n \setminus Z_a(g_2)$ and with values in $\mathbb{K}$.

By contrast, given a projective space $\mathbb{P}(V)$, an element $g \in S(V^*)$ cannot be considered as function on $\mathbb{P}(V)$ with values in $\mathbb{K}$, even if g is homogeneous. On the other hand, if we take homogeneous elements $G_1, G_2 \in H(S(V^*))$ of the same degree and $P = [\underline{p}] \in U_p(G_2) :=\mathbb{P}(V) \setminus Z_p(G_2)$, where $\underline{p} \in V$, it makes sense to consider

$$\frac{G_1}{G_2}(P) := \frac{G_1(\underline{p})}{G_2(\underline{p})} = \frac{G_1(t\underline{p})}{G_2(t\underline{p})}, \quad \forall t \in \mathbb{K} \setminus \{0\}.$$

Thus, as in the affine case, $\frac{G_1}{G_2}$ can be considered as a function defined on the principal open subset $U_p(G_2)$ and with values in $\mathbb{K}$. We sometimes will drop the indices a and p in $U_a(g_2), U_p(G_2)$, respectively, when this does not create ambiguity.

Remark 5.2.1. In what follows, the field of fractions $\mathcal{Q}(A^{(n)})$ will be simply denoted by $\mathcal{Q}^{(n)}$ and its elements will be simply called *rational functions* in the indeterminates $x_1, \ldots, x_n$ (or even *rational functions on* $\mathbb{A}^n$). Similarly, the field $\mathcal{Q}(S(V^*))$ contains the subfield $\mathcal{Q}(V^*)$ which consists of all fractions $\frac{G_1}{G_2}$ such that $G_1, G_2 \in H(S(V^*))$ are of the same degree, where $G_2 \neq 0$. These will be called *degree-zero rational functions*. If we introduce in $\mathbb{P}(V)$ homogeneous coordinates $[X_0, \ldots, X_n]$, then $\mathcal{Q}(V^*)$ identifies with the subfield $\mathcal{Q}_0(\mathcal{S}^{(n)})$ of $\mathcal{Q}^{(n+1)}$ consisting of all fractions $\frac{G_1}{G_2}$ where G_1, G_2 are homogeneous polynomials of the same degree in the indeterminates $X_0, \ldots, X_n$, with G_2 non-zero (recall Definition 1.10.10). This sub-field of $\mathcal{Q}^{(n+1)}$ will be simply denoted by $\mathcal{Q}_0^{(n)}$.

Let now $Y \subseteq \mathbb{P}(V)$ be a locally closed subset. Let $f : Y \to \mathbb{K}$ be a function and let $P \in Y$ be a point. Then f is said to be *regular at* the point P if there exist an open neighborhood $U \subseteq Y$ of P and $G_1, G_2 \in H(S(V^*))$ of the same degree, such that $Z_p(G_2) \cap U = \emptyset$, i.e. $U \subseteq U_p(G_2)$, and such that the restriction of f to U coincides with the restriction of $\frac{G_1}{G_2}$ to U. A function $f : Y \to \mathbb{K}$ is said to be *regular on* Y, if it is regular at any point of Y (in the above sense).

Remark 5.2.2. For $f : Y \to \mathbb{K}$ being regular at a point P is both a *local* and *open* property, i.e. it only depends on a (Zariski) open neighborhood of P and it holds for all points in the (Zariski) open neighborhood of P.

If, as customary, we identify $\mathbb{A}^n$ with the affine open set U_0 of $\mathbb{P}^n$, a straightforward application of maps in Definition 1.10.16 gives the following.

Lemma 5.2.3. *Let $Y \subseteq \mathbb{A}^n$ be any locally closed subset and let $f : Y \to \mathbb{K}$ be a function. Then f is regular at $P \in Y$ (in the sense of the above definition) if and only if there exist a (Zariski) open neighborhood $U \subseteq Y$ of P and polynomials $g_1, g_2 \in A^{(n)}$, such that $Z_a(g_2) \cap U = \emptyset$ and the restriction of f to U coincides with the restriction of $\frac{g_1}{g_2}$ to U.*

The same strategy also shows that

$$Q_0^{(n)} = Q_0 \left(S^{(n)} \right) \cong Q^{(n)} = Q \left(A^{(n)} \right). \tag{5.1}$$

Let therefore Y be any algebraic variety. For any non-empty, open subset $U \subseteq Y$ we denote by $\mathcal{O}_Y(U)$ the set of regular functions on U. Constant functions on Y are obviously regular functions on any open set $U \subseteq Y$. If moreover $f, g \in \mathcal{O}_Y(U)$, then the functions

$$f + g : P \in U \to f(P) + g(P), \quad f \cdot g : P \in U \to f(P) \cdot g(P) \tag{5.2}$$

are also regular on U. Thus, for any algebraic variety Y and for any non-empty open subset $U \subseteq Y$, $\mathcal{O}_Y(U)$ is a $\mathbb{K}$-algebra with respect to the operations $(+, \cdot)$ as in (5.2), which is called the *algebra of regular functions on U*. If moreover $U, U' \subseteq Y$ are non-empty open subsets, with $U' \subseteq U$, one has an obvious *restriction map*

$$\rho^U_{U'} : f \in \mathcal{O}_Y(U) \to f|_{U'} \in \mathcal{O}_Y(U'), \tag{5.3}$$

which is well-defined and which clearly is a $\mathbb{K}$-algebra homomorphism with respect to the operations (5.2). We will sometimes abuse notation and denote $\rho^U_{U'}(f)$ simply by $f|_{U'}$.

Since $\mathbb{K}$-algebras $\mathcal{O}_Y(U)$ and homomorphisms (5.3) satisfy conditions (F1) and (F2) in Definition 5.1.1, we will denote by $\mathcal{O}_Y$ the *pre-sheaf of regular functions* on Y as the datum of these $\mathbb{K}$-algebras and morphisms (in Remark 5.2.5 below we will see that it is a pre-sheaf of integral $\mathbb{K}$-algebras and in Proposition 5.3.4 we will more precisely prove that $\mathcal{O}_Y$ is actually a sheaf).

For any non-empty open subset $U \subseteq Y$, if $f \in \mathcal{O}_Y(U)$ is non-zero we will denote by $Z_U(f)$ the *zero locus* of f in U, i.e. the set

$$Z_U(f) := f^{-1}(0). \tag{5.4}$$

Proposition 5.2.4. *Let Y be any algebraic variety.*

(i) *Any $f \in \mathcal{O}_Y(Y)$ is continuous, if f is considered as a function with values in $(\mathbb{A}^1_{\mathbb{K}}, \mathfrak{Zar}^1_a)$.*

(ii) *If $f, g \in \mathcal{O}_Y(Y)$ are such that there exists a non-empty open subset U of Y such that $f|_U = g|_U$, then $f = g$ on Y.*

Proof.

(i) One has to show that $Z_Y(f - a)$ is closed for all $a \in \mathbb{K}$. By replacing f with $f - a$, it suffices to verify that $Z_Y(f)$ is closed; this can be done locally. If $Z_Y(f) \neq \emptyset$, let $P \in Z_Y(f)$ be any point. Since f is regular at P, there exists an open neighborhood $U_P \subseteq Y$ of P such that $f|_{U_P} = \frac{G_1}{G_2}|_{U_P}$, for suitable G_1, $G_2 \in H(\mathcal{S}^{(n)})$ of the same degree, $G_2 \neq 0$, such that $U_P \subseteq U_p(G_2) = Z_p(G_2)^c$. Thus,

$$Z_Y(f) \cap U_P = Z_p(G_1) \cap U_P,$$

where $Z_p(G_1) \subset \mathbb{P}^n$ a projective hypersurface. In other words we have shown that, for any $P \in Z_Y(f)$, there exists an open neighborhood U_P of P in Y s.t. $Z_Y(f) \cap U_P$ is closed in U_P. This implies that $Z_Y(f)$ is closed in Y. Indeed, denote by $\overline{Z_Y(f)}$ the closure in Y of $Z_Y(f)$; if we had $Z_Y(f) \subsetneq \overline{Z_Y(f)}$, any point $Q \in \overline{Z_Y(f)} \setminus Z_Y(f)$ would give that, for any open neighborhood V of Q in $\overline{Z_Y(f)}$, $V \cap Z_Y(f) \neq \emptyset$ but $Q \notin V \cap Z_Y(f)$. On the other hand, all the open sets of $Z_Y(f)$ are induced by opens sets of $\overline{Z_Y(f)}$ thus we would contradict that $Z_Y(f) \cap U_P$ is closed in U_P, for some $P \in V \cap Z_Y(f)$.

(ii) Note that $Z_Y(f - g)$ is closed in Y and contains U, which is dense (see Proposition 4.1.2). $\qquad \square$

Remark 5.2.5.

(a) Proposition 5.2.4-(i) more generally holds for any locally closed subset in $\mathbb{P}(V)$, on the other hand we will not use it.

(b) If $f \in \mathcal{O}_Y(Y)$ is non-zero, then $Z_Y(f)$ is a proper closed subset of Y. Therefore $U_Y(f) := Y \setminus Z_Y(f)$, also denoted by $U(f)$ for brevity, is a non-empty open subset of Y which is called the *principal open subset* of

Y associated to f. Note that, by construction, in $U(f)$ one can consider the function $\frac{1}{f}$ which is regular therein.

(c) A proof similar to that of Proposition 5.2.4-(ii) shows that, for any non-empty open subset $U \subseteq Y$, the $\mathbb{K}$-algebra $\mathcal{O}_Y(U)$ is *integral*. Indeed if $0 = f \cdot g \in \mathcal{O}_Y(U)$ then $Z_U(f \cdot g) = U$, i.e. $U = Z_U(f) \cup Z_U(g)$. Since U is irreducible, then either $U = Z_U(f)$ or $U = Z_U(g)$, i.e. either $f = 0$ on U or $g = 0$ on U.

(d) If $f : Y \to \mathbb{K}$ is a function which is regular at a point $P \in Y$, then the value $f(P)$ depends only on f and P, i.e. it does not depend on the choice of the open neighborhood $U = U_P$ of P and on the choice of representative homogeneous polynomials G_1, G_2 of the same degree such that $f|_U = \frac{G_1}{G_2}|_U$.

(e) For any point $P \in Y$, let $\Phi_P := \frac{G_1}{G_2} \in \mathcal{Q}_0^{(n)}$ be a rational function and let $U_P \subseteq Y$ be an open neighborhood of P s.t. $U_P \subseteq U_p(G_2) = Z_p(G_2)^c$. Then $\Phi_P|_{U_P} \in \mathcal{O}_Y(U_P)$. For any pair of points $P \neq Q \in Y$, and for any choice of open neighborhoods U_P and U_Q, respectively, one has $U_P \cap U_Q \neq \emptyset$ since Y is irreducible. Thus, if one has $\Phi_P|_{U_P \cap U_Q} = \Phi_Q|_{U_P \cap U_Q}$, the datum

$$\{(U_P, \Phi_P|_{U_P}), (U_Q, \Phi_Q|_{U_Q})\}$$

defines an element in $\mathcal{O}_Y(U)$, where $U := U_P \cup U_Q$ open set in Y.

5.3   Rational Functions

For Y any algebraic variety, we set

$$\mathcal{H}(Y) := \{(U, f) \mid U \subseteq Y \text{ non-empty, open subset, } f \in \mathcal{O}_Y(U)\}$$

and we define a relation $\mathcal{R}$ on $\mathcal{H}(Y)$ in this way:

$$(U, f) \, \mathcal{R} \, (U', f') \Leftrightarrow f|_{U \cap U'} = f'|_{U \cap U'}.$$

Since Y is irreducible, $\mathcal{R}$ is a well-defined equivalence relation.

Definition 5.3.1. The quotient set $\frac{\mathcal{H}(Y)}{\mathcal{R}}$ will be denoted by $K(Y)$ and the equivalence class of (U, f) will be denoted by $[U, f]$ (sometimes also by Φ_f) and called a *rational function* on Y.

Lemma 5.3.2. $K(Y)$ *is endowed with a structure of a field, which gives a field extension of the ground field* $\mathbb{K}$.

Proof. One poses

(i) $[U, f] + [V, g] := [U \cap V, f + g]$,

(ii) $[U, f] \cdot [V, g] := [U \cap V, fg]$,

(iii) one has an injective map $\mathbb{K} \hookrightarrow K(Y)$ defined by $\lambda \to [Y, \lambda]$, for any $\lambda \in \mathbb{K}$,

(iv) if $[V, g]$ is such that $g \neq 0$, then $[V, g]^{-1} := [V \setminus Z_V(g), \frac{1}{g}] = [U_V(g), \frac{1}{g}]$, as in Remark 5.2.5.

It is straightforward to check that the above conditions are well-defined, i.e. they do not depend on the representatives, and moreover that in (ii) one has $[U \cap V, fg] \neq 0$ if and only if $[U, f]$ and $[V, g]$ are both non-zero (cf. Remark 5.2.5-(c)). $\qquad\square$

For this reason, $K(Y)$ is called the *field of rational functions on* Y.

Lemma 5.3.3. *For any non-empty, open subset* $U \subseteq Y$, *one has an injective homomorphism of integral* $\mathbb{K}$-*algebras*

$$j_U : \mathcal{O}_Y(U) \hookrightarrow K(Y), \text{ defined by } f \xrightarrow{j_U} [U, f]. \tag{5.5}$$

Proof. Obvious. $\qquad\square$

In particular, one has

$$\mathcal{O}_Y(Y) \subseteq K(Y). \tag{5.6}$$

Note that the previous inclusion in general is strict: e.g. $\frac{1}{x} \in K(\mathbb{A}^1)$ is not regular on $\mathbb{A}^1$ so $\frac{1}{x} \notin \mathcal{O}_{\mathbb{A}^1}(\mathbb{A}^1)$. On the other hand, if we replace $\mathbb{A}^1$ with the open subset $W := \mathbb{A}^1 \setminus \{0\} \subset \mathbb{A}^1$, which is also an algebraic variety, then $\frac{1}{x} \in \mathcal{O}_W(W)$.

Proposition 5.3.4. *For any algebraic variety* Y, $\mathcal{O}_Y$ *is a sheaf (of integral* $\mathbb{K}$-*algebras) which is called the* structural sheaf *of* Y.

Proof. We already know that $\mathcal{O}_Y$ is a pre-sheaf of integral $\mathbb{K}$-algebras on Y. We need to show that condition (F3) in Definition 5.1.2 holds. Let $\{U_i\}_{i \in I}$ be any family of non-empty, open subsets of Y and correspondingly, consider any collection of regular functions $(f_i) \in \Pi_{i \in I} \mathcal{O}_Y(U_i)$ s.t.

$$\rho^{U_i}_{U_i \cap U_j}(f_i) = \rho^{U_j}_{U_i \cap U_j}(f_j), \quad \forall\, i, j \in I.$$

This condition implies $j_{U_i}(f_i) = j_{U_j}(f_j)$, i.e. there exists a rational function $\Phi_f \in K(Y)$ such that $\Phi_f = [U_i, f_i] = [U_j, f_j]$, On the other hand, since its restriction to any U_i is $f_i \in \mathcal{O}_Y(U_i)$, for any $i \in I$, then $\Phi_f|_U = f \in \mathcal{O}_Y(U)$, where $U = \cup_{i \in I} U_i$.

The fact that $f \in \mathcal{O}_Y(U)$ is uniquely determined follows from the fact that restriction maps $\rho^U_{U_i}$ are injective, for any $i \in I$; indeed, assume there exist $f, g \in \mathcal{O}_Y(U)$ s.t., for any $i \in I$, one has

$$\rho^U_{U_i}(f) = f|_{U_i} = f_i = g|_{U_i} = \rho^U_{U_i}(g).$$

Since f and g are both regular on U and since $U_i \subseteq U$ is open in U, then one concludes by Proposition 5.2.4-(ii). $\square$

Definition 5.3.5. For any rational function $\Phi \in K(Y)$, we consider the open set $U_\Phi := \cup_i U_i$, where the union is over all the representatives $[U_i, f_i]$ of Φ in $K(Y)$. This is called the *domain of* Φ (or even *the open-set of definition of* Φ), denoted also by $\mathrm{Dom}(\Phi)$, and it is the biggest open subset of Y over which Φ is regular.

With this set-up, for any open set $U \subseteq Y$, one simply has

$$\mathcal{O}_Y(U) = \{\Phi \in K(Y) \mid U \subseteq U_\Phi\}.$$

Definition 5.3.6. Let Y be any algebraic variety. Any locally-closed and irreducible subset $W \subseteq Y$ is said to be an (*algebraic*) *subvariety* of Y.

In other words, W is an algebraic variety on its own which is a locally-closed subset of Y.

Remark 5.3.7.

(i) Take any algebraic variety $Y \subseteq \mathbb{A}^n$ and consider its (affine) coordinate ring $A(Y)$, as in Definition 4.1.7. For any subvariety $W \subseteq Y$, the ideal

$$I_{a,Y}(W) := \frac{I_a(W)}{I_a(Y)}, \tag{5.7}$$

defined as the image of the ideal $I_a(W) \subseteq A^{(n)}$ via the canonical quotient morphism $A^{(n)} \twoheadrightarrow A(Y)$, is a prime ideal of $A(Y)$ (when otherwise $W = Y$, one obviously has $I_{a,Y}(W) = (0)$).

(ii) Similarly, let $Y \subseteq \mathbb{P}^n$ be any algebraic variety and let $\mathcal{S}(Y)$ be its (homogeneous) coordinate ring. For any subvariety $W \subseteq Y$, the homogeneous ideal

$$I_{p,Y}(W) := \frac{I_p(W)}{I_p(Y)}, \tag{5.8}$$

defined as the image of the homogeneous ideal $I_p(W) \subseteq \mathcal{S}^{(n)}$ via the surjective homomorphism $\mathcal{S}^{(n)} \twoheadrightarrow \mathcal{S}(Y)$, is a homogeneous prime ideal of $\mathcal{S}(Y)$.

Definition 5.3.8. Let W be a subvariety of an algebraic variety Y. A rational function $\Phi \in K(Y)$ is said to be *defined in* W if there exists a representative (U, f) of $\Phi = [U, f]$ s.t. $U \cap W \neq \emptyset$, i.e. Φ is regular in an open, dense subset of W. The set of rational functions defined in W will be denoted by $\mathcal{O}_{Y,W}$; in symbols $\mathcal{O}_{Y,W} := \{\Phi \in K(Y) \mid U_\Phi \cap W \neq \emptyset\}$.

Operations (i)–(iii) as in the proof of Lemma 5.3.2 endow $\mathcal{O}_{Y,W}$ with an integral $\mathbb{K}$-algebra structure, for any subvariety $W \subseteq Y$. The following properties are easy consequences of the previous definitions.

Proposition 5.3.9. *Let Y be an algebraic variety and let $W \subseteq Y$ be a subvariety. Then*

(i) $\mathcal{O}_{Y,Y} = K(Y)$;
(ii) $\mathcal{O}_{Y,W}$ *is a subring of the field $K(Y)$;*
(iii) $\mathcal{O}_{Y,W}$ *contains $\mathcal{O}_Y(Y)$ as a subring;*
(iv) *the set $\mathfrak{m}_{Y,W} := \{\Phi \in \mathcal{O}_{Y,W} \mid \Phi(P) = 0, \, \forall \, P \in U_\Phi \cap W\}$ is an ideal of $\mathcal{O}_{Y,W}$.*

When in particular $W = \{P\}$ is a point, $\mathcal{O}_{Y,P}$ is called the *ring of germs of regular functions at P* and $\mathfrak{m}_{Y,P}$ is the *ideal of germs of regular functions vanishing at P*. For any point $P \in Y$, one has

$$\mathcal{O}_Y(Y) \subseteq \mathcal{O}_{Y,P} \subseteq K(Y). \tag{5.9}$$

Thus, for any non-empty, open set $U \subseteq Y$,

$$\mathcal{O}_Y(U) = \{\Phi \in K(Y) \mid \Phi \text{ regular at } P, \, \forall \, P \in U\} = \bigcap_{P \in U} \mathcal{O}_{Y,P}, \tag{5.10}$$

whereas

$$K(Y) = \{\Phi \in K(Y) \mid \Phi \text{ regular at some point } P \in Y\} = \bigcup_{P \in Y} \mathcal{O}_{Y,P}. \tag{5.11}$$

Recalling definitions as in Section 1.11, we can prove the following result.

Theorem 5.3.10. *Let Y be an algebraic variety and let W be any subvariety of Y. Then $(\mathcal{O}_{Y,W}, \mathfrak{m}_{Y,W})$ is a local ring of residue field $K(W)$.*

Proof. Take any $\Phi := [U, f] \in \mathcal{O}_{Y,W} \setminus \mathfrak{m}_{Y,W}$ where, up to a change of representative (U, f) for Φ, we can assume $U \cap W \neq \emptyset$. Since $f \in \mathcal{O}_Y(U)$, $Z_U(f)$ is a proper closed subset of U (cf. the proof of Proposition 5.2.4). Moreover, since $f \notin \mathfrak{m}_{Y,W}$, then also $Z_U(f) \cap W$ is a proper, closed subset of $U \cap W$. Let $U' := U_U(f) = U \setminus Z_U(f)$ be the principal open set in

U associated to f. From Remark 5.2.5-(b), one has $f' := \frac{1}{f} \in \mathcal{O}_Y(U')$; moreover

$$U' \cap W = (U \cap W) \setminus (Z_U(f) \cap W) \neq \emptyset.$$

Thus, Φ is invertible with $\Phi^{-1} := [U', f'] \in \mathcal{O}_{Y,W}$. From Proposition 1.11.10 we conclude that $(\mathcal{O}_{Y,W}, \mathfrak{m}_{Y,W})$ is a local ring.

Let now $K_{Y,W} := \frac{\mathcal{O}_{Y,W}}{\mathfrak{m}_{Y,W}}$ be the residue field of $(\mathcal{O}_{Y,W}, \mathfrak{m}_{Y,W})$. Consider the map

$$\begin{array}{rccc} \varphi_{Y,W}: & K_{Y,W} & \longrightarrow & K(W) \\ & [U, f] + \mathfrak{m}_{Y,W} & \longrightarrow & [U \cap W, f|_{U \cap W}]. \end{array} \tag{5.12}$$

This is well-defined and it is easy to check that it is a (non-zero) homomorphism of $\mathbb{K}$-algebras; since $K_{Y,W}$ and $K(W)$ are both fields, $\varphi_{Y,W}$ is injective. We now prove that $\varphi_{Y,W}$ is also surjective. Let $[U', f'] \in K(W)$ and let $P \in U'$ be any point. The fact that f' is regular at P implies there exists an open neighborhood U'' of P in U' and two homogeneous polynomials G_1, $G_2 \in H(S(V^*))$ of the same degree, with $G_2 \neq 0$, such that

$$f'|_{U''} = \frac{G_1}{G_2}|_{U''} \tag{5.13}$$

and $U'' \subseteq U_p(G_2)$ (recall definitions as in Section 5.2). One can always find an open subset $\widetilde{U} \subset \mathbb{P}(V)$ s.t. $\widetilde{U} \cap W = U''$ and $\widetilde{U} \cap Z_p(H) = \emptyset$ (otherwise one can always replace $\widetilde{U}$ with $\widetilde{U} \cap U_p(H)$). Let $U := \widetilde{U} \cap Y$. By the choice of $\widetilde{U}$, one has $U \subseteq U_p(H)$. Define

$$f := \frac{G_1}{G_2}|_U \in \mathcal{O}_Y(U) \subseteq K(Y). \tag{5.14}$$

Since $U \cap W = \widetilde{U} \cap W = U'' \subset W$, then $[U, f] \in \mathcal{O}_{Y,W}$. From (5.13) and (5.14) it follows that $[U'', f|_{U''}] \, \mathcal{R} \, [U', f']$, i.e.

$$\varphi_{Y,W}([U, f] + \mathfrak{m}_{Y,W}) = [U', f'],$$

proving the surjectivity of $\varphi_{Y,W}$. □

Definition 5.3.11. For any algebraic variety Y and any subvariety $W \subseteq Y$, the ring $(\mathcal{O}_{Y,W}, \mathfrak{m}_{Y,W})$ is called the *local ring of W in Y*.

For any open subset $U \subseteq Y$ s.t. $U \cap W \neq \emptyset$ the map j_U as in (5.5) induces an injective $\mathbb{K}$-algebra homomorphism

$$\gamma_U : \mathcal{O}_Y(U) \hookrightarrow \mathcal{O}_{Y,W}, \text{ defined by } f \xrightarrow{\gamma_U} [U, f]. \tag{5.15}$$

Definition 5.3.12. The ideal

$$\mathfrak{I}_U(W) := \gamma_U^{-1}(\mathfrak{m}_{Y,W}) = \{f \in \mathcal{O}_Y(U) \mid Z_U(f) \supseteq U \cap W\}$$

will be called the *ideal of W in U*.

Lemma 5.3.13. *Let Y be any algebraic variety and let $U \subseteq Y$ be any non-empty, open subset.*

(i) *For any subvariety $W \subseteq Y$ s.t. $W' := U \cap W \neq \emptyset$, one has $\mathcal{O}_{U,W'} \cong \mathcal{O}_{Y,W}$.*

(ii) *In particular, $K(U) \cong K(Y)$.*

Proof. It is clear that (ii) is a particular case of (i), taking $W = Y$. Therefore, we only need to prove (i). Let $[U', f'] \in \mathcal{O}_{U,W'}$, where $U' \subseteq U$ an open set s.t. $U' \cap W' = U' \cap U \cap W \neq \emptyset$. Since U' is open in Y, $f' \in \mathcal{O}_Y(U')$ and $U' \cap W \neq \emptyset$, then $[U', f'] \in \mathcal{O}_{Y,W}$. Thus, one has an injective $\mathbb{K}$-algebra homomorphism

$$\psi : \mathcal{O}_{U,W'} \hookrightarrow \mathcal{O}_{Y,W}, \text{ defined by } [U', f'] \to [U', f'].$$

Note that ψ is also surjective; indeed for any $[U'', f''] \in \mathcal{O}_{Y,W}$, where the representative (U'', f'') is s.t. $U'' \cap W \neq \emptyset$, one has

$$[U'', f''] = \psi([U'' \cap U, f''|_{U'' \cap U}])$$

since $U'' \cap U \cap W \neq \emptyset$. $\qquad \square$

Recalling notation as in Remarks 1.11.11, 5.3.7, we can prove the following fundamental result.

Theorem 5.3.14 (Fundamental theorem on regular and rational functions). *Let $Y \subseteq \mathbb{A}^n$ be any affine variety. Then:*

(a) $\mathcal{O}_Y(Y) = A(Y)$.

(b) *For any subvariety $W \subseteq Y$, the ideal $\mathfrak{I}_Y(W)$ (cf. Definition 5.3.12) is a prime ideal in $\mathcal{O}_Y(Y)$ which is isomorphic to the prime ideal $I_{a,Y}(W)$ of $A(Y)$ as in (5.7). Conversely any prime ideal of $\mathcal{O}_Y(Y)$ is of this type; furthermore, $\mathfrak{I}_Y(W)$ is a maximal ideal if and only if W is a point $P \in Y$.*

(c) *For any subvariety $W \subseteq Y$ one has*

$$\mathcal{O}_{Y,W} \cong \mathcal{O}_Y(Y)_{\mathfrak{J}_Y(W)} \cong A(Y)_{I_{a,Y}(W)}.$$

(d) $K(Y) = \mathfrak{Q}(A(Y))$ *whereas, for any subvariety $W \subseteq Y$ one has*

$$K(W) \cong \frac{\mathcal{O}_Y(W)_{\mathfrak{J}_Y(W)}}{\mathfrak{J}_Y(W)\mathcal{O}_Y(W)_{\mathfrak{J}_Y(W)}} \cong \frac{A(Y)_{I_{a,Y}(W)}}{I_{a,Y}(W)A(Y)_{I_{a,Y}(W)}}.$$

Let $Y \subseteq \mathbb{P}^n$ be any projective variety. Then
(e) $\mathcal{O}_Y(Y) = \mathbb{K}$.
(f) *For any subvariety $W \subseteq Y$, the ideal $I_{p,Y}(W)$ as in (5.8), is a homogeneous, prime ideal of $\mathcal{S}(Y)$. Conversely any homogeneous, non-irrelevant, prime ideal of $\mathcal{S}(Y)$ is of this type.*
(g) *For any subvariety $W \subseteq Y$, one has*

$$\mathcal{O}_{Y,W} \cong \mathcal{S}(Y)_{(I_{p,Y}(W))}.$$

(h) $K(Y) = \mathcal{S}(Y)_{((0))}$ *whereas, for any subvariety $W \subseteq Y$,*

$$K(W) \cong \frac{\mathcal{S}(Y)_{(I_{p,Y}(W))}}{I_{p,Y}(W)\mathcal{S}(Y)_{(I_{p,Y}(W))}}.$$

Proof. Let us first focus on the affine case.

(a) By definition of $A(Y)$, there is a natural injective $\mathbb{K}$-algebra homomorphism $\alpha : A(Y) \hookrightarrow \mathcal{O}_Y(Y)$. From the Hilbert "Nullstellensatz"-weak form in $\mathbb{A}^n$ (cf. Theorem 2.2.1) and (5.7), there is a bijective correspondence between points $P \in Y$ and maximal ideals $\mathfrak{m}_Y(P) := I_{a,Y}(P)$ of $A(Y)$. Identifying $A(Y)$ with its image $\alpha(A(Y))$, we can therefore interpret $\mathfrak{m}_Y(P) = \{f \in A(Y) \subseteq \mathcal{O}_Y(Y) \mid f(P) = 0\} \subset \mathcal{O}_Y(Y)$, i.e. $\mathfrak{m}_Y(P) \subseteq \mathfrak{J}_Y(P)$, where the latter ideal is as in Definition 5.3.12.

Claim 5.3.15. For any $P \in Y$, one has $A(Y)_{\mathfrak{m}_Y(P)} \cong \mathcal{O}_{Y,P}$.

Proof. For any $\frac{f}{g} \in A(Y)_{\mathfrak{m}_Y(P)}$, the set $U_Y(g) := Y \backslash Z_Y(g)$ is a non-empty, open subset of Y containing P, indeed $g \in A(Y) \backslash \mathfrak{m}_Y(P)$ by definition of $A(Y)_{\mathfrak{m}_Y(P)}$ so $\frac{f}{g}$ is regular over $U_Y(g)$ (cf. the proof of Proposition 5.2.4-(i) and Remark 5.2.5-(b)). We can therefore consider the following map:

$$\alpha_P : A(Y)_{\mathfrak{m}_Y(P)} \longrightarrow \mathcal{O}_{Y,P}, \text{ defined by } \frac{f}{g} \xrightarrow{\alpha_P} \left[U_Y(g), \frac{f}{g}\right]. \qquad (5.16)$$

This is a $\mathbb{K}$-algebra homomorphism. Note moreover that α_P is an isomorphism: first $\left[U_Y(g), \frac{f}{g}\right] = \left[U_Y(g'), \frac{f'}{g'}\right]$ if and only if on $U_Y(g) \cap U_Y(g') \neq \emptyset$

one has $fg' - f'g = 0$. Since $U_Y(g) \cap U_Y(g')$ is an open, dense subset of Y, by Proposition 5.2.4-(ii), one has $fg' - f'g = 0$ on Y, i.e. $fg' - f'g = 0 \in A(Y)$. Since $g, g' \in A(Y) \setminus \mathfrak{m}_Y(P)$, from (1.35) we get that $\frac{f}{g} = \frac{f'}{g'} \in A(Y)_{\mathfrak{m}_Y(P)}$, so α_P is injective. At last, the surjectivity of α_P directly follows from Lemma 5.2.3. $\qquad\square$

By construction, α_P is a *local isomorphism of local rings*, i.e. the maximal ideals $\mathfrak{m}_Y(P)A(Y)_{\mathfrak{m}_Y(P)}$ and $\mathfrak{m}_{Y,P} = \mathfrak{I}_Y(P)$ bijectively correspond under α_P. In particular, this proves points (b), (c) and (d) for the case $W = \{P\}$. From (5.10) and Claim 5.3.15 we get

$$A(Y) \cong \alpha(A(Y)) \subseteq \mathcal{O}_Y(Y) = \bigcap_{P \in Y} \mathcal{O}_{Y,P} \cong \bigcap_{\mathfrak{m}_Y(P) \in \mathrm{Specm}(A(Y))} A(Y)_{\mathfrak{m}_Y(P)},$$

where

$$\mathrm{Specm}(A(Y)) := \{\text{maximal ideals of } A(Y)\}. \tag{5.17}$$

Since $A(Y)$ is integral, any $A(Y)_{\mathfrak{m}_Y(P)}$ is a subring of $\mathcal{Q}(A(Y))$ and, from the definition of localization (cf. also Matsumura, 1980, Lemma 2, p. 8) one has $\bigcap_{\mathfrak{m}_Y(P) \in \mathrm{Specm}(A(Y))} A(Y)_{\mathfrak{m}_Y(P)} = A(Y)$, proving (a).

(b) Since $I_a(W) = I_a(\overline{W})$, we can consider W a closed subvariety. By definition of $A(Y)$, we have a bijective correspondence between prime ideals $\overline{\mathfrak{p}} \subseteq A(Y)$ and prime ideals $\mathfrak{p} \subset A^{(n)}$ containing $I_a(Y)$. From Proposition 2.1.11-(i) and (2.15), any such prime ideal $\mathfrak{p}$ corresponds to a closed subvariety $W := Z_a(\mathfrak{p}) \subseteq Z_a(I_a(Y)) = Y \subseteq \mathbb{A}^n$. Since $\mathfrak{p}$ is prime, it is radical so $\mathfrak{p} = I_a(W)$ and $\overline{\mathfrak{p}} = I_{a,Y}(W)$. Using (a), the prime ideal $I_{a,Y}(W)$ by definition coincides with the ideal $\mathfrak{I}_Y(W) \subset \mathcal{O}_Y(Y)$. At last, from the proof of part (a), $I_{a,Y}(W) \cong \mathfrak{I}_Y(W)$ is maximal if and only if W is a point.

(c) On the one hand, for any subvariety $W \subseteq Y$, the ring $\mathcal{O}_{Y,W}$ is local, with maximal ideal $\mathfrak{m}_{Y,W}$ (cf. Theorem 5.3.10). On the other hand, from part (a) above, we have $A(Y) \cong \mathcal{O}_Y(Y)$ and $I_{a,Y}(W) \cong \mathfrak{I}_Y(W)$. Thus, one has that $A(Y)_{I_{a,Y}(W)} \cong \mathcal{O}_Y(Y)_{\mathfrak{I}_Y(W)}$ is a local ring of maximal ideal $I_{a,Y}(W)A(Y)_{I_{a,Y}(W)} \cong \mathfrak{I}_Y(W)\mathcal{O}_Y(Y)_{\mathfrak{I}_Y(W)}$ (cf. Proposition 1.11.12). Since $A(Y)_{I_{a,Y}(W)} \subseteq K(Y)$, as in the proof of Claim 5.3.15, one constructs a map $\alpha_W : A(Y)_{I_{a,Y}(W)} \longrightarrow \mathcal{O}_{Y,W}$ which is a local isomorphism of local $\mathbb{K}$-algebras.

(d) From Proposition 5.3.9-(i) and from (c), we get $K(Y) = \mathcal{O}_{Y,Y} \cong A(Y)_{(0)} = \mathcal{Q}(A(Y))$, the last equality following from Remark 1.11.11.

Similarly, from Theorem 5.3.10, Proposition 1.11.12-(i) and from (c) above, one gets the assertions on $K(W)$.

We now consider the case of $Y \subseteq \mathbb{P}^n$ a projective variety. Recalling notation as in (1.40) and (3.14), a preliminary step is given by the following result.

Lemma 5.3.16. *Let* $i \in \{0, \ldots, n\}$ *be any integer for which* $Y_i := Y \cap U_i \neq \emptyset$. *Then* $A(Y_i) \cong S(Y)_{([X_i])}$, *where* $X_i \in S(Y)$ *denotes the image via the quotient morphism* $S^{(n)} \twoheadrightarrow S(Y)$ *of the indeterminate* $X_i \in S^{(n)}$.

Proof. Assuming that the affine variety $Y_0 \neq \emptyset$, for simplicity we consider only the case $i = 0$, the other cases being similar. One has a natural $\mathbb{K}$-algebra isomorphism

$$A^{(n)} \xrightarrow{\varphi_0} S^{(n)}_{([X_0])}, \text{ defined by } f(x_1, \ldots, x_n) \xrightarrow{\varphi_0} f\left(\frac{X_1}{X_0}, \ldots, \frac{X_n}{X_0}\right).$$

If $f \in I_a(Y_0)$, by (1.32) and the definition of $I_p(Y)$, we get that $\mathfrak{h}_0(f) = X_0^{\deg(f)} \varphi_0(f) \in I_p(Y)$, i.e. $\varphi_0(f) \in I_p(Y)S^{(n)}_{([X_0])}$. Note that $I_p(Y)S^{(n)}_{([X_0])}$ is a proper ideal of $S^{(n)}_{([X_0])}$: indeed if we denote by S the multiplicative system given by 1 and by all the powers of the indeterminate X_0, by the fact that $Y_0 \neq \emptyset$ it follows that $I_p(Y) \cap S = \emptyset$, so one concludes by Proposition 1.11.6-(ii). In particular, φ_0 induces a homomorphism of $\mathbb{K}$-algebras

$$\overline{\varphi}_0 : A(Y) \longrightarrow \frac{S^{(n)}_{([X_0])}}{I_p(Y)S^{(n)}_{([X_0])}} \cong S(Y)_{([X_0])},$$

where the isomorphism on the right directly follows from the fact that $S(Y) = \frac{S^{(n)}}{I_p(Y)}$ and from the definition of localization with respect to the multiplicative system $S = \{1, X_0, X_0^2, X_0^3, \ldots\}$. Note that $\overline{\varphi}_0$ is an isomorphism: it is injective since $\overline{\varphi}_0(f) = 0$ if and only if $\mathfrak{h}_0(f) = 0$, i.e. if and only if $f = 0$; the surjectivity follows from the fact that, for any positive integer n and for any $F \in S(Y)_n$, there always exists a polynomial $f \in A(Y)$ such that $\frac{F}{X_0^n} = \overline{\varphi}_0(f)$. $\qquad \square$

We can proceed with the proof of the "projective" part of the statement.

(g) Let $W \subseteq Y$ be any (closed) subvariety and let $i \in \{0, \ldots, n\}$ be any integer for which $Y_i \neq \emptyset$ and $W_i := W \cap U_i \neq \emptyset$. For simplicity, assume that it occurs for, e.g. $i = 0$. From Lemma 5.3.13-(i), we have $\mathcal{O}_{Y,W} \cong \mathcal{O}_{Y_0,W_0}$. Since Y_0 and W_0 are affine varieties, from

(c) above we have $\mathcal{O}_{Y_0,W_0} \cong \mathcal{O}_{Y_0}(Y_0)_{\mathcal{I}_{Y_0}(W_0)}$. Moreover, from (a), one has also $\mathcal{O}_{Y_0}(Y_0) \cong A(Y_0)$ and $\mathcal{I}_{Y_0}(W_0) \cong I_{a,Y_0}(W_0)$. From the proof of Lemma 5.3.16, the isomorphism $\overline{\varphi}_0$ bijectively maps the ideal $I_{a,Y_0}(W_0)$ to the ideal $I_{p,Y}(W)\mathcal{S}(Y)_{([X_0])}$ so we get $\mathcal{O}_{Y,W} \cong A(Y_0)_{I_{a,Y_0}(W_0)} \cong \left(\mathcal{S}(Y)_{([X_0])}\right)_{I_{p,Y}(W)\mathcal{S}(Y)_{([X_0])}}$. Since $X_0 \notin I_{p,Y}(W)$ (because $W_0 \neq \emptyset$ by assumption), reasoning as in Lemma 5.2.3, one easily notices that $\left(\mathcal{S}(Y)_{([X_0])}\right)_{I_{p,Y}(W)\mathcal{S}(Y)_{([X_0])}} \cong \mathcal{S}(Y)_{(I_{p,Y}(W))}$, proving (g).

(h) The statement about $K(Y)$ follows from Proposition 5.3.9-(i) and from (g) above, whereas statement about $K(W)$ follows from Theorem 5.3.10, Proposition 1.11.12-(ii) and from (g) above.

(f) As in the affine case, it directly follows from bijective correspondence as in Corollary 4.1.12 for $\mathbb{P}^n$.

(e) Let $f \in \mathcal{O}_Y(Y)$ be any non-zero element. Since for any algebraic variety we have $\mathcal{O}_Y(Y) \subseteq K(Y)$, from (h) above any such f can be considered as an element of $K(Y) \cong \mathcal{S}(Y)_{((0))} = \mathcal{Q}_0(\mathcal{S}(Y)) \subset \mathcal{Q}(\mathcal{S}(Y))$, where $\mathcal{Q}_0(\mathcal{S}(Y))$ as in Definition 1.10.10. Notice first that an index $j \in \{0,\dots,n\}$ is such that $Y_j = \emptyset$ if and only if $Y \subseteq Z_p(X_j) = H_j$, where $H_j \cong \mathbb{P}^{n-1}$ the hyperplane at infinity of the affine open set U_j; this occurs if and only if $X_j \in I_p(Y)$, in which case $\mathcal{S}(Y) = \frac{\mathcal{S}^{(n)}}{I_p(Y)} \cong \frac{\mathcal{S}^{(n-1)}}{I_{p,H_j}(Y)}$, where $\mathcal{S}^{(n-1)} \cong \frac{\mathcal{S}^{(n)}}{(X_j)}$ and $I_{p,H_j}(Y) = \frac{I_p(Y)}{(X_j)}$. In other words, for any such j, we can replace $\mathcal{S}^{(n)}$ with $\mathcal{S}^{(n-1)}$, i.e. $Y \subset \mathbb{P}^n$ is degenerate as it is contained into the hyperplane $H_j \cong \mathbb{P}^{n-1}$ and one can work directly therein.

From now on we will therefore assume $Y_i \neq \emptyset$, for any $0 \leqslant i \leqslant n$. Since $f \in \mathcal{O}_Y(Y)$ then $f|_{Y_i} \in \mathcal{O}_{Y_i}(Y_i)$, for any i. Moreover, as Y_i is affine, from (a) and Lemma 5.3.16, $f|_{Y_i} \in \mathcal{O}_{Y_i}(Y_i) \cong A(Y_i) \cong \mathcal{S}(Y)_{([X_i])}$. Thus, for any $i \in \{0,\dots,n\}$, there exist a positive integer N_i and a polynomial $G_i \in \mathcal{S}(Y)_{N_i}$ such that $f|_{Y_i} := \frac{G_i}{X_i^{N_i}}$, i.e. $X_i^{N_i} f|_{Y_i} \in \mathcal{S}(Y)_{N_i}$, for any $0 \leqslant i \leqslant n$. Since $f \in \mathcal{O}_Y(Y) \subset K(Y) \cong \mathcal{S}(Y)_{((0))} = \mathcal{Q}_0(\mathcal{S}(Y)) \subset \mathcal{Q}(\mathcal{S}(Y))$ and $f|_{Y_i} \in \mathcal{S}(Y)_{([X_i])} \subset \mathcal{Q}_0(\mathcal{S}(Y)) \subset \mathcal{Q}(\mathcal{S}(Y))$, for any $0 \leqslant i \leqslant n$, considering f as an element of $\mathcal{Q}(\mathcal{S}(Y))$, we have

$$X_i^{N_i} f \in \mathcal{S}(Y)_{N_i}, \ \forall\, i \in \{0,\dots,n\}. \tag{5.18}$$

Let N be any positive integer such that $N \geqslant \Sigma_{i=1}^n N_i$. Since $\mathcal{S}(Y)_N$ is generated as a $\mathbb{K}$-vector space by monomials $X_0^{\alpha_0} X_1^{\alpha_1} \cdots X_n^{\alpha_n}$ such that

$\Sigma_{i=0}^n \alpha_i = N$, from (5.18), we get

$$\mathcal{S}(Y)_N f \subseteq \mathcal{S}(Y)_N. \tag{5.19}$$

Recursively applying (5.19), for any positive integer q one has $\mathcal{S}(Y)_N f^q \subseteq \mathcal{S}(Y)_N$. In particular, $X_0^N f^q \in \mathcal{S}(Y)_N$, for any integer $q \geqslant 1$, i.e.

$$\mathcal{S}(Y)[f] \subseteq \mathcal{S}(Y) \cdot \frac{1}{X_0^N}. \tag{5.20}$$

Since $\mathcal{S}(Y)$ is a noetherian ring and since $\mathcal{S}(Y) \cdot \frac{1}{X_0^N}$ is a finitely generated $\mathcal{S}(Y)$-module, then $\mathcal{S}(Y)[f]$ is a finitely generated $\mathcal{S}(Y)$-module (cf. Atiyah and McDonald, 1969, Propositions 6.2, 6.5). From Proposition 1.6.3, it follows that $f \in \mathcal{Q}_0(\mathcal{S}(Y))$ is therefore integral over $\mathcal{S}(Y)$, i.e. there exist $a_1, \ldots, a_m \in \mathcal{S}(Y)$ s.t.

$$f^m + a_1 f^{m-1} + \cdots + a_m = 0. \tag{5.21}$$

Since $f \in \mathcal{Q}_0(\mathcal{S}(Y))$ there exist homogeneous polynomials $G_1, G_2 \in H(\mathcal{S}(Y))$ of the same degree such that $f = \frac{G_1}{G_2}$, so (5.21) becomes $G_1^m + a_1 G_1^{m-1} G_2 + \cdots + a_m G_2^m = 0$.

Let $a_j = A_{j0} + A_{j1} + \cdots + A_{j\ell_j}$ be the decomposition of the coefficients $a_j \in \mathcal{S}(Y)$ in their homogeneous components, $1 \leqslant j \leqslant n$. Decomposing the previous equality into homogeneous pieces we get in particular $G_1^m + A_{10} G_1^{m-1} G_2 + \cdots + A_{m0} G_2^m = 0$, i.e.

$$f^m + A_{10} f^{m-1} + \cdots + A_{m0} = 0, \tag{5.22}$$

where $A_{j0} \in \mathcal{S}(Y)_0 = \mathbb{K}$, for any $1 \leqslant j \leqslant m$. In particular, $f \in K(Y)$ is algebraic over $\mathbb{K}$. Since $\mathbb{K}$ is algebraically closed, $f \in \mathbb{K}$ proving (e). $\qquad \square$

From Theorem 5.3.14-(a) and (e) it follows that for Y either an affine or a projective variety, $\mathcal{O}_Y(Y)$ is an integral $\mathbb{K}$-algebra of finite type. This has already been observed to more generally hold for any algebraic variety Y (cf. Remark 5.2.5-(c)).

Corollary 5.3.17. *Let $Y \subset \mathbb{P}^n$ be any projective variety and let $\mathcal{S}(Y)$ be its homogeneous coordinate ring. Let $\mathcal{Q}(\mathcal{S}(Y))$ be the quotient field of the integral $\mathbb{K}$-algebra $\mathcal{S}(Y)$. Let $x_0, \ldots, x_n$ be the images via the canonical epimorphism $\pi : \mathcal{S}^{(n)} \twoheadrightarrow \mathcal{S}(Y)$ of the homogeneous indeterminates $X_0, X_1, \ldots, X_n \in \mathcal{S}^{(n)}$. Then, for any $0 \leqslant j \leqslant n$, one has:*

(i) $\mathcal{Q}(\mathcal{S}(Y)) \cong K(Y)(x_j)$;

(ii) *x_j is transcendental over $K(Y)$.*

Proof. It suffices to prove the statements for $j = 0$. Let us consider any non-zero element $\frac{f}{g} \in \mathcal{Q}(\mathcal{S}(Y))$; thus one can write $f = F_0 + F_1 + \cdots + F_h$ and $g = G_0 + G_1 + \cdots + G_k$, where each F_i, respectively, G_j, is either zero or homogeneous of degree i, respectively, j, for $1 \leqslant i \leqslant h$, $0 \leqslant j \leqslant k$. Therefore,

$$\frac{f}{g} = \frac{F_0 + F_1 + \cdots + F_h}{G_0 + G_1 + \cdots + G_k} = \frac{\sum_{i=0}^{h} F_i\left(1, \frac{x_1}{x_0}, \ldots, \frac{x_n}{x_0}\right) x_0^i}{\sum_{j=0}^{k} G_j\left(1, \frac{x_1}{x_0}, \ldots, \frac{x_n}{x_0}\right) x_0^j}.$$

Thus, $\frac{f}{g} \in \mathcal{S}(Y)_{((0))}(x_0)$. Since Y is projective, from Theorem 5.3.14-(h), we know that $K(Y) \cong \mathcal{S}(Y)_{((0))}$ so (i) is proved.

Assume now $\sum_s f_s\left(\frac{x_1}{x_0}, \ldots, \frac{x_n}{x_0}\right) x_0^s = 0$, where $f_s \in K(Y)$ with $d_s = \deg(f_s)$. If we set $d := \max_s(d_s)$ and $F_s(x_0, \ldots, x_n) := f_s\left(\frac{x_1}{x_0}, \ldots, \frac{x_n}{x_0}\right) x_0^{d_s}$, then we get

$$0 = \sum_s F_s(x_0, \ldots, x_n) x_0^{s-d_s} = x_0^{-d} \sum_s F_s(x_0, \ldots, x_n) x_0^{d+s-d_s}$$

$$= x_0^{-d} \sum_s G_s(x_0, \ldots, x_n) x_0^s,$$

with $G_s = x_0^{d-d_s} F_s \in \mathcal{S}(Y)_d$, i.e homogeneous of degree d, for any s. Since $\mathcal{S}(Y)$ is an integral domain, the previous equality imposes $\sum_s G_s(x_0, \ldots, x_n) x_0^s = 0$. Moreover, since $\mathcal{S}(Y)$ is a graded ring, it then follows that $G_s(x_0, \ldots, x_n) = 0$ for any s. This implies that $f_s = 0$ for any s, i.e. that x_0 is transcendental over $K(Y)$. $\qquad\square$

5.3.1   *Consequences of the fundamental theorem on regular and rational functions*

We want to discuss some fundamental consequences of Theorem 5.3.14.

Corollary 5.3.18. *For any integer $n \geqslant 0$, one has*

$$\mathcal{O}_{\mathbb{A}^n}(\mathbb{A}^n) = A^{(n)}, \quad \mathcal{O}_{\mathbb{P}^n}(\mathbb{P}^n) = \mathbb{K}, \quad \text{and} \quad K(\mathbb{A}^n) \cong K(\mathbb{P}^n) \cong \mathcal{Q}^{(n)}.$$

Proof. The first two equality are given by (a) and (e) of Theorem 5.3.14. The fact that $K(\mathbb{A}^n) \cong \mathcal{Q}^{(n)}$ is point (d) of the same result. The fact that also $K(\mathbb{P}^n) \cong \mathcal{Q}^{(n)}$ follows either from Lemma 5.3.13-(ii) or from point (h) of Theorem 5.3.14, Proposition 1.11.12-(ii) and isomorphism (5.1), namely one has $K(\mathbb{P}^n) \cong \mathcal{S}^{(n)}_{((0))} \cong \mathcal{Q}_0(\mathcal{S}^{(n)}) \cong \mathcal{Q}(A^{(n)}) = \mathcal{Q}^{(n)}$. $\qquad\square$

Corollary 5.3.19 (cf. also Mumford, 1988, Proposition 2, pp. 20–21). *If Y is any affine variety, the ring $A(Y)$ completely determines the ringed space $(Y, \mathcal{O}_Y)$. In other words, $A(Y)$ determines the topological space Y since:*

- *as a set of points one has $Y = \operatorname{Specm}(A(Y))$, where $\operatorname{Specm}(A(Y))$ as in (5.17), whereas,*
- *all irreducible closed subsets of Y are in bijective correspondence with elements in $\operatorname{Spec}(A(Y))$, where $\operatorname{Spec}(A(Y))$ defined in (1.43), moreover,*
- *all closed subsets of Y are unions of finitely many irreducible closed subsets,*
- *finally, for any open set $U \subseteq Y$ and for any point $P \in Y$, all integral $\mathbb{K}$-algebras $\mathcal{O}_Y(U)$ and $\mathcal{O}_{Y,P}$ giving rise to the structural sheaf of Y are suitable localizations of the affine coordinate ring $A(Y)$.*

Proof. From the bijective correspondence of Theorem 5.3.14-(b) between points $P \in Y$ and maximal ideals of $A(Y)$, as a set of points Y can be identified with $\operatorname{Specm}(A(Y))$. In this identification, the correspondence between prime ideals $\mathfrak{p}$ of $A(Y)$ and irreducible closed subvarieties $Z_Y(\mathfrak{p}) \subseteq Y$ in Theorem 5.3.14-(b), reads as

$$Z_Y(\mathfrak{p}) = \{\mathfrak{m} \in \operatorname{Specm}(A(Y)) \mid \mathfrak{p} \subseteq \mathfrak{m}\}. \tag{5.23}$$

Since Y is a noetherian topological space, any arbitrary closed subset Z is a finite union of its irreducible components, say $Z_1, \ldots, Z_k$. From Proposition 2.2.8 applied to ideals $I_{a,Y}(Z_i) = \frac{I_a(Z_i)}{I_a(Y)}$ of $A(Y)$, we get $I_{a,Y}(Z) = \cap_{i=1}^k I_{a,Y}(Z_i)$, so identification (5.23) can be extended to any radical ideal of $A(Y)$. This implies that $A(Y)$ recover the whole topology of Y, since it determines the family of all its closed subsets. Finally, from $\mathcal{O}_{Y,P} \cong A(Y)_{\mathfrak{m}_{Y,P}}$ and from (5.10), where any $\mathcal{O}_Y(U)$ and any $\mathcal{O}_{Y,P}$ is viewed as a subring of $K(Y)$ as in (5.11), one realizes that $A(Y)$ determines also the structural sheaf $\mathcal{O}_Y$. □

Corollary 5.3.20. *Let Y be any algebraic variety, and let $U \subset Y$ be any non-empty, open subset. Then $\mathcal{O}_U \cong \mathcal{O}_Y|_U$ (cf. Remark 5.1.4). In particular, if Y is any affine variety and if $\overline{Y}$ denotes its projective closure, then for any $P \in Y$ one has $\mathcal{O}_{Y,P} \cong \mathcal{O}_{\overline{Y},P}$ and $\mathfrak{m}_{Y,P} \cong \mathfrak{m}_{\overline{Y},P}$.*

Proof. From Definition 5.3.5 and (5.10) one has that, for any non-empty, open set $V \subseteq U \subset Y$, $\mathcal{O}_U(V) = \{\Phi \in K(U) \mid V \subset U_\Phi\}$ and $\mathcal{O}_Y(V) = \{\Phi \in K(Y) \mid V \subset U_\Phi\}$. Then, one concludes by Lemma 5.3.13-(ii). □

Recalling Definition 1.5.5, one has also the following fundamental result.

Corollary 5.3.21. *Let Y be any algebraic variety. Then $K(Y)$ is a finitely generated field extension of the base field $\mathbb{K}$.*

Proof. From Lemma 5.3.13-(ii), we can assume Y to be projective. Thus, from Theorem 5.3.14-(h), $K(Y) \cong \mathcal{S}(Y)_{((0))} = \mathcal{Q}_0(\mathcal{S}(Y))$. Since $\mathcal{S}(Y)$ is a quotient of $\mathcal{S}^{(n)}$, it is a $\mathbb{K}$-algebra of finite type generated by the images of the indeterminates $X_0, \ldots, X_n$ via the quotient morphism. Thus, $\mathcal{Q}(\mathcal{S}(Y))$ is a finitely generated field extension of $\mathbb{K}$. Since $\mathcal{Q}_0(Y)$ is the subfield of degree-zero fractions in $\mathcal{Q}(\mathcal{S}(Y))$, one can conclude. $\qquad\square$

5.3.2   Examples

We discuss here some applications of the previous results.

Example 5.3.22 (Points).

(i) If $Y = \{P\}$ is a point, from the first part of Theorem 5.3.14 we get that $\mathcal{O}_P(P) \cong K(P) \cong \mathbb{K}$.

   More generally, if Y is an algebraic variety and $P \in Y$ is a point then $K(P) \cong \mathbb{K}$, i.e. the field of rational functions of a point is of *intrinsic nature* for P and it is independent from the inclusion of P as a subvariety of Y. To prove this, with no loss of generality we may assume Y to be affine. Indeed, if Y is any quasi-projective variety, consider $\overline{Y} \subseteq \mathbb{P}^n$ its projective closure and, for any index $i \in \{0, \ldots, n\}$ s.t. $\overline{Y} \cap U_i \neq \emptyset$, take the affine variety $Y_i := \overline{Y} \cap U_i$; from Lemma 5.3.13-(ii), $K(Y) = K(\overline{Y}) = K(Y_i)$ so we can replace Y with Y_i.

   Thus, since Y can be considered affine, from Theorem 5.3.14-(c), $\mathcal{O}_{Y,P} \cong A(Y)_{\mathfrak{m}_{Y,P}}$ where $\mathfrak{m}_{Y,P}$ denotes both the ideal of germs of regular functions vanishing at P and the ideal $\frac{\mathfrak{m}_P}{I_a(Y)}$, via identification in Theorem 5.3.14-(a). From Theorem 5.3.14-(d) and Proposition 1.11.12-(i), we therefore get $K(P) \cong \frac{A(Y)_{\mathfrak{m}_{Y,P}}}{\mathfrak{m}_{Y,P} A(Y)_{\mathfrak{m}_{Y,P}}} \cong \mathcal{Q}\left(\frac{A(Y)}{\mathfrak{m}_{Y,P}}\right) \cong \mathbb{K}$ since $\frac{A(Y)}{\mathfrak{m}_{Y,P}} \cong \mathbb{K}$.

(ii) If $Y \subset \mathbb{P}^n$ is any projective variety and if $P \in Y$ is any point, from Theorem 5.3.14-(f), one has that $I_{p,Y}(P)$ is a homogeneous prime ideal of $\mathcal{S}(Y)$ which does not coincide with the irrelevant ideal of $\mathcal{S}(Y)$; in particular it is not maximal (as it occurs also for $I_p(P) \subset \mathcal{S}^{(n)}$, cf. Section 3.3.1). The ideal $I_{p,Y}(P)$ is the prime ideal generated by all homogeneous elements $F \in H(\mathcal{S}(Y))$ vanishing at P.

Example 5.3.23 (Irreducible conics).

(i) Let $Y := Z_a(x_1 x_2 - 1) \subset \mathbb{A}^2$, which we call a *hyperbola* (as in the real case). Since $x_1 x_2 - 1 \in A^{(2)}$ is irreducible, $I_a(Y) \subset A^{(2)}$ is a prime ideal. The (affine) coordinate ring $A(Y)$ is s.t. $A(Y) \cong \mathbb{K}[x_1][x_1^{-1}] = A_{x_1}^{(1)}$ (cf. notation as in (1.39)). From Theorem 5.3.14, $\mathcal{O}_Y(Y) \cong A_{x_1}^{(1)}$. In particular $\mathcal{O}_{\mathbb{A}^1}(\mathbb{A}^1) \cong A^{(1)}$ is identified with a (proper) subring of $A(Y)$. On the other hand Theorem 5.3.14 ensures also $K(\mathbb{A}^1) \cong K(Y) \cong \mathbb{Q}^{(1)}$. Geometric motivations about these function rings are suggested in Exercise 5.1 and discussed in full details in its solution.

(ii) Let $V := Z_a(x_2 - x_1^2) \subset \mathbb{A}^2$ be a *parabola*. This conic has "more similarity" with $\mathbb{A}^1$ than what the hyperbola above does. Indeed $A(V) \cong \mathbb{K}[x_1] = A^{(1)} = A(\mathbb{A}^1)$ and, from Theorem 5.3.14, $K(V) \cong \mathbb{K}(x_1) \cong \mathbb{Q}^{(1)} \cong K(\mathbb{A}^1)$, the latter as occurred for the hyperbola. For geometric motivations, see Exercise 5.2 and its solution.

(iii) At last, consider $Z := Z_a(x_1^2 + x_2^2 - 1) \subset \mathbb{A}^2$, with e.g. $\mathbb{K} = \mathbb{C}$, and $\mathbb{A}^2$ considered as the *complexification* of a real affine plane. Thus, classification of conics is up to affine transformations with coefficients from the subfield $\mathbb{R}$; then Z is a non-degenerate *ellipse*, since its points at infinity are the *cyclic points* $[0, 1, i]$ and $[0, 1 - i]$. Similar computations as in Remark 2.1.16-(iii) show that one has a rational parametrization

$$
\begin{array}{ccccc}
\mathbb{A}^1 \supset & W & \xrightarrow{\phi_3} & Z & \subset \mathbb{A}^2 \\
 & t & \to & (\frac{t^2+1}{2it}, \frac{t^2-1}{2t}) & \\
x_2 - i x_1 & \leftarrow & (x_1, x_2) & &
\end{array}
$$

where $i^2 = -1$ and $W := \mathbb{A}^1 \setminus \{0\}$. The map ϕ_3^{-1} is given by the pencil of (parallel) lines in $\mathbb{A}^2$, $\ell_t : x_2 - i x_1 - t = 0$, $t \in W$. This is actually given by the pencil of (projective) lines through one of the cyclic points, namely from $[0, i, 1]$. One then computes that $A(Z) \cong A_{x_1}^{(1)} \cong A(W)$ and $K(Z) = K(W) = K(\mathbb{A}^1) \cong \mathbb{Q}^{(1)}$, as it occurs for the hyperbola above.

Example 5.3.24 (Semi-cubic parabola). Let $Y := Z_a(x_1^3 - x_2^2) \subset \mathbb{A}^2$, which is called a *semi-cubic parabola*, or even a *cuspidal plane cubic* (cf. Figure 5.1). In this case, Y has a polynomial parametrization as in (3.31) given by:

$$
\begin{array}{ccc}
\mathbb{A}^1 & \xrightarrow{\phi} & Y \subset \mathbb{A}^2 \\
t & \to & (t^2, t^3) ,
\end{array}
$$

and $A(Y) \cong \mathbb{K}[t^2, t^3] \subset \mathbb{K}[t]$.

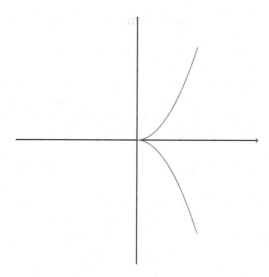

Fig. 5.1 Semi-cubic parabola or cuspidal plane cubic (real part).

In particular, $A(Y)$ is not isomorphic to $A(\mathbb{A}^1)$; on the other hand, from Theorem 5.3.14 one has $K(Y) \cong \mathbb{K}(t) \cong K(\mathbb{A}^1)$. The pencil of lines $x_2 = t \, x_1$, $t \in \mathbb{A}^1 \cong \mathbb{K}$, through the origin gives a bijective correspondence between Y and $\mathbb{A}^1$ by the following rule:

$$Y \xrightarrow{\ \psi\ } \mathbb{A}^1$$
$$(0,0) \neq (x_1, x_2) \ \rightarrow\ \tfrac{x_2}{x_1}$$
$$(0,0) = (x_1, x_2) \ \rightarrow\ 0.$$

The map ψ is bijective so it is a homeomorphism between $(Y, \mathfrak{Zar}^2_{a,Y})$ and $(\mathbb{A}^1, \mathfrak{Zar}^1_a)$; moreover $\psi^{-1} = \phi$, i.e. ϕ is a homeomorphism. In Example 6.2.5-(v), we will show that ϕ, on the other hand, is not an isomorphism of algebraic varieties.

Corollary 5.3.21 ensures that, for any algebraic variety Y, the field extension $\mathbb{K} \subseteq K(Y)$ is finitely generated. Note that all fields of rational functions appearing in Examples 5.3.23 and 5.3.24 are more precisely isomorphic to $\mathbb{K}(t)$, where t an indeterminate, so they all are purely transcendental extensions of $\mathbb{K}$ with transcendence degree 1 (cf. (1.15)). The following example shows that this is not always the case, i.e. not all algebraic varieties have fields of rational functions which are purely transcendental extension of $\mathbb{K}$.

Example 5.3.25 (Some elliptic affine plane cubics). Let $\mathbb{K} = \mathbb{C}$ and

$$Y_a := Z_a(x_2^2 - x_1(x_1 - 1)(x_1 - a)) \subset \mathbb{A}^2, \quad a \in \mathbb{C}.$$

When a is either 0 or 1, Y_a has a *node* (i.e. an ordinary double point) at the origin and for this reason it has a rational parametrization. To see this consider for simplicity the case $a = 0$, the other case being similar. In such a case, Y_0 is an affine rational cubic with polynomial parametrization

$$\mathbb{A}^1 \xrightarrow{\phi} Y_0, \text{ defined by } t \xrightarrow{\phi} (t^2 + 1, t^3 + t).$$

Note that the map ϕ is injective only on $W := \mathbb{A}^1 \setminus \{i, -i\}$, $i^2 = -1$, whereas $\phi(i) = \phi(-i) = (0, 0)$. The non-injectivity of ϕ produces the nodal singularity of Y_0 at the origin. Conversely, the pencil of lines $x_2 = tx_1$, $t \in \mathbb{C}$, through the origin gives bijective correspondence between $Y_0 \setminus \{(0, 0)\}$ and W. Using Theorem 5.3.14 one can compute that $K(Y_0) \cong \mathbb{K}(t)$, as we expected from the existence of the polynomial parametrization ϕ and the fact that $\phi|_W$ will be an isomorphism of algebraic varieties (cf. Chapter 6). To sum up, in terms of fields of rational functions, Y_0 and Y_1 behave exactly as the previous examples of conics and cuspidal plane cubics.

Let us consider therefore the case $a \neq 0, 1$. Since Y_a is affine, from Theorem 5.3.14-(a) we have that $\mathcal{O}_{Y_a}(Y_a) \cong A(Y_a)$, where

$$A(Y_a) \cong \{f + x_2 g \mid f, g \in A^{(1)} \text{ and } x_2^2 = x_1(x_1 - 1)(x_1 - a)\},$$

where $A^{(1)} = \mathbb{C}[x_1]$. From Theorem 5.3.14-(d), we get

$$K(Y_a) \cong \{\phi + x_2 \psi \mid \phi, \psi \in \mathcal{Q}^{(1)} \text{ and } x_2^2 = x_1(x_1 - 1)(x_1 - a)\},$$

where $\mathcal{Q}^{(1)} = \mathbb{C}(x_1)$; indeed, any element of $\mathcal{Q}(A(Y_a))$ is of the form $\frac{a_0 + x_2 a_1}{b_0 + x_2 b_1}$, with $a_0, a_1, b_0, b_1 \in A^{(1)}$ and $x_2^2 = x_1(x_1 - 1)(x_1 - a)$, which can be therefore written as $\frac{a_0 + x_2 a_1}{b_0 + x_2 b_1} \frac{b_0 - x_2 b_1}{b_0 - x_2 b_1} = \phi(x_1) + x_2 \psi(x_2)$, for some $\phi, \psi \in \mathcal{Q}^{(1)}$, as claimed. Thus, $x_2 \in K(Y_a)$ is a root of the polynomial $P_{x_2}(t) := t^2 - x_1(x_1 - 1)(x_1 - a) \in \mathcal{Q}^{(1)}[t]$, where t is an indeterminate over $\mathcal{Q}^{(1)}$. A key remark is now the following.

Lemma 5.3.26 (cf. also Reid, 1988, § 2.2). *Let $\mathbb{K}$ be a field of characteristic different from 2 and let $\lambda \in \mathbb{K}$, with $\lambda \neq 0, 1$. Let t be an indeterminate and $\Phi, \Psi \in \mathbb{K}(t)$ rational functions s.t. $\Phi^2 = \Psi(\Psi - 1)(\Psi - \lambda)$. Then $\Phi, \Psi \in \mathbb{K}$.*

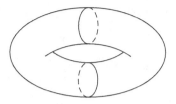

Fig. 5.2   Complex Torus or elliptic plane cubic.

The proof of this result is of pure algebraic nature and uses the fact that $\mathbb{K}[t]$ is a UFD and *Fermat's method of infinite descent*. The interested reader is referred to Reid (1988, Chapter 2, § 2.2). In our set-up, Lemma 5.3.26 ensures it cannot exist a non-constant map from a non-empty open set $U \subseteq \mathbb{A}^1$ to Y_a given by rational functions in the coordinate $t \in U$ (i.e. for $a \neq 0, 1$, Y_a admits no rational parametrization). Note moreover that it also implies that the polynomial $P_{x_2}(t)$ is irreducible over $\mathbb{Q}^{(1)}$ (otherwise, it would admit a root in $\mathbb{Q}^{(1)} = \mathbb{C}(x_1)$ contradicting Lemma 5.3.26). Thus, for $a \neq 0, 1$, the rational function $x_2 \in K(Y_a)$ is algebraic over $\mathbb{Q}^{(1)}$ (but not in $\mathbb{Q}^{(1)}$) and $K(Y_a) \cong \frac{\mathbb{Q}^{(1)}[t]}{(P_{x_2}(t))}$ is an algebraic extension of degree 2 of $\mathbb{Q}^{(1)}$, i.e. $K(Y_a)$ is a finitely generated field extension of $\mathbb{C}$ of transcendental degree 1 over $\mathbb{C}$, but not purely transcendental over $\mathbb{C}$.

Identifying the complex affine plane with the real affine four-dimensional space, in the Euclidean topology any Y_a with $a \neq 0, 1$ has a form as in Figure 5.2.

Exercises

Exercise 5.1. Let $\mathbb{K}$ be an algebraically closed field. Let $Y := Z_a(x_1 \, x_2 - 1) \subset \mathbb{A}^2$ be the *hyperbola* as in Example 5.3.23-(i). Consider the quasi-affine variety $W := \mathbb{A}^1 \setminus \{0\}$.

Show that the map

$$W \xrightarrow{\phi_1} Y,$$

defined as

$$t \xrightarrow{\phi_1} \left(t, \frac{1}{t} \right)$$

is a homeomorphism of algebraic varieties. Compare the ring $\mathcal{O}_W(W)$ and the field $K(W)$ with $\mathcal{O}_Y(Y)$ and $K(Y)$. Show that the map ϕ_1 is the restriction of a homeomorphism

$$\Phi_1 : \mathbb{P}^1 \to \overline{Y},$$

where $\overline{Y}$ denotes the projective closure in $\mathbb{P}^2$ of Y, when $\mathbb{A}^2$ is identified with the affine chart U_0 of $\mathbb{P}^2$.

Exercise 5.2. Let $\mathbb{K}$ be an algebraically closed field. Let $V := Z_a(x_2 - x_1^2) \subset \mathbb{A}^2$ be the *parabola* as in Example 5.3.23-(ii). Taking the map

$$\mathbb{A}^1 \xrightarrow{\ \phi_2\ } V,$$

defined as

$$t \xrightarrow{\ \phi_2\ } \left(t, t^2 \right),$$

answer to the same questions as those in Exercise 5.1.

Exercise 5.3. Let $\mathbb{K}$ be an algebraically closed field. Let $d \geqslant 2$ be any integer and let $V := \{(t, t^2, \ldots, t^d) \mid t \in \mathbb{A}^1\} \subset \mathbb{A}^d$. Show that V is an affine variety in $\mathbb{A}^d$. Determine $\mathcal{O}_V(V)$, $K(V)$ and $\operatorname{trdeg}_{\mathbb{K}}(K(V))$.

Exercise 5.4. Let $\mathbb{K}$ be an algebraically closed field. Consider the affine variety $V := Z_a(x_1 x_4 - x_2 x_3) \subset \mathbb{A}^4$ and let $f := \frac{x_1}{x_2} \in K(V)$. Describe $\operatorname{Dom}(f)$ as defined in Definition 5.3.5 (cf. Reid, 1988, Ex. 4.9, p. 83).

Exercise 5.5. Let $\mathbb{K}$ be an algebraically closed field. Consider $Z := Z_p(X_1^2 - X_0 X_2) \subset \mathbb{P}^3$. Prove that Z is a projective variety and describe $K(Z)$.

Chapter 6

Morphisms of Algebraic Varieties

In this chapter, we will study suitable maps which preserve the structure of algebraic varieties.

6.1 Morphisms

Let V and W be algebraic varieties.

Definition 6.1.1. A map $\varphi : V \to W$ is said to be a *morphism of algebraic varieties*, or simply a *morphism*, if:

(i) φ is a continuous map between the topological spaces V and W, and
(ii) for any non-empty open set $U \subseteq W$ and for any $f \in \mathcal{O}_W(U)$, one has
$$f_\varphi := f \circ \varphi \in \mathcal{O}_V(\varphi^{-1}(U)).$$

The set of all morphisms from V to W will be denoted by $\mathrm{Morph}(V, W)$.

For any algebraic variety V, one has $\mathrm{Id}_V \in \mathrm{Morph}(V, V)$. Moreover, if V, W Z are algebraic varieties and if $\varphi \in \mathrm{Morph}(V, W)$ and $\psi \in \mathrm{Morph}(W, Z)$, then $\psi \circ \varphi \in \mathrm{Morph}(V, Z)$.

Definition 6.1.2. $\varphi \in \mathrm{Morph}(V, W)$ is said to be an *isomorphism* if there exists $\psi \in \mathrm{Morph}(W, V)$ such that $\varphi \circ \psi = \mathrm{Id}_W$ and $\psi \circ \varphi = \mathrm{Id}_V$. If such a ψ exists, it is uniquely determined and will be denoted by φ^{-1}. The set of all isomorphisms from V to W will be denoted by $\mathrm{Isom}(V, W)$.

If $\mathrm{Isom}(V, W) \neq \emptyset$, then V and W are said to be *isomorphic* algebraic varieties, which will be denoted by $V \cong W$. The set $\mathrm{Isom}(V, V)$ will be denoted by $\mathrm{Aut}(V)$ and called the set of *automorphisms* of V.

149

Remark 6.1.3.

(i) For any $\varphi \in \mathrm{Morph}(V, W)$ and for any non-empty open subset $U \subseteq W$, from the previous definitions one has a natural map

$$\varphi^U : \mathcal{O}_W(U) \longrightarrow \mathcal{O}_V(\varphi^{-1}(U))$$
$$f \longrightarrow f_\varphi := f \circ \varphi, \tag{6.1}$$

which is a homomorphism of integral $\mathbb{K}$-algebras.

(ii) When $U = W$, the morphism φ^W will be simply denoted by

$$\varphi^{\#} : \mathcal{O}_W(W) \longrightarrow \mathcal{O}_V(V). \tag{6.2}$$

(iii) If $\varphi \in \mathrm{Isom}(V, W)$ then, for any non-empty open subset $U \subseteq W$, φ^U is an isomorphism of $\mathbb{K}$-algebras; indeed $(\varphi^U)^{-1} := (\varphi^{-1})^{\varphi^{-1}(U)}$. In particular

$$V \cong W \ \Rightarrow \ \mathcal{O}_V(V) \cong \mathcal{O}_W(W). \tag{6.3}$$

Remark 6.1.4. Recalling Example 5.3.23 and Exercises 5.1, 5.2, isomorphisms in (6.3) imply that the hyperbola $Y = Z_a(x_1 x_2 - 1)$ and the ellipse $Z = Z_a(x_1^2 + x_2^2 - 1)$ cannot be isomorphic to $\mathbb{A}^1$. Below (cf. Example 6.2.5-(iv)), we will show that they both are isomorphic to the quasi-affine variety $W := \mathbb{A}^1 \setminus \{0\}$ whereas the parabola $V = Z_a(x_2 - x_1^2)$ is instead isomorphic to the whole $\mathbb{A}^1$.

Definition 6.1.5. If $\varphi \in \mathrm{Morph}(V, W)$ is such that $\overline{\mathrm{Im}(\varphi)} = W$, then φ is said to be a *dominant morphism*, or simply *dominant*.

Parametrizations ϕ_1, in Exercise 5.1, and ϕ_3, in Example 5.3.23-(iii), are dominant (but not surjective) morphisms with target $\mathbb{A}^1$.

Remark 6.1.6. If $\varphi \in \mathrm{Morph}(V, W)$ is dominant then, for any non-empty open subset $U \subseteq W$, one has $U \cap \mathrm{Im}(\varphi) \neq \emptyset$. Indeed, if otherwise there existed a non-empty open set $U_1 \subset W$ such that $U_1 \cap \mathrm{Im}(\varphi) = \emptyset$ one would have $\mathrm{Im}(\varphi) \subseteq U_1^c \subset W$, with U_1^c a proper, non-empty, closed subset of W, contradicting the dominance of φ. Later on (cf. Corollary 11.1.2), we will more precisely show that any dominant morphism $\varphi \in \mathrm{Morph}(V, W)$ is such that $\mathrm{Im}(\varphi)$ always contains a non-empty open subset of the target variety W.

Proposition 6.1.7. *If $\varphi \in \mathrm{Morph}(V, W)$ is dominant, then $K(W) \subseteq K(V)$ is a field extension. In particular*

$$V \cong W \ \Rightarrow \ K(V) \cong K(W). \tag{6.4}$$

Proof. Recalling notation as in Definition 5.3.1 and taking into account Remark 6.1.6, the morphism φ induces a map

$$\mathcal{H}(W) \longrightarrow \mathcal{H}(V)$$
$$(U, f) \longrightarrow (\varphi^{-1}(U), f_\varphi),$$

which is compatible with the equivalence relation $\mathcal{R}$. Thus, it defines a map

$$\varphi^* : K(W) \longrightarrow K(V)$$
$$[U, f] \longrightarrow [\varphi^{-1}(U), f_\varphi], \tag{6.5}$$

which is a $\mathbb{K}$-algebra homomorphism. Since φ^* is a non-zero field homomorphism it is therefore injective, proving the first part of the statement. When in particular $\varphi \in \mathrm{Isom}(V, W)$, φ^* is a field isomorphism between $K(W)$ and $K(V)$. $\qquad\square$

Remark 6.1.8. The converse of (6.4) does not hold: take, e.g. the hyperbola $Y = Z_a(x_1 x_2 - 1)$, for which we found $K(Y) \cong K(\mathbb{A}^1)$ (cf. Example 5.3.23-(i)), even if Y cannot be isomorphic to $\mathbb{A}^1$ as observed in Remark 6.1.4.

Definition 6.1.9. Let V be an algebraic variety and let $W \subseteq V$ be any subvariety. The natural inclusion map $\iota_W : W \hookrightarrow V$ is a morphism which is called an *open* (*closed, locally-closed*, respectively) *immersion* if W is an open (closed, locally closed, respectively) subvariety of V.

Remark 6.1.10. If V is an algebraic variety and $\phi \in \mathrm{Morph}(V, W)$ then, in general, $\phi(V)$ is neither open nor closed in W. In general $\phi(V)$ is a *constructible set*, i.e. a union of finitely many locally closed subsets of W (cf. Corollary 11.1.3). To have an example, take, e.g. the *hyperbolic paraboloid* $V := Z_a(x_1 x_3 - x_2) \overset{\iota_V}{\hookrightarrow} \mathbb{A}^3$. Let $\pi : \mathbb{A}^3 \to \mathbb{A}^2$ be the projection defined by $\pi(x_1, x_2, x_3) = (x_1, x_2)$, which is easily seen to be a morphism (cf. also Example 6.2.5-(i)). Then $\phi := \pi \circ \iota_V \in \mathrm{Morph}(V, \mathbb{A}^2)$ and $\phi(V) = (\mathbb{A}^2 \setminus Z_a(x_1)) \cup \{(0, 0)\}$, which is neither closed nor open in $\mathbb{A}^2$ but it is a constructible set in $\mathbb{A}^2$.

In the next sections, we shall give some useful criteria ensuring that a given map between algebraic varieties is actually a morphism.

6.2 Morphisms with (Quasi) Affine Target

In this section, we focus on cases where the target of $\varphi : V \to W$ is a quasi-affine (or affine) variety. We start with the following basic result.

Proposition 6.2.1. *Let V be an algebraic variety. There is a bijective correspondence between* $\mathrm{Morph}(V, \mathbb{A}^1)$ *and* $\mathcal{O}_V(V)$.

Proof. Any $f \in \mathcal{O}_V(V)$ is a continuous map from V to $\mathbb{A}^1$, as it follows from Proposition 5.2.4-(i). Let now $U \subseteq \mathbb{A}^1$ be any non-empty open set. From Lemma 5.3.13-(ii) and Corollary 5.3.18, one has $K(U) \cong K(\mathbb{A}^1) = \mathbb{Q}^{(1)} \cong \mathbb{K}(t)$, where t an indeterminate over $\mathbb{K}$. Let $\frac{c(t)}{b(t)} \in \mathcal{O}_{\mathbb{A}^1}(U)$ be any regular function on U, i.e. $c(t), b(t) \in A^{(1)} = \mathbb{K}[t]$ and $U \subseteq U_a(b) = Z_a(b)^c$. Using notation as in (6.1), one has

$$f^U\left(\frac{c(t)}{b(t)}\right) = \frac{c(t)}{b(t)} \circ f = \frac{c(f)}{b(f)} \in \mathcal{O}_V(f^{-1}(U)),$$

i.e. f defines a morphism from V to $\mathbb{A}^1$. Conversely, any $\varphi \in \mathrm{Morph}(V, \mathbb{A}^1)$ gives rise to $\varphi^{\#}(\mathrm{Id}_{\mathbb{A}^1}) = (\mathrm{Id}_{\mathbb{A}^1})_\varphi \in \mathcal{O}_V(V)$ and, from the previous step, it turns out that φ is the morphism defined by this regular function. $\square$

Since $\mathbb{A}^n$ is affine, from Theorem 5.3.14-(a), any indeterminate $x_i \in A^{(n)} = A(\mathbb{A}^n) = \mathcal{O}_{\mathbb{A}^n}(\mathbb{A}^n)$, $1 \leqslant i \leqslant n$, can be identified with the regular function

$$\mathbb{A}^n \xrightarrow{x_i} \mathbb{A}^1, \quad (a_1, a_2, \ldots, a_n) \to a_i \tag{6.6}$$

which, in this interpretation, is called the *projection π_i of $\mathbb{A}^n$ onto its ith-axis*. If $W \subseteq \mathbb{A}^n$ is any quasi-affine variety, by abuse of notation, we will always denote by $x_i \in A(W)$ the restriction of the regular function (6.6) to the subvariety W.

Proposition 6.2.2. *Let V be an algebraic variety and let $W \subseteq \mathbb{A}^n$ be a quasi-affine variety. Then*

$$\varphi \in \mathrm{Morph}(V, W) \;\Leftrightarrow\; \varphi_i := x_i \circ \varphi \in \mathcal{O}_V(V), \quad \forall\, 1 \leqslant i \leqslant n.$$

In particular, for any $n \geqslant 1$, there is a bijective correspondence between $\mathrm{Morph}(V, \mathbb{A}^n)$ *and* $\mathcal{O}_V(V)^{\oplus n}$.

Proof. ($\Rightarrow$) Since $x_i \in A(W) \subseteq \mathcal{O}_W(W)$, by composition of morphisms we have $\varphi_i = x_i \circ \varphi \in \mathrm{Morph}(V, \mathbb{A}^1) = \mathcal{O}_V(V)$, as it follows from Lemma 6.2.1.

($\Leftarrow$) Let $\varphi : V \to W$ be any map s.t. $\varphi_i = x_i \circ \varphi \in \mathcal{O}_V(V)$, for any $1 \leqslant i \leqslant n$. Then, for any $f \in A(\mathbb{A}^n) = A^{(n)}$, one has $f' := f(\varphi_1, \ldots, \varphi_n) \in \mathcal{O}_V(V)$, since f' is a polynomial expression in the

φ_i's. From the proof of Proposition 5.2.4-(i), $Z_V(f') \subseteq V$ is a closed subset of V and moreover $\varphi^{-1}(Z_a(f) \cap W) = Z_V(f')$, i.e. φ is continuous. Let us show that φ is also a morphism. Take any non-empty open subset $U \subseteq W$ and any $f \in \mathcal{O}_W(U)$; by definition of regularity and by Lemma 5.2.3, for any $P' \in U$ there exist an open neighborhood $U_{P'} \subseteq U$ and two polynomials $g_{P'}, h_{P'} \in A^{(n)}$ such that $h_{P'} \neq 0$, $U_{P'} \subseteq U_a(h_{P'}) = Z_a(h_{P'})^c$, such that $f|_{U_{P'}} = \frac{g_{P'}}{h_{P'}}|_{U_{P'}}$ and moreover, for any $Q' \in U$ different from P' one has

$$(*) \quad \frac{g_{P'}}{h_{P'}}|_{U_{P'} \cap U_{Q'}} = f|_{U_{P'} \cap U_{Q'}} = \frac{g_{Q'}}{h_{Q'}}|_{U_{P'} \cap U_{Q'}}.$$

Consider $\varphi^U(f) = f \circ \varphi$ as in (6.1); we need to show that $\varphi^U(f)$ is regular on the open set $\varphi^{-1}(U) \subseteq V$. From the facts that: regularity is a local condition, $\{U_{P'}\}_{P' \in U}$ is an open covering of U, $\{\varphi^{-1}(U_{P'})\}_{P' \in U}$ is an open covering of $\varphi^{-1}(U)$ and from $(*)$, it suffices to check that for any $P' \in U$ one has $\varphi^{U_{P'}}(f) \in \mathcal{O}_V(\varphi^{-1}(U_{P'}))$. Note that

$$\varphi^{U_{P'}}(f) = f \circ \varphi|_{\varphi^{-1}(U_{P'})} = \frac{g_{P'}(\varphi_1, \ldots, \varphi_n)}{h_{P'}(\varphi_1, \ldots, \varphi_n)}|_{\varphi^{-1}(U_{P'})}.$$

Since $\varphi_i \in \mathcal{O}_V(V)$ for any $1 \leqslant i \leqslant n$ and since $U_{P'} \subseteq U_a(h_{P'})$, one has $\frac{g_{P'}(\varphi_1,\ldots,\varphi_n)}{h_{P'}(\varphi_1,\ldots,\varphi_n)} \in \mathcal{O}_V(\varphi^{-1}(U_{P'}))$, as desired.

For what concerns $\mathrm{Morph}(V, \mathbb{A}^n)$, the case $n = 1$ is Lemma 6.2.1 whereas the case $n \geqslant 2$ follows from the first part applied to $W = \mathbb{A}^n$. $\square$

Similarly to (5.4), for any $f_1, \ldots, f_n \in \mathcal{O}_V(V)$ one can define

$$Z_V(f_1, \ldots, f_n) := \{P \in V \mid f_1(P) = f_2(P) = \cdots = f_n(P) = 0\} \subseteq V. \quad (6.7)$$

Thus, from Proposition 6.2.2, for any $\varphi \in \mathrm{Morph}(V, \mathbb{A}^n)$ one can define

$$Z_V(\varphi) := \mathbb{Z}_V(\varphi_1, \ldots, \varphi_n) = \varphi^{-1}((0, \ldots, 0)).$$

In the sequel, for any $\mathbb{K}$-algebras R and S, we will denote by

$$\mathrm{Hom}_\mathbb{K}(R, S) \qquad (6.8)$$

the set of all $\mathbb{K}$-algebra homomorphisms from R to S.

Recalling (6.2), one has:

Proposition 6.2.3. *Let V be an algebraic variety and let $W \subseteq \mathbb{A}^n$ be an affine variety. The map*

$$\mathrm{Morph}(V, W) \xrightarrow{\alpha} \mathrm{Hom}_\mathbb{K}(A(W), \mathcal{O}_V(V))$$
$$\varphi \longrightarrow \varphi^\# \qquad\qquad (6.9)$$

is bijective.

Proof. By Theorem 5.3.14-(i), one has $\mathcal{O}_W(W) = A(W) = \frac{A^{(n)}}{I_a(W)}$. Take now any $h \in \mathrm{Hom}_{\mathbb{K}}(A(W), \mathcal{O}_V(V))$ and let

$$\xi_i := h(x_i) \in \mathcal{O}_V(V), \qquad (6.10)$$

for any $x_i \in A(W)$, $1 \leqslant i \leqslant n$. Consider the map $V \xrightarrow{\varphi_h} \mathbb{A}^n$, defined by $P \longrightarrow (\xi_1(P), \ldots, \xi_n(P))$. From Proposition 6.2.2, $\varphi_h \in \mathrm{Morph}(V, \mathbb{A}^n)$. If we show that $\varphi_h(V) \subseteq W$, we will get that $\varphi_h \in \mathrm{Morph}(V, W)$.

To prove this, consider any polynomial $f \in I_a(W)$. For any $P \in V$ we have $\varphi_h(f)(P) = (f \circ \varphi_h)(P) = f(\xi_1(P), \ldots, \xi_n(P))$, which means

$$\varphi_h(f) = f \circ \varphi_h = f(h(x_1), \ldots, h(x_n)).$$

Since $h \in \mathrm{Hom}_{\mathbb{K}}(A(W), \mathcal{O}_V(V))$, the latter equals $h(f(x_1, \ldots, x_n)) = h(0) = 0$, as $x_i \in A(W)$ whereas $f \in I_a(W)$. Thus $\varphi_h(V) \subseteq W$, as desired. At last, the map

$$\mathrm{Hom}_{\mathbb{K}}(A(W), \mathcal{O}_V(V)) \longrightarrow \mathrm{Morph}(V, W)$$
$$h \qquad \longrightarrow \qquad \varphi_h$$

is the inverse of α, which is therefore a bijection as desired. $\qquad\square$

Corollary 6.2.4. *Let V and W be affine varieties. Then $V \cong W$ as affine varieties if and only if $A(V) \cong A(W)$ as integral $\mathbb{K}$-algebras of finite type.*

Proof. One implication is (6.3), the other follows from Proposition 6.2.3. $\qquad\square$

Example 6.2.5.

(i) Let $1 \leqslant m < n$ be integers and consider the set inclusion $I := \{i_1, i_2, \ldots, i_m\} \subset \{1, 2, \ldots, n\}$. The injective, $\mathbb{K}$-algebra homomorphism $\mathbb{K}[x_{i_1}, \ldots, x_{i_m}] \xrightarrow{\pi_I^{\#}} \mathbb{K}[x_1, \ldots, x_n]$ corresponds to the surjective morphism

$$\pi_I : \mathbb{A}^n \twoheadrightarrow \mathbb{A}^m, \quad (a_1, \ldots, a_n) \xrightarrow{\pi_I} (a_{i_1}, \ldots, a_{i_m}), \qquad (6.11)$$

which is called the *projection of $\mathbb{A}^n$ onto the coordinates* $I = \{i_1, i_2, \ldots, i_m\}$.

(ii) Let $(b_1, \ldots, b_n) \neq (0, \ldots, 0)$ be non-negative integers. The morphism $\varphi : \mathbb{A}^1 \to \mathbb{A}^n$, defined by $t \xrightarrow{\varphi} (t^{b_1}, \ldots, t^{b_n})$, corresponds to the $\mathbb{K}$-algebra homomorphism

$$\varphi^{\#} : \mathbb{K}[x_1, \ldots, x_n] \to \mathbb{K}[t], \ x_i \xrightarrow{\varphi^{\#}} t^{b_i}, \quad 1 \leqslant i \leqslant n.$$

The image of φ is an affine rational curve with polynomial parametrization (cf. (3.31)).

(iii) In particular, for any non-negative integer b, $\varphi : \mathbb{A}^1 \to \mathbb{A}^2$, $t \xrightarrow{\varphi} (t, t^b)$ defines an isomorphism between $\mathbb{A}^1$ and the affine plane curve $V = Z_a(x_2 - x_1^b)$ as φ corresponds to the $\mathbb{K}$-algebra isomorphism $A(V) = \frac{\mathbb{K}[x_1,x_2]}{(x_2-x_1^b)} \xrightarrow{\varphi^\#} \mathbb{K}[t] = A(\mathbb{A}^1)$, defined by $\varphi^\#(x_1) = t$ and $\varphi^\#(x_2) = t^b$ (cf. Corollary 6.2.4). Specifically, the inverse morphism φ^{-1} is given by $\pi_1|_V$, where π_1 the projection of the affine plane $\mathbb{A}^2$ onto the x_1-axis, which is a morphism since it is a restriction to V of a morphism from $\mathbb{A}^2$ to $\mathbb{A}^1$.

(iv) The map ϕ_2 in Example 5.3.23-(ii) and Exercise 5.2 is an isomorphism. Indeed the parametrization

$$\phi_2 : \mathbb{A}^1 \to V, \ t \xrightarrow{\phi_2} (t, t^2)$$

of the parabola $V = Z_a(x_2 - x_1^2) \subset \mathbb{A}^2$ is given by the regular functions $t, t^2 \in \mathcal{O}_{\mathbb{A}^1}(\mathbb{A}^1)$ so, from Proposition 6.2.2, ϕ_2 is a morphism (cf. Figure 6.1). The map ϕ_2^{-1} is given by the restriction to V of the projection

$$\pi_1 : \mathbb{A}^2 \to \mathbb{A}^1, \ (x_1, x_2) \to x_1$$

onto the first coordinate as in (6.6), which is therefore a morphism. Since V and $\mathbb{A}^1$ are both affine varieties, Corollary 6.2.4 gives intrinsic motivation of $A(V) \cong A(\mathbb{A}^1)$ as observed in Example 5.3.23-(ii) and in Exercise 5.2.

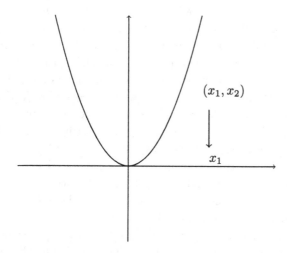

Fig. 6.1 Projection of the parabola $Z_a(x_2 - x_1^2)$ onto the x_1-axis.

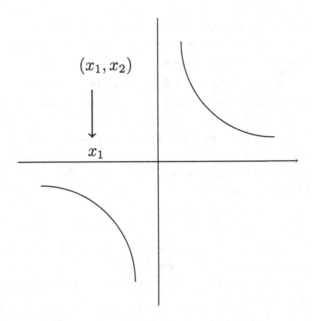

Fig. 6.2   Projection of the hyperbola $Z_a(x_1 x_2 - 1)$ to the x_1-axis.

Different situation occurs for, e.g. the hyperbola $Y = Z_a(x_1 x_2 - 1)$ as in Example 5.3.23-(i) and in Exercise 5.1 (cf. Figure 6.2).

From Proposition 6.2.2 the parametrization $\phi_1 : \mathbb{A}^1 \setminus \{0\} := W \to Y$ is an isomorphism, i.e. Y is isomorphic to the (quasi-affine) variety $W \subset \mathbb{A}^1$. From (6.3) it follows that $\mathcal{O}_W(W) \cong \mathcal{O}_Y(Y) \cong A(Y)$, the latter isomorphism coming from the fact that Y is affine (cf. Theorem 5.3.14). Recall that in Example 5.3.23-(i) we showed that $A(Y) \cong A_{x_1}^{(1)}$; the fact that ϕ_1 is an isomorphism in particular proves that $\mathcal{O}_W(W) \cong A_{x_1}^{(1)}$ (as stated without proof in Example 5.3.23-(i) and in Exercise 5.1). In particular, this is another way to prove that $K(W) \cong \mathcal{Q}^{(1)}$.

Considerations about the hyperbola hold verbatim for the ellipse $Z := Z_a(x_1^2 + x_2^2 - 1)$ with the map ϕ_3 as in Example 5.3.23-(iii), namely Z turns out to be isomorphic to $W = \mathbb{A}^1 \setminus \{0\}$.

(v) By definition of isomorphism, any $\phi \in \mathrm{Isom}(V, W)$ is also a homeomorphism between V and W as topological spaces; on the other hand, the converse does not hold. Consider, e.g. the semi-cubic parabola $Y = Z_a(x_1^3 - x_2^2)$ in Example 5.3.24. The map $\phi : \mathbb{A}^1 \to Y$ therein is a morphism, as it follows from Proposition 6.2.2. In Example 5.3.24, we proved that ϕ is also a homeomorphism. On the other hand, we also showed that $A(Y) \cong \mathbb{K}[t, t^2]$, which is a (proper) subring of $A(\mathbb{A}^1) \cong \mathbb{K}[t]$. Since Y and $\mathbb{A}^1$ are both affine varieties, from Corollary 6.2.4, ϕ cannot be an isomorphism. In fact that the map ψ constructed in Example 5.3.24 is the inverse of ϕ as a homeomorphism, but it is not a *morphism*, as ψ is not a *polynomial map*, i.e. a map given by a collection of polynomials, as prescribed by Proposition 6.2.2. Indeed, for any open set $U \subseteq \mathbb{A}^1$ containing 0, one has $t \in \mathcal{O}_{\mathbb{A}^1}(U)$ but $\psi^{-1}(t) = \frac{x_2}{x_1} \notin \mathcal{O}_Y(\psi^{-1}(U))$ as $(0,0) \in \psi^{-1}(U)$.

Interesting consequences of previous results are the following.

Corollary 6.2.6. *If V is an affine variety which is isomorphic to a projective variety, then V is a point.*

Proof. From Theorem 5.3.14-(a), if V is an affine variety, then $\mathcal{O}_V(V) \cong A(V)$. On the other hand, since V is isomorphic to a projective variety, by (6.3) and Theorem 5.3.14-(e), we have also $\mathcal{O}_V(V) \cong \mathbb{K}$. Thus, for some non-negative integer n, one has that $I_a(V) \subset A^{(n)}$ must be a maximal ideal. By Theorem 2.2.1, V is a point. $\qquad\square$

Corollary 6.2.7. *Any morphism $\varphi : V \to W$, where V a projective variety and W an affine variety, is constant.*

Proof. Since V is projective, $\mathcal{O}_V(V) = \mathbb{K}$ (cf. Theorem 5.3.14-(e)). At the same time, since W is affine then it is a closed, irreducible subset of some affine space $\mathbb{A}^n$, for some non-negative integer n, so $A(W) = \frac{A^n}{I_a(W)}$. For any $\varphi \in \mathrm{Morph}(V, W)$, consider $\varphi^\# \in \mathrm{Hom}_\mathbb{K}(A(W), \mathcal{O}_V(V))$. As in (6.10), for any generator $x_i \in A(W)$ one has $\xi_i = \varphi^\#(x_i) := a_i \in \mathbb{K}$, for any $1 \leqslant i \leqslant n$. This implies that for any point $P \in V$ one has $\xi_i(P) := a_i$ so $\varphi(P) = (a_1, \ldots, a_n)$, i.e. φ is constant. $\qquad\square$

Corollary 6.2.8. *Let V and W be affine varieties and $\varphi \in \mathrm{Morph}(V, W)$. Then, φ is dominant if and only if the map $\varphi^\# \in \mathrm{Hom}_\mathbb{K}(A(W), A(V))$ is injective.*

Proof. ($\Rightarrow$) If by contradiction $\mathrm{Ker}(\varphi^\#) \neq 0$, for any non-zero polynomial $f \in \mathrm{Ker}(\varphi^\#)$ one has $\varphi^\#(f) = 0 \in A(V)$, i.e. $\varphi^\#(f) \in I_a(V)$. Since $f \in A(W) = \mathcal{O}_W(W)$, by the proof of Proposition 5.2.4-(i), $Z_W(f)$ is a proper closed subset of W (as $0 \neq f \in A(W)$). Now, $\varphi^\#(f) \in I_a(V)$ is equivalent to $f \circ \varphi$ identically zero on W, i.e. $\varphi(V) \subseteq Z_W(f) \subsetneq W$, contradicting the dominance of φ.

($\Leftarrow$) Assume by contradiction that φ is not dominant, i.e. $\overline{\varphi(V)} = K \subsetneq W$, where K is an irreducible, proper closed subset of W (recall Corollary 4.1.6 and Proposition 4.1.2-(iii)). Then, with notation as in (5.7), for any non-zero $f \in I_{a,W}(K)$ we would have $\overline{\varphi(V)} \subseteq Z_W(f)$. Any such f determines a non-zero regular function $f \in A(W) = \mathcal{O}_W(W)$ s.t. $f \circ \varphi(P) = 0$ for any $P \in V$, i.e. a non-zero element of $\mathrm{Ker}(\varphi^{\#})$, contradicting the injectivity assumption. $\square$

The previous results are also consequences of more general facts which will be discussed later on (cf. Remark 9.2.2).

Example 6.2.9. Morphisms ϕ_1^{-1} and ϕ_3^{-1} in Examples 5.3.23-(i), (iii) and 6.2.5-(iv) and in Exercise 5.1 are dominant but not surjective morphisms with target $\mathbb{A}^1$. Indeed, in both cases we have the injection $A(\mathbb{A}^1) = A^{(1)} \hookrightarrow A^{(1)}_{x_1} \cong A(Y) \cong A(Z)$, where Y the hyperbola and Z the ellipse, respectively.

The use of isomorphisms allows to extend some previous definitions.

Definition 6.2.10. Let V be any algebraic variety. V is said to be an *affine variety* if V is isomorphic to an irreducible closed subset in some affine space. If $U \subseteq V$ is an open subset which is an affine variety, then U will be called an *affine open set* of V.

Lemma 6.2.11. *For any* $0 \leqslant i \leqslant n$, *the principal open set* $U_i \subset \mathbb{P}^n$ *is an affine open set of* $\mathbb{P}^n$ *isomorphic to* $\mathbb{A}^n$; *in other words the homeomorphism* ϕ_i *as in Proposition 3.3.2 is actually an isomorphism, for any* $0 \leqslant i \leqslant n$. *More generally, any projective variety* $V \subseteq \mathbb{P}^n$ *has a finite open covering* $\{V_i\}_{0 \leqslant i \leqslant n}$, *where* $V_i := V \cap U_i$ *is an affine open set of* V.

Proof. One is reduced to showing that the map ϕ_i as in (3.15) is an isomorphism, for any $i \in \{0, \ldots, n\}$. From Proposition 3.3.2, any such ϕ_i is a homeomorphism. The fact that ϕ_i is a morphism follows from the expression of ϕ_i in (3.15) and from Proposition 6.2.2. To prove that ϕ_i^{-1} is a morphism too, one uses the definition of regular function and Lemma 5.2.3. $\square$

Another example of affine open set of an algebraic variety is given by $W := \mathbb{A}^1 \setminus \{0\} \subset \mathbb{A}^1$: W is isomorphic to, e.g. the hyperbola $Y = Z_a(x_1 x_2 - 1) \subset \mathbb{A}^2$ as in Example 6.2.5-(iv). This is a particular case of the following more situation.

Lemma 6.2.12. *For any positive integer n and for any non-constant polynomial $f \in A^{(n)}$, let $Z := Z_a(f) \subset \mathbb{A}^n$ be the affine hypersurface defined by f. Then $U := \mathbb{A}^n \setminus Z \subset \mathbb{A}^n$ is an affine open set, being isomorphic to the irreducible hypersurface $Z' := Z_a(x_{n+1}f - 1) \subset \mathbb{A}^{n+1}$. Moreover, one has $\mathcal{O}_U(U) \cong A_f^{(n)}$.*

Proof. First note that $(x_{n+1}f - 1) \in A^{(n+1)}$ is irreducible so Z' is an affine variety: this follows from $A(Z') = \frac{A^{(n+1)}}{(x_{n+1}f-1)} \cong A_f^{(n)} \subset \mathbb{Q}(A^{(n)})$, which therefore has to be an integral domain, and from Proposition 4.1.10. Consider now the map

$$\psi : Z' \to \mathbb{A}^n \setminus Z, \quad (a_1, a_2, \ldots, a_{n+1}) \xrightarrow{\psi} (a_1, a_2, \ldots, a_n).$$

Since $\psi = \pi_I|_{Z'}$, for the multi-index $I := (1, 2, \ldots, n)$, then ψ is a morphism. Conversely, the map

$$\varphi : \mathbb{A}^n \setminus Z \to Z', \quad (a_1, a_2, \ldots, a_n) \xrightarrow{\varphi} \left(a_1, a_2, \ldots, a_n, \frac{1}{f(a_1, \ldots, a_n)}\right)$$

is a morphism by Proposition 6.2.2 which is the inverse of the morphism ψ, proving that Z' and $U = \mathbb{A}^n \setminus Z$ are isomorphic. At last, since Z' is an affine hypersurface, by Theorem 5.3.14-(a) and by (6.3) we have $\mathcal{O}_U(U) \cong A(Z') \cong A_f^{(n)}$, where the last isomorphism has been proved above. $\qquad\square$

Corollary 6.2.13. $\mathbb{A}^n$ *admits a basis for the topology $\mathfrak{Zar}_a^n$ consisting of affine open sets.*

Proof. It directly follows from Lemma 2.1.21 and the previous result. $\quad\square$

The next example shows that there actually exist quasi-affine varieties which cannot be affine varieties.

Example 6.2.14. From Lemma 6.2.12, if $Z \subset \mathbb{A}^2$ is any curve then $\mathbb{A}^2 \setminus Z$ is an affine open set of $\mathbb{A}^2$. On the contrary, for any point $P \in \mathbb{A}^2$, we now show that $\mathbb{A}^2 \setminus \{P\}$ is a quasi-affine variety which cannot be affine. With the use of affine transformations, it is not a restriction to consider P to be the origin $O = (0,0)$. Then $W := \mathbb{A}^2 \setminus \{O\} = Z_a((x_1, x_2))^c$ is quasi-affine. Let $\iota_W : W \hookrightarrow \mathbb{A}^2$ be the open immersion of W as a subvariety, so $\iota_W \in \operatorname{Morph}(W, \mathbb{A}^2)$. Since $\mathbb{A}^2$ is affine, from Proposition 6.2.3, ι_W corresponds to the $\mathbb{K}$-algebra homomorphism $\iota_W^\# \in \operatorname{Hom}_{\mathbb{K}}(A(\mathbb{A}^2), \mathcal{O}_W(W))$. Since $A(\mathbb{A}^2) = A^{(2)}$, $\iota_W^\#$ simply sends any polynomial $f \in A^{(2)}$ (viewed as a regular function on $\mathbb{A}^2$) to its restriction to W. Since W is a dense open set in $\mathbb{A}^2$, it is clear that $\iota_W^\#$ is an injective homomorphism.

Claim 6.2.15. $\iota_W^{\#}$ *is an isomorphism.*

Proof. From above, we need to show that $\iota_W^{\#}$ is surjective. From Lemma 5.3.13 and Theorem 5.3.14-(d), we have $\mathcal{O}_W(W) \subset K(W) \cong K(\mathbb{A}^2) = \mathcal{Q}(A^{(2)}) = \mathcal{Q}^{(2)}$, i.e. any regular function on W can be identified with a rational function $\phi = \frac{g}{h} \in \mathcal{Q}^{(2)}$ such that $g, h \in A^{(2)}$, $h \neq 0$ and $W \subseteq U_a(h) = Z_a(h)^c$. Thus, to prove the surjectivity of $\iota_W^{\#}$ is equivalent to showing that any $\phi \in \mathcal{O}_W(W)$ is actually a polynomial, i.e. that $h \in \mathbb{K}^*$. Take therefore any $\phi = \frac{g}{h} \in \mathcal{O}_W(W)$ and assume by contradiction that $h \in A^{(2)} \setminus \mathbb{K}$. Since $W \subseteq U_a(h)$, then one has either $Z_a(h) = \emptyset$ or $Z_a(h) = \{O\}$. From Theorem 2.2.3, in the first case we would have $\sqrt{(h)} = (1)$ whereas in the second case $\sqrt{(h)} = \mathfrak{m}_O = (x_1, x_2)$, a contradiction in both cases. This concludes the proof of Claim 6.2.15. $\qquad\square$

Thus, if W were affine, from Theorem 5.3.14-(a) we would have $\mathcal{O}_W(W) \cong A(W)$ so, from Claim 6.2.15 and Corollary 6.2.4, ι_W would be an isomorphism which is a contradiction, since ι_W is injective but not surjective onto $\mathbb{A}^2$.

6.3   Morphisms with (Quasi) Projective Target

We start with the following general result.

Proposition 6.3.1. *Let n, m be non-negative integers and let $V \subseteq \mathbb{P}^n$ and $W \subseteq \mathbb{P}^m$ be quasi-projective varieties. A non-constant map $\psi : V \to W$ is a morphism if and only if, for any $P \in V$, there exist an open neighborhood U_P of P in V, a non-negative integer k and $m+1$ homogeneous polynomials $F_0, \ldots, F_m \in \mathcal{S}_k^{(n)}$ such that:*

(a) *there exists $i \in \{0, \ldots, m\}$ for which $F_i(P') \neq 0$, for any $P' \in U_P$, and*
(b) *$\psi(P') = [F_0(P'), \ldots, F_m(P')]$, for any $P' \in U_P$.*

Proof. ($\Rightarrow$) For $P \in V$, let $Q := \psi(P) \in W \subseteq \mathbb{P}^m$. Since $W \neq \emptyset$, there exists an index $i \in \{0, \ldots, m\}$ for which $W_i := W \cap U_i \neq \emptyset$, U_i the fundamental affine open set of $\mathbb{P}^m$. For simplicity, assume this occurs for $i = 0$. Thus, W_0 is an open set of W, so $U_P := \psi^{-1}(W_0) \subseteq V$ is an open neighborhood of P in V, being ψ continuous. The map $\psi' := \psi|_{U_P} : U_P \to W_0$ is a morphism; up to composing with the inclusion $W_0 \overset{\iota_{W_0}}{\hookrightarrow} U_0$ and with the isomorphism $U_0 \overset{\phi_0}{\longrightarrow} \mathbb{A}^m$, ψ' has quasi-affine target W_0. From Proposition 6.2.2, ψ' is therefore given by a m-tuple of regular functions over $U_P \subseteq V \subseteq \mathbb{P}^n$. Up to suitably restricting the open neighborhood U_P, this means there exist m non-negative integers d_i, $1 \leqslant i \leqslant m$, and m pairs of homogeneous polynomials

$$(F_1, F_{1,0}), \ (F_2, F_{2,0}), \dots, (F_m, F_{m,0}),$$

where $F_i, F_{i,0} \in \mathcal{S}_{d_i}^{(n)}$, $1 \leqslant i \leqslant m$, such that $U_P \subseteq \cap_{i=1}^m U_p(F_{i,0}) = \cap_{i=1}^m Z_p(F_{i,0})^c$ (i.e. $F_{i,0}(P') \neq 0$ for any $P' \in U_P$ and for any $i \in \{1, \dots, m\}$) and

$$\psi'(P') = \left(\frac{F_1(P')}{F_{1,0}(P')}, \frac{F_2(P')}{F_{2,0}(P')}, \dots, \frac{F_m(P')}{F_{m,0}(P')} \right), \quad \forall \, P' \in U_P.$$

If we put

$$\widehat{F} := F_{1,0} F_{2,0} \cdots F_{m,0} \text{ and } \widehat{F}_i := F_{1,0} F_{2,0} \cdots F_{i-1,0} \, F_i \, F_{i+1,0} \cdots F_{m,0},$$

for any $i \in \{1, \dots m\}$ we have $\widehat{F}, \widehat{F}_i \in \mathcal{S}_k^{(n)}$, where $k := \Sigma_{i=1}^m d_i$. Moreover, for any $P' \in U_P$, we have $\widehat{F}(P') \neq 0$ and

$$\psi'(P') = \left(\frac{\widehat{F}_1(P')}{\widehat{F}(P')}, \frac{\widehat{F}_2(P')}{\widehat{F}(P')}, \dots, \frac{\widehat{F}_m(P')}{\widehat{F}(P')} \right).$$

Thus, we can assume that $F_0 := F_{1,0} = F_{2,0} = \cdots = F_{m,0} \in \mathcal{S}_k^{(n)}$, so $F_1, F_2, \dots, F_m \in \mathcal{S}_k^{(n)}$ and that for any $P' \in U_P$ one has $F_0(P') \neq 0$ and

$$\psi'(P') = \left(\frac{F_1(P')}{F_0(P')}, \frac{F_2(P')}{F_0(P')}, \dots, \frac{F_m(P')}{F_0(P')} \right).$$

Using the isomorphism $\phi_0^{-1} : \mathbb{A}^m \to U_0$, the previous equality reads as

$$\psi'(P') = \left[1, \frac{F_1(P')}{F_0(P')}, \frac{F_2(P')}{F_0(P')}, \dots, \frac{F_m(P')}{F_0(P')} \right] = [F_0(P'), F_1(P'), \dots, F_m(P')],$$

for any $P' \in U_P$, as desired.

($\Leftarrow$) Take $P \in V$ and assumptions as in the statement. Composing with the isomorphism $\phi_i : U_i \to \mathbb{A}^m$, from Proposition 6.2.2 the map $\psi|_{U_P}$ is a morphism with target $W_i := W \cap U_0$ a quasi-affine variety. Since $\{U_P\}_{P \in V}$ is an open covering of V, it follows that ψ is a morphism. $\qquad \square$

Example 6.3.2. For any integer $n \geqslant 1$, consider the projective space $\mathbb{P}^n$.

(i) Let $d \geqslant 1$ be an integer and let $F_0, \dots, F_r \in \mathcal{S}_d^{(n)}$ be homogeneous polynomials. Put

$$B := Z_p(F_0, \dots, F_r) \subset \mathbb{P}^n \tag{6.12}$$

and

$$U := \mathbb{P}^n \setminus B. \tag{6.13}$$

B is a proper (possibly empty) closed subset of $\mathbb{P}^n$, whereas U is an open (so dense and irreducible) subset of $\mathbb{P}^n$. Define the map

$$\nu : U \longrightarrow \mathbb{P}^r, \quad [\underline{P}] = [p_0, \dots, p_n] \overset{\nu}{\longrightarrow} [F_0(\underline{P}), \dots, F_r(\underline{P})]. \tag{6.14}$$

From Proposition 6.3.1, ν is a morphism. The open set U is called the *domain* of ν, i.e. the set where ν is defined, whereas B is called the *indeterminacy locus* of ν.

Take homogeneous coordinates $[Y_0, \dots, Y_r]$ in the target and let $H_i := Z_p(Y_i) \subset \mathbb{P}^r$ be the fundamental hyperplane. Note that $\text{Im}(\nu)$ is not contained in any of the hyperplanes H_i's, $0 \leqslant i \leqslant r$: indeed $\nu(U) \subseteq H_i$, for some i, means that $U \subseteq Z_p(F_i)$; since $U \subset \mathbb{P}^n$ is open and dense, the latter condition would imply $F_i = 0$ as a polynomial, contradicting that any F_i is homogeneous of some degree $d > 0$. Note moreover that, for any $0 \leqslant i \leqslant r$, one has

$$\nu^{-1}(H_i) = U \cap Z_p(F_i) \subset \mathbb{P}^n, \tag{6.15}$$

i.e. the hypersurface $U \cap Z_p(F_i) = Z_U(F_i)$ is the pre-image via the map ν of $H_i \cap \text{Im}(\nu)$, which is called a *hyperplane section* of $\text{Im}(\nu)$.

The polynomials $F_0, \dots, F_r$ are linearly independent in $\mathcal{S}_d^{(n)}$ if and only if $\text{Im}(\nu)$ is non-degenerate in $\mathbb{P}^r$: any possible hyperplane

$$H := Z_p \left(\sum_{i=0}^r a_i Y_i \right) \subset \mathbb{P}^r$$

for which $\text{Im}(\nu) \subseteq H$ would give $U \subseteq Z_p(\Sigma_{i=0}^r a_i F_i) \subset \mathbb{P}^n$. In such a case $\Sigma_{i=0}^r a_i F_i$ is identically zero which, by the linear independence of the F_i's, implies $(a_0, \dots, a_r) = (0, \dots, 0)$. If otherwise $F_0, \dots, F_r$ are linearly dependent, any non-trivial linear combination $\Sigma_{i=0}^r a_i F_i = 0$ corresponds, via ν, to a hyperplane of $\mathbb{P}^r$ containing $\text{Im}(\nu)$.

Thus, if $F_0, \dots, F_r$ are linearly independent in $\mathcal{S}_d^{(n)}$, we pose $L :=$ $\text{Span}\{F_0, \dots, F_r\}$ and denote by

$$\nu_L : U \longrightarrow \mathbb{P}^r \tag{6.16}$$

the associated morphism as in (6.14). The target $\mathbb{P}^r$ of ν_L can be identified with the linear subspace $\mathbb{P}(L)$ of $\mathbb{P}(\mathcal{S}_d^n)$; this linear subspace will be called *linear system of hypersurfaces of degree d in $\mathbb{P}^n$*, of *(projective) dimension $r = \dim(L) - 1$*.

When $L = \mathcal{S}_d^{(n)}$ the associated map will be simply denoted by $\nu_{n,d}$ (cf. also Section 6.5) and $\mathbb{P}(\mathcal{S}_d^{(n)})$ will be called the *complete linear system of hypersurfaces of degree d in $\mathbb{P}^n$*; it has projective dimension $N(n,d) := \mathbf{b}(n,d) - 1$, where $\mathbf{b}(n,d)$ the binomial coefficient as in (1.26).

(ii) Consider $\mathbb{P}^1$ and take $d = 2$. If we take the complete linear system $\mathbb{P}(\mathcal{S}_2^{(1)})$, with $F_0 = X_0^2, F_1 = X_0 X_1, F_2 = X_1^2$ the canonical basis of $\mathcal{S}_2^{(1)}$, then $B = \emptyset$, $U = \mathbb{P}^1$ and $\nu_{1,2} : \mathbb{P}^1 \to \mathbb{P}^2$ is such that $\mathrm{Im}(\nu_{1,2}) = Z_p(Y_0 Y_2 - Y_1^2) \subset \mathbb{P}^2$, where $[Y_0, Y_1, Y_2]$ are homogeneous coordinates for $\mathbb{P}^2$. Any linear section of the conic $Z_p(Y_0 Y_2 - Y_1^2)$ is given by $Z_p(Y_0 Y_2 - Y_1^2, a_0 Y_0 + a_1 Y_1 + a_2 Y_2)$, where $(a_0, a_1, a_2) \neq (0,0,0)$, and it bijectively corresponds to the two roots (counted with multiplicity) in $\mathbb{P}^1$ of the homogeneous polynomial $a_0 X_0^2 + a_1 X_0 X_1 + a_2 X_1^2 \in \mathcal{S}_2^{(1)}$ in the sense of Proposition 1.10.19, i.e. to a pair (counted with multiplicity) of points in $\mathbb{P}^1$. Note that $\nu_{1,2}$ coincides with the map Φ_2 considered in Example 5.3.23-(ii) and Exercise 5.2, which is an isomorphism from $\mathbb{P}^1$ onto the conic $Z_p(Y_0 Y_2 - Y_1^2)$, i.e. the isomorphism ϕ_2 between affine varieties in Example 6.2.5-(iv) naturally extends to an isomorphism between their projective closures.

(iii) With notation as in (ii), if otherwise we take $L = \mathrm{Span}\{F_0 = X_0^2, F_1 = X_0 X_1\}$ a proper subspace of $\mathcal{S}_2^{(1)}$, then $B = \{[0,1]\}$ is a *base point* of the linear system $\mathbb{P}(L)$, $U = \mathbb{P}^1 \setminus \{[0,1]\} = U_0 \subset \mathbb{P}^1$ and the morphism ν_L is given by $U_0 \xrightarrow{\nu_L} \mathbb{P}^1$, $[X_0, X_1] \to [X_0^2, X_0 X_1]$. Since $X_0 \neq 0$, on points of U_0 one has $[X_0^2, X_0 X_1] = [X_0, X_1]$ so ν_L extends to the identity of $\mathbb{P}^1$ (this follows from more general facts which will be discussed later on; cf. Example 8.1.12).

(iv) If we consider instead $d = 3$, $L = \mathcal{S}_3^{(1)}$, F_0, F_1, F_2, F_3 the canonical basis of $\mathcal{S}_3^{(1)}$, one easily deduces that $\mathrm{Im}(\nu_{1,3})$ is the (standard) projective twisted cubic as in Section 3.3.12 and that $\nu_{1,3}$ is an isomorphism onto its image.

Example 6.3.3. With notation as above, consider the case of $F_0, \ldots, F_r$ linearly independent linear forms in $\mathcal{S}_1^{(n)}$.

(i) If $r = n$, then $F_0, \ldots, F_n$ is a basis of $\mathcal{S}_1^{(n)}$ and the morphism ν_L is simply a projectivity of $\mathbb{P}^n$.

(ii) If otherwise $r < n$ then $B = Z_p(F_0, \ldots, F_r) \subset \mathbb{P}^n$ is a linear subspace of dimension $n - r - 1$, the open set of definition is $U := \mathbb{P}^n \setminus B$ and the morphism $\nu_L : U \to \mathbb{P}^r$ is called *projection of $\mathbb{P}^n$ to $\mathbb{P}^r$ with center the linear subspace B*. A geometric interpretation of the morphism ν_L is given by the following construction: the target $\mathbb{P}^r$ is identified with any r-dimensional linear subspace $\Lambda \subset \mathbb{P}^n$ which is skew with respect to the center of projection B and, for any point $P \in U$, $\nu_L(P) \in \Lambda$ is given by the intersection of Λ with the linear subspace $P \vee B \subset \mathbb{P}^n$ (recall Section 3.3.6, in particular (3.23)). For more details on projections of a projective space from a non-empty linear subspace, see Example 8.1.14.

6.4 Local Properties of Morphisms: Affine Open Coverings of an Algebraic Variety

Before treating some particularly interesting examples (from the geometric point of view) of morphisms with projective target (cf., e.g. Section 6.5), we give here two fundamental results which will be frequently used later on (cf., e.g. Theorem 6.5.1 as well as Chapters 7 and 8).

The first result gives a general criterion to verify when a continuous map between general algebraic varieties is actually a morphism.

Proposition 6.4.1. *Let V and W be any algebraic varieties and let $\varphi : V \to W$ be any continuous map. Then φ is a morphism if and only if there exists an open covering $\{W_j\}_{j \in J}$ of W such that, for any $j \in J$, there exists an open covering $\{V_{ij}\}_{i \in I(j)}$ of $\varphi^{-1}(W_j)$ such that the continuous map $\varphi_{ij} : V_{ij} \to V_j$ induced by φ is a morphism, for any $j \in J$ and any $i \in I(j)$.*

The previous proposition states that for a continuous map φ between algebraic varieties to be a morphism is a *local property*, i.e. φ globally is a morphism if and only if it locally is a morphism on open sets of suitable open coverings of the algebraic varieties which are the domain and the target of φ, respectively.

Proof. It is clear that if φ is a morphism, then each $\varphi_{ij} = \iota_{W_i}^{-1} \circ \varphi \circ \iota_{V_{ij}}$ is a morphism, being a composition of morphisms. Conversely, assume that any φ_{ij} is a morphism and let $U \subseteq W$ be any non-empty open set and let $f \in \mathcal{O}_W(U)$ be any regular function on it. Let f_{ij} be the restriction of the continuous map $f \circ \varphi$ to the open set $\varphi^{-1}(U) \cap V_{ij}$ of V. Note that f_{ij} coincides with the restriction to $\varphi^{-1}(U) \cap V_{ij}$ of $f|_{U \cap W_j} \circ \varphi_{ij}$, i.e. $f_{ij} \in \mathcal{O}_V(\varphi^{-1}(U) \cap V_{ij})$. Since $\{\varphi^{-1}(U) \cap V_{ij}\}_{i \in I(j)}$ is an open covering of $\varphi^{-1}(U)$, it follows that $\varphi^U(f)$ is regular of $\varphi^{-1}(U)$, i.e. φ is a morphism. $\qquad\square$

The second result extends Corollary 6.2.13 to any algebraic variety.

Proposition 6.4.2 (Affine open coverings of an algebraic variety). *On any algebraic variety V there exists a basis for the topology $\mathfrak{Zar}_V$ consisting of open affine subsets.*

Proof. We must show that, for any point $P \in V$ and for any open set $U \subset V$ containing P, there exists an open affine subset U_P with $P \in U_P \subseteq U$. Since U is also a variety, we may assume $V = U$. Moreover, since any variety can be covered by quasi-affine varieties, we can assume V to be quasi-affine in $\mathbb{A}^n$, for some non-negative integer n.

Let $Z := \overline{V}^a \setminus V \neq \emptyset$, where $\overline{V}^a$ the closure of V which is quasi-affine in $\mathbb{A}^n$. Then Z is a proper closed subset of $\overline{V}^a$ and let $I_a(Z) \subset A^{(n)}$ be its ideal. Then, since Z is closed and $P \notin Z$, there exists $f \in I_a(Z)$ such that $f(P) \neq 0$. Now $Z \subseteq Z_a(f)$ and $P \notin Z_a(f)$, so $P \in U_P := V \setminus (V \cap Z_a(f))$ which is an open subset of V, since $V \cap Z_a(f)$ is a closed subset of V by induced topology; in particular U_P is irreducible. On the other hand

$$U_P = \overline{V}^a \cap (\mathbb{A}^n \setminus Z_a(f)) = (V \cup Z) \cap (\mathbb{A}^n \setminus Z_a(f)) = V \cap (\mathbb{A}^n \setminus Z_a(f)),$$

the latter equality following from $Z \subseteq Z_a(f)$; thus U_P is also a closed subset of $\mathbb{A}^n \setminus Z_a(f)$ which, by Lemma 6.2.12, is an affine open set of $\mathbb{A}^n$. Hence, U_P is affine, as desired. $\qquad\square$

Corollary 6.4.3. *Let V be any algebraic variety and let $W \subseteq Y$ be any subvariety. Then:*

(i) *there is a bijective correspondence between prime ideals in $\mathcal{O}_{Y,W}$ and closed subvarieties of Y containing W.*

(ii) $\mathcal{Q}(\mathcal{O}_{Y,W}) = K(Y)$.

Proof. Recall that, by its very definition, $\mathcal{O}_{V,W} \subseteq K(V)$. Moreover, from Lemma 5.3.13-(ii), for any non-empty open subset $U \subseteq V$ one has $K(V) \cong K(U)$ and $W' := W \cap U$ is a subvariety of U. Therefore, by Proposition 6.4.2, up to replacing V with any of its affine open set, we may assume V to be affine. Since V is affine, statement (i) directly follows Theorem 5.3.14-(b) and (c) and from Remark 1.11.9.

As for (ii), by Theorem 5.3.14-(a) and (d) one has $\mathcal{O}_Y(Y) = A(Y) \subseteq \mathcal{O}_{Y,W} \subseteq K(Y)$ and $\mathcal{Q}(A(Y)) = K(Y)$, which concludes the proof. $\qquad\square$

6.5   Veronese Morphism: Divisors and Linear Systems

Keep notation as in Example 6.3.2-(i) and, for simplicity, set $N := N(n, d)$. If we let $F_0, \ldots, F_N$ be any basis of the $\mathbb{K}$-vector space $\mathcal{S}_d^{(n)}$, we can consider the induced complete linear system of hypersurfaces of degree d in $\mathbb{P}^n$. By Theorem 3.2.4, in such a case the base locus is

$$B = Z_p(F_0, \ldots, F_N) = Z_p((\mathcal{S}_+^{(n)})^d) = Z_p(\mathcal{S}_+^{(n)}) = \emptyset.$$

Therefore, from Proposition 6.3.1 and from (6.14), the map

$$\nu_{n,d} : \mathbb{P}^n \to \mathbb{P}^N, \quad [X_0, \ldots, X_n] \overset{\nu_{n,d}}{\longmapsto} [F_0(X_0, \ldots, X_n), \ldots, F_N(X_0, \ldots, X_n)] \tag{6.17}$$

is a morphism, which will be called *Veronese morphism of indexes n and d* induced by the chosen basis $F_0, \ldots, F_N$ of $\mathcal{S}_d^{(n)}$. We also pose $V_{n,d} := \mathrm{Im}(\nu_{n,d})$, which we call *Veronese variety of indexes n and d*. In Theorem 6.5.1 we will justify the use of the term "variety" for such an image.

In view of the proof of Lemma 1.10.12, one can restrict the analysis by focusing on a more specific case which, in a certain sense, "generates" all Veronese varieties of given indexes n and d. Precisely, let $\underline{X}$ denote the row–vector of the homogeneous coordinates of $\mathbb{P}^n$, i.e. $\underline{X} = (X_0, \ldots, X_n)$, and, for any multi-index $I := (i_0, \ldots, i_n) \in \mathbb{Z}_{\geqslant 0}^{n+1}$ such that $|I| := \Sigma_{j=0}^n i_j = d$, set $\underline{X}^I := X_0^{i_0} X_1^{i_1} \cdots X_n^{i_n}$, which is a homogeneous, degree-d monomial in $\mathcal{S}_d^{(n)}$ in the given homogeneous indeterminates. Letting $I \in \mathbb{Z}_{\geqslant 0}^{n+1}$ vary among all the multi-indexes such that $|I| = d$, with lexicographic ordering among the multi-indexes, the collection

$$\{\underline{X}^I\}_{I \in \mathbb{Z}_{\geqslant 0}^{n+1}, \, |I| = d} \tag{6.18}$$

is the *canonical* (or *standard*) basis of the $\mathbb{K}$-vector space $\mathcal{S}_d^{(n)}$, which is formed by all homogeneous, degree-d monomials in the $(n+1)$-homogeneous indeterminates $X_0, \ldots, X_n$. The Veronese morphism induced by the standard basis (6.18) will be denoted by

$$\nu_{n,d}^{\mathrm{st}} : \mathbb{P}^n \to \mathbb{P}^N, \quad \underline{X} \overset{\nu_{n,d}^{\mathrm{st}}}{\longmapsto} [\cdots, \underline{X}^I, \cdots]$$

and called the *standard Veronese morphism*. Similarly its image $\mathrm{Im}(\nu_{n,d}^{\mathrm{st}}) \subset \mathbb{P}^N$ will be denoted by $V_{n,d}^{\mathrm{st}}$ and called the *standard Veronese variety of indexes n and d*.

Any basis $F_0, \ldots, F_N$ of $\mathcal{S}_d^{(n)}$ arises from the standard basis (6.18) via a linear transformation in $\mathrm{GL}(\mathcal{S}_d^{(n)})$ which in turn induces, up to scalar multiplication, an automorphism of $\mathbb{P}^N$, i.e. a projectivity in $\mathrm{PGL}(N+1, \mathbb{K})$ giving rise to a change of homogeneous coordinates of $\mathbb{P}^N$. Thus, any Veronese morphism $\nu_{n,d}$ is such that $\nu_{n,d} = \Phi \circ \nu_{n,d}^{\mathrm{st}}$, where Φ a suitable projective transformation of the ambient space $\mathbb{P}^N$ and, consequently, any Veronese variety $V_{n,d}$ is projectively equivalent to the standard Veronese variety $V_{n,d}^{\mathrm{st}}$, in the sense of Definition 3.3.12. Thus, to study Veronese varieties $V_{n,d} \subset \mathbb{P}^N$ one can reduce to standard models $V_{n,d}^{\mathrm{st}} \subset \mathbb{P}^N$; moreover we can obviously avoid the study of the case $d = 1$ since, from Example 6.3.3, any map $\nu_{n,1}$ is simply a projectivity of $\mathbb{P}^n$.

With this set-up, we are now in position to prove the following.

Theorem 6.5.1. *For any integers $n \geqslant 1$ and $d \geqslant 2$, any $V_{n,d}$ is a non-degenerate, projective variety in $\mathbb{P}^N$ and any Veronese morphism $\nu_{n,d}$ is an isomorphism; in particular $V_{n,d}$ is isomorphic to $\mathbb{P}^n$. Furthermore, the homogeneous ideal $I_p(V_{n,d})$ is generated by homogeneous quadratic polynomials, i.e. any $V_{n,d}$ is cut-out by finitely many quadric hypersurfaces of $\mathbb{P}^N$.*

Proof. From the previous observation, we can reduce to prove the statement for the standard cases $\nu_{n,d}^{\mathrm{st}}$ and $V_{n,d}^{\mathrm{st}}$. Since $\nu_{n,d}^{\mathrm{st}}$ is a morphism, irreducibility of $V_{n,d}^{\mathrm{st}}$ directly follows from Corollary 4.1.6. Moreover $V_{n,d}^{\mathrm{st}}$ is non-degenerate in $\mathbb{P}^N$, since $\nu_{n,d}^{\mathrm{st}}$ is defined by a basis of $\mathcal{S}_d^{(n)}$ (cf. Example 6.3.2-(i)). Observe now that the homogeneous coordinates of $\mathbb{P}^N$ are indexed by all homogeneous, degree-d monomials $\underline{X}^I$ in $\mathcal{S}_d^{(n)}$, taken with the lexicographic order; correspondingly denote by Z_I the homogeneous coordinates of $\mathbb{P}^N$. If $I = (i_0, \ldots, i_n)$ and $J = (j_0, \ldots, j_n)$ are two multi-indexes such that $|I| = |J| = d$, then set $I + J := (i_0 + j_0, \ldots, i_n + j_n)$.

If we moreover take multi-indexes I, J, K, L such that $|I| = |J| = |K| = |L| = d$ and that $I + J = K + L$, the quadric hypersurface defined by the homogeneous equation

$$Z_I Z_J - Z_K Z_L = 0$$

contains $V_{n,d}^{\mathrm{st}}$, as $\underline{X}^I \underline{X}^J - \underline{X}^K \underline{X}^L = \underline{X}^{I+J} - \underline{X}^{K+L} = 0$. Thus,

$$V_{n,d}^{\mathrm{st}} \subseteq W := Z_p \Big(Z_I Z_J - Z_K Z_L \mid I, J, K, L \in \mathbb{Z}_{\geqslant 0}^{n+1},$$
$$|I| = |J| = |K| = |L| = d \text{ and } I + J = K + L \Big).$$

We need to show that W is irreducible and that the other inclusion holds, namely that $\nu_{n,d}^{\mathrm{st}}$ surjects onto W.

To prove the irreducibility of W, consider the $\mathbb{K}$-algebra homomorphism

$$\theta_{n,d} : \mathbb{K}[\cdots, Z_I, \cdots] \to \mathbb{K}[\underline{X}], \quad f(\cdots, Z_I, \cdots) \overset{\theta_{n,d}}{\longmapsto} f(\cdots, \underline{X}^I, \cdots),$$

where $\mathbb{K}[\cdots, Z_I, \cdots] = \mathcal{S}^{(N)}$ is interpreted as the homogeneous coordinate ring of $\mathbb{P}^N$ whereas $\mathbb{K}[\underline{X}] = \mathcal{S}^{(n)}$ that of $\mathbb{P}^n$. The image of $\theta_{n,d}$ is the subring of $\mathcal{S}^{(n)}$ defined as

$$\mathcal{S}^{(n)}(d) := \bigoplus_{k \geqslant 0} \mathcal{S}^{(n)}(d)_k, \tag{6.19}$$

where, as in (1.33), one sets $\mathcal{S}^{(n)}(d)_k := \mathcal{S}_{d+k}^{(n)}$ for any $k \geqslant 0$. Note that $\mathcal{S}^{(n)}(d)$ is a graded integral domain, $\theta_{n,d}$ is a homogeneous homomorphism of degree d and $\ker(\theta_{n,d}) = I_p(W)$. Since $I_p(W)$ is a homogeneous prime ideal, this shows that $W \subset \mathbb{P}^N$ is irreducible.

To prove that $W \subseteq V_{n,d}^{\mathrm{st}}$, we claim that for any point of W at least one of the coordinates $Z_{(d,0,\ldots,0)}, Z_{(0,d,\ldots,0)}, \ldots, Z_{(0,0,\ldots,d)}$ is non-zero: indeed assume by contradiction they are all zero, on the other hand since $W \subset \mathbb{P}^N$ there must be a coordinate $Z_{(i_0,\ldots,i_n)} \neq 0$ which, with no loss of generality, we may assume such that $i_0 > 0$ with the index i_0 maximal w.r.t. this property, namely $Z_{j_0,\ldots,j_n} = 0$ for all $j_0 > i_0$. Note that, by assumption, one has $i_0 < d$ so there is another index, say i_1, such that $0 < i_1 < d$. From the equations $Z_I Z_J = Z_K Z_L$ defining W, one gets

$$0 \neq Z_{(i_0,i_1,\ldots,i_n)}^2 = Z_{(i_0+1,i_1-1,\ldots,i_n)} Z_{(i_0-1,i_1+1,\ldots,i_n)},$$

i.e. $Z_{(i_0+1,i_1-1,\ldots,i_n)} \neq 0$ contradicting the maximality of i_0. With this for granted, let therefore

$$H_0 := Z_p(Z_{(d,0,\ldots,0)}), \quad H_1 := Z_p(Z_{(0,d,\ldots,0)}), \ldots, H_n := Z_p(Z_{(0,0,\ldots,d)})$$

be fundamental hyperplanes in $\mathbb{P}^N$ and let $U_i := \mathbb{P}^N \setminus H_i$, $0 \leqslant i \leqslant n$, be the corresponding affine open sets of $\mathbb{P}^N$. Then, the open sets $W_i := W \cap U_i$ of W are not empty, dense (since W is irreducible) and cover W. For any $0 \leqslant i \leqslant n$, we can consider the map $W_i \overset{\phi_i}{\longrightarrow} \mathbb{P}^n$ defined as

$$[\ldots, Z_I, \ldots] \overset{\phi_i}{\longrightarrow} [Z_{(1,0,\ldots,d-1^i,\ldots,0)}, Z_{(0,1,\ldots,d-1^i,\ldots,0)}, \ldots, Z_{(0,0,\ldots,d-1^i,\ldots,1)}],$$

where the symbol $d - 1^i$ stands for "the integer $d - 1$ is at the ith position of the corresponding multi-index" and where the right-side-member equals

$$[X_0 X_i^{d-1}, X_1 X_i^{d-1}, \ldots, X_n X_i^{d-1}] = [X_0, X_1, \ldots, X_n],$$

the second equality following from the fact that on W_i one has $X_i^d \neq 0$ which therefore implies $X_i \neq 0$. From Proposition 6.3.1, it follows that ϕ_i is a morphism, for any $0 \leqslant i \leqslant n$. Note also that, for $0 \leqslant i \neq j \leqslant n$, one has $W_i \cap W_j \neq \emptyset$ by the irreducibility of W and, from the equations defining W,

$$Z_{(0,0,\ldots,1^t,\ldots,d-1^i,\ldots,0)} = \frac{Z_{(0,0,\ldots,d^i,\ldots,0)}}{Z_{(0,0,\ldots,1^i,\ldots,d-1^j,\ldots,0)}} Z_{(0,0,\ldots,1^t,\ldots,d-1^j,\ldots,0)}$$

holds true, for any $0 \leqslant i \neq j \neq t \leqslant n$. Therefore, on $W_i \cap W_j$ one has

$$[Z_{(1,0,\ldots,d-1^i,\ldots,0)}, Z_{(0,1,\ldots,d-1^i,\ldots,0)}, \ldots, Z_{(0,0,\ldots,d-1^i,\ldots,1)}]$$
$$= [Z_{(1,0,\ldots,d-1^j,\ldots,0)}, Z_{(0,1,\ldots,d-1^j,\ldots,0)}, \ldots, Z_{(0,0,\ldots,d-1^j,\ldots,1)}],$$

namely the morphisms ϕ_i and ϕ_j agree on the overlaps $W_i \cap W_j$, for any $0 \leqslant i \neq j \leqslant n$. From Proposition 6.4.1, these maps patch together to define a global morphism $\phi : W \to \mathbb{P}^n$ which clearly is the inverse of the morphism $\nu_{n,d}^{\mathrm{st}}$. This last assertion not only shows that $\nu_{n,d}^{\mathrm{st}}$ surjects onto W, i.e. $V_{n,d}^{\mathrm{st}} = W$ is a projective variety in $\mathbb{P}^N$, but also that $\nu_{n,d}^{\mathrm{st}}$ is actually an isomorphism, thus $V_{n,d}^{\mathrm{st}}$ is isomorphic to $\mathbb{P}^n$, and also that $I_p(V_{n,d}^{\mathrm{st}}) = I_p(W)$ is generated by homogeneous quadratic polynomials. $\qquad\square$

Remark 6.5.2. The fact that $V_{n,d}$ is Zariski closed in $\mathbb{P}^N$ will more easily follow from some general properties related to *completeness* of projective varieties (cf. Theorem 9.2.1 and Remark 9.2.2-(iii)).

From the fact that $\nu_{n,d}$ is an isomorphism onto its image, one has the following.

Corollary 6.5.3. *Let $n \geqslant 1$ and $d \geqslant 2$ be integers. If $Y \subset \mathbb{P}^n$ is a projective variety, then $\nu_{n,d}(Y)$ is a subvariety of $V_{n,d}$ and so of $\mathbb{P}^N$.*

The importance and the significance of the Veronese morphisms $\nu_{n,d}$ allows us to specify in more details Corollary 6.5.3. If $F := \Sigma_I a_I \underline{X}^I \in \mathcal{S}_d^{(n)}$, the hypersurface $Y := Z_p(F) \subset \mathbb{P}^n$ is mapped by $\nu_{n,d}$ to the intersection of $V_{n,d} = \nu_{n,d}(\mathbb{P}^n)$ with the hyperplane $H_F := Z_p(\Sigma_I a_I Z_I) \subset \mathbb{P}^N$, namely the Veronese morphism allows to reduce the study of some problems concerning hypersurfaces of given degree $d \geqslant 2$ in $\mathbb{P}^n$ to hyperplane sections of the projective variety $V_{n,d} \subset \mathbb{P}^N$. If more generally $F \in \mathcal{S}_{kd}^{(n)}$, for some integer $k \geqslant 2$, then $Y = Z_p(F) \subset \mathbb{P}^n$ is mapped by $\nu_{n,d}$ to the intersection of $V_{n,d}$ with the degree-k hypersurface $\nu_{n,d}(Y) \subset \mathbb{P}^N$ obtained by the equation of F where one replaces the monomials $\underline{X}^I$ with Z_I. Similarly, for $F \in \mathcal{S}_m^{(n)}$, where m not a multiple of d, note that one can always find a positive integer a such that $X_i^{a-m} F \in \mathcal{S}_{kd}^{(n)}$, for some k and for any $0 \leqslant i \leqslant n$. Since $Y = Z_p(F) = Z_p(X_0^{a-m}F, \, X_1^{a-m}F, \ldots, X_n^{a-m}F) \subset \mathbb{P}^n$ (cf. Exercise 3.5) then $\nu_{n,d}(Y)$ is cut-out on $V_{n,d}$ by the corresponding $n+1$ hypersurfaces of $\mathbb{P}^N$. This extends also to the most general case of $Y = Z_p(F_1, \ldots, F_r) \subset \mathbb{P}^n$, for some $F_i \in \mathcal{S}_{m_i}^{(n)}$, $1 \leqslant i \leqslant r$ (cf. also Harris, 1995, Example 2.7, p. 24).

Below we discuss in more details some particular cases.

Example 6.5.4 (Rational normal curves of degree d). This is the case $V_{1,d}$, for any $d \geqslant 2$. From Theorem 6.5.1, any $V_{1,d}$ is a non-degenerate curve in $\mathbb{P}^d$ isomorphic to $\mathbb{P}^1$. Such a curve is called a *rational normal curve of degree d in $\mathbb{P}^d$*. Up to a projective transformation of $\mathbb{P}^d$, any such curve is projectively equivalent to $V_{1,d}^{\mathrm{st}}$, so we can focus on standard cases.

We have already met some standard rational normal curves; the conic $Z_p(X_0 X_2 - X_1^2) \subset \mathbb{P}^2$ in Example 6.3.2-(ii) is nothing but $V_{1,2}^{\mathrm{st}} \subset \mathbb{P}^2$, whereas the projective twisted cubic in Section 3.3.12 is $V_{1,3}^{\mathrm{st}} \subset \mathbb{P}^3$. Any $V_{1,d}^{\mathrm{st}}$ is irreducible and non-degenerate in $\mathbb{P}^d$; by the correspondence among hyperplane sections of $V_{1,d}^{\mathrm{st}}$ and hypersurfaces of degree d in $\mathbb{P}^1$, $V_{1,d}^{\mathrm{st}}$ turns out to be a curve of degree d in $\mathbb{P}^d$. If we restrict the Veronese morphism $\nu_{1,d}^{\mathrm{st}}$ to the affine chart $U_0 \subset \mathbb{P}^1$, we get the morphism

$$\nu_{1,d}^0 : \mathbb{A}^1 \to V_{1,d}^0 := V_{1,d}^{\mathrm{st}} \cap U_0 \subset U_0 \cong \mathbb{A}^d, \quad t \to (t, t^2, t^3, \ldots, t^d).$$

Therefore $V_{1,d}^0$ is an affine rational curve with polynomial parametrization, as in (3.31). With no use of Theorem 6.5.1, one directly gets that $V_{1,d}^0$ is irreducible and isomorphic to $\mathbb{A}^1$, since $\nu_{1,d}^0$ is given by a d-tuple of regular functions on $\mathbb{A}^1$ (cf. Proposition 6.2.2) and its inverse morphism simply is the restriction to $V_{1,d}^0$ of the first projection $\pi_1 : \mathbb{A}^d \to \mathbb{A}^1$. As for the projective twisted cubic in (3.34), $V_{1,d}^{\mathrm{st}}$ is the projective closure in $\mathbb{P}^d$ of $V_{1,d}^0$. Moreover, the isomorphism $\nu_{1,d}^0$ in the affine chart extends to $\nu_{1,d}^{\mathrm{st}}$ as an isomorphism between $\mathbb{P}^1$ and $V_{1,d}$. As done in Section 3.3.12 for the projective twisted cubic, by the proof of Theorem 6.5.1, $V_{1,d}^{\mathrm{st}}$ is given by the zero-locus of all the quadratic polynomials coming from the maximal minors of the matrix of linear forms

$$A := \begin{pmatrix} X_0 \ X_1 \ X_2 \ \ldots \ X_{d-1} \\ X_1 \ X_2 \ X_3 \ \ldots \ X_d \end{pmatrix}, \tag{6.20}$$

which also generate $I_p(V_{1,d}^{\mathrm{st}})$.

For further readings, see, e.g. Harris (1995, Examples 1.14 and 1.15, pp. 10–11).

Example 6.5.5 (Veronese surface). Another well-known classical example of Veronese variety is given by $V_{2,2}$, which is called *Veronese surface*. This is a non-degenerate, projective surface in $\mathbb{P}^5$, which is isomorphic to $\mathbb{P}^2$. Furthermore one deduces that its degree is 4, indeed by using the bijective correspondence among conics in $\mathbb{P}^2$ and hyperplane sections of $V_{2,2}$, two general such hyperplane sections intersect along four distinct points as two general plane conics do. The two-dimensional complete linear system of lines in $\mathbb{P}^2$ gives rise to a two-dimensional family of conics contained in $V_{2,2}$, whereas the five-dimensional complete linear system of conics in $\mathbb{P}^2$ (cf. Exercise 4.5) gives rise to the five-dimensional linear system of hyperplane section curves in $V_{2,2}$ whose general element (i.e. the image of any conic in $\mathcal{C} \setminus \mathcal{D} \cong \mathbb{P}^5 \setminus \Sigma_3$ in notation of Exercise 4.5) is a rational normal quartic curve in $\mathbb{P}^4$ and so on (see also, e.g. Harris, 1995, Exercise 2.5, p. 23).

6.5.1 *Veronese morphism and consequences*

One can use the Veronese isomorphism $\nu_{n,d}$ to prove some interesting general properties of arbitrary algebraic varieties. For example, the next result is the extension to the projective case of what proved in Lemma 6.2.12 for $\mathbb{A}^n$.

Proposition 6.5.6. *For any non-negative integers n, d, let $W \subset \mathbb{P}^n$ be any hypersurface of degree d. Then $\mathbb{P}^n \setminus W$ is an affine variety.*

Proof. Using the Veronese isomorphism $\nu_{n,d}$, $\mathbb{P}^n \setminus W \cong V_{n,d} \setminus (H_W \cap V_{n,d})$, where $H_W \subset \mathbb{P}^N$ denotes the hyperplane which cuts out the hyperplane section $H_W \cap V_{n,d}$ isomorphic to the hypersurface $W \subset \mathbb{P}^n$ of degree d. One concludes by observing that $V_{n,d} \setminus (H_W \cap V_{n,d}) \cong V_{n,d} \cap (\mathbb{P}^n \setminus H_W) \cong V_{n,d} \cap \mathbb{A}^N$. $\qquad\square$

Corollary 6.5.7. *If $V \subset \mathbb{P}^n$ is a projective variety, which is not a point, and $W \subset \mathbb{P}^n$ is any hypersurface then $V \cap W \neq \emptyset$. In particular, any two projective plane curves intersect.*

Proof. If by contradiction one had $V \cap W = \emptyset$, choosing $d := \deg(W)$, from the proof of Proposition 6.5.6, we would get $V \cong \nu_{n,d}(V) \subset Z_p(H_W)^c \cong \mathbb{A}^N$. Then V would be isomorphic to an irreducible, closed subset in $\mathbb{A}^N$ i.e. to an affine variety. From Corollary 6.2.6, V would be a point against the assumptions. $\qquad\square$

From Corollary 6.2.7, we already know that $\mathbb{P}^2$ and $\mathbb{A}^2$ cannot be isomorphic; the second part of Corollary 6.5.7 more generally shows that they are not even homeomorphic, as in $\mathbb{A}^2$ there actually exist curves with empty intersections.

Similarly to Example 6.2.14, with the use of the Veronese isomorphism $\nu_{n,d}$ we can show that there exist quasi-projective varieties which cannot be either projective, or affine or quasi-affine. To do this, consider first the following preliminary result.

Lemma 6.5.8. *For any integer $n \geqslant 2$, let $\Lambda \subset \mathbb{P}^n$ be a linear subspace s.t. $\dim(\Lambda) \leqslant n - 2$. Let $W := \mathbb{P}^n \setminus \Lambda$. Then $\mathcal{O}_W(W) \cong \mathbb{K}$.*

Proof. Since $W \subset \mathbb{P}^n$ is an open dense subset of $\mathbb{P}^n$, from Lemma 5.3.13 and Theorem 5.3.14 (h) we have $K(W) \cong K(\mathbb{P}^n) = \mathcal{S}^{(n)}_{((0))}$. Since $\mathcal{O}_W(W) \subset K(W)$, then any regular function $\phi \in \mathcal{O}_W(W)$ is of the form $\phi = \frac{G_1}{G_2}$, for some G_1, $G_2 \in \mathcal{S}^{(n)}_d$, with $d \geqslant 0$ an integer, $G_2 \neq 0$ and $W \subset U_p(G_2) = Z_p(G_2)^c$. Since $\dim(\Lambda) \leqslant n - 2$, reasoning as in the proof of Claim 6.2.15, we get that $G_2 \in \mathbb{K}^*$ and so also $G_1 \in \mathbb{K}$. $\square$

Example 6.5.9. We can now show that, for any point $P \in \mathbb{P}^2$, the algebraic variety $W := \mathbb{P}^2 \setminus \{P\}$ is a quasi-projective variety which is neither projective, nor affine nor quasi-affine. By using projective transformations, with no loss of generality we may assume P to be the fundamental point $[1, 0, 0]$. We first note that W cannot be a projective variety: otherwise, for any line ℓ not passing through P, $W \setminus \ell$ would be affine from Proposition 6.5.6, on the other hand $W \cap (\mathbb{P}^2 \setminus \ell) \cong \mathbb{A}^2 \setminus \{(0,0)\}$, contradicting Example 6.2.14. By Lemma 6.5.8, W can be neither an affine variety: otherwise $A(W) = \mathcal{O}_W(W) \cong \mathbb{K}$, i.e. $I_a(W)$ would be maximal i.e. W would be a point, a contradiction. Similarly, W cannot be quasi-affine: otherwise, if $\overline{W}$ denotes its affine closure, always by Lemma 6.5.8, we would have $A(\overline{W}) \subseteq \mathcal{O}_W(W) \cong \mathbb{K}$ and we can argue as before.

We conclude this section by observing that, from the proof of Theorem 6.5.1, another fundamental difference among affine and projective varieties enter explicitly into the game.

Remark 6.5.10. Corollary 6.2.4 establishes that the coordinate ring $A(V)$ of any affine variety V is invariant for the isomorphism class represented by V; in other words, if $V \subset \mathbb{A}^n$ and $W \subset \mathbb{A}^m$ are isomorphic affine varieties, then $A(V) \cong A(W)$, no matter the embedding in different affine spaces. Instead, projective varieties behave differently from this point of view; namely for $V \subset \mathbb{P}^n$ any projective variety, the homogeneous coordinate ring $\mathcal{S}(V)$ in general is not invariant under isomorphism.

To be more precise, if $V, W \subseteq \mathbb{P}^n$ are projective varieties which are isomorphic via a projectivity $\Phi \in \mathrm{PGL}(n+1, \mathbb{K})$ sending V to W (as, e.g. it occurs between any Veronese variety $V_{n,d}$ and its standard model $V_{n,d}^{\mathrm{st}}$) then, by Definition 3.2.1, Φ gives rise to an affinity $\Phi \in \mathrm{GL}(n+1, \mathbb{K})$ of $\mathbb{A}^{n+1}$ inducing an isomorphism $C_a(V) \cong C_a(W)$. Since these are both affine varieties, by Corollary 6.2.4 we have $A(C_a(V)) \cong A(C_a(W))$. Thus, since $A^{(n+1)} \cong \mathcal{S}^{(n)}$ as integral $\mathbb{K}$-algebras and since, from (3.8), one has that for any projective variety $Z \subseteq \mathbb{P}^n$

$$\mathcal{S}(Z) = \frac{\mathcal{S}^{(n)}}{I_p(Z)} \cong \frac{A^{(n+1)}}{I_a(C_a(Z))} = A(C_a(Z))$$

holds, one deduces that $\mathcal{S}(V) \cong \mathcal{S}(W)$, as it occurs for coordinate rings of affine varieties.

On the contrary, if $V \subseteq \mathbb{P}^n$ and $W \subseteq \mathbb{P}^m$ are projective varieties, embedded in different ambient projective spaces, but which are isomorphic as algebraic varieties (as it occurs for $\mathbb{P}^n$ and $V_{n,d} \subset \mathbb{P}^N$, with $d \geqslant 2$) $\mathcal{S}(V)$ is in general not isomorphic to $\mathcal{S}(W)$. This has been implicitly encountered in the proof of Theorem 6.5.1. To let things be easily understood, we discuss in full details the following example.

Example 6.5.11. For any $d \geqslant 2$, we know that any rational normal curve $V_{1,d} \subset \mathbb{P}^d$ is isomorphic to $\mathbb{P}^1$. On the one hand, we have $\mathcal{S}(\mathbb{P}^1) = \mathcal{S}^{(1)} :=S$; on the other, since any $V_{1,d}$ is projectively equivalent inside $\mathbb{P}^d$ to the standard rational normal curve $V_{1,d}^{\mathrm{st}}$, we have that $\mathcal{S}(V_{1,d}) = \frac{\mathcal{S}^{(d)}}{I_p(V_{1,d})}$ is isomorphic to the homogeneous coordinate ring $\mathcal{S}(V_{1,d}^{\mathrm{st}})$. Now, by definition of the standard Veronese morphism $\nu_{1,d}^{\mathrm{st}}$, the graded ring $\mathcal{S}(V_{1,d}^{\mathrm{st}}) := R$ is the image of the homogeneous homomorphism

$$\theta_{1,d} : \mathcal{S}^{(d)} \longrightarrow \mathcal{S}^{(1)} = S, \ Z_{(d-i,i)} \overset{\theta_{1,d}}{\longrightarrow} X_0^{d-i} X_1^i, \quad 0 \leqslant i \leqslant d,$$

where $Z_{d,0}, \ldots, Z_{0,d}$ the indeterminates in $\mathcal{S}^{(d)}$ whereas X_0, X_1 those in $\mathcal{S}^{(1)}$, and the kernel of the previous homomorphism is exactly the ideal generated by the maximal minors of the matrix of linear forms (6.20). Note that $\theta_{1,d}$ is not surjective: focusing on the graded summands we get that $S_0 = R_0$ but $R_1 \cong S_d$ and more generally, for any $k \geqslant 1$, $R_k \cong S_{kd}$. In other words,

R is isomorphic to the graded subring of $S = \mathcal{S}^{(1)}$ given by

$$\mathcal{S}^{(1)}(d) := \bigoplus_{k \geqslant 0} \left(\mathcal{S}^{(1)}(d)\right)_k,$$

where, as in (1.33), each graded summand is defined as $\left(\mathcal{S}^{(1)}(d)\right)_k := \mathcal{S}^{(1)}_{kd}$.

6.5.2 Divisors and linear systems

The study of Veronese morphisms suggests a more general approach. Let W be either an affine or a projective space.

We will denote by $\mathrm{Div}(W)$ the free abelian group generated by the set Ω of all irreducible hypersurfaces in W. Any element of $\mathrm{Div}(W)$ is of the form $D := \Sigma_{Y \in \Omega} r_Y Y$, where $r_Y \in \mathbb{Z}$ are integers which are almost all zero except for finitely many of them. Such a D is called a *divisor* of W, the integer r_Y is called the *multiplicity* of Y in D and the hypersurfaces for which $r_Y \neq 0$ are called the *irreducible components of D*. The hypersurface $\mathrm{supp}(D) := \cup_{r_Y \neq 0} Y$ is called the *support* of D. If $r_Y \in \{-1, 0, 1\}$ for any Y, then D is said to be a *reduced divisor*; if $r_Y \geqslant 0$ for any Y in D, then D is said to be an *effective divisor*; if $D = Y$, where Y an irreducible hypersurface, then D is said to be an *irreducible divisor*.

For any $D \in \mathrm{Div}(W)$ one defines the *degree of D* as $\deg(D) := \Sigma_{Y \in \Omega} r_Y \deg(Y)$. If $f_Y = 0$ is an equation of Y in Ω (in a chosen system of coordinates of W and moreover homogeneous if W a projective space), given $D = \Sigma_{Y \in \Omega} r_Y Y$, one sets $f_D := \Pi_{Y \in \Omega} f_Y^{r_Y}$ so that $f_D = 0$ is an *equation* of D (it is homogeneous in the projective case). By the Hilbert "Nullstellensatz", two equations of a divisor D differ by a non-zero constant multiplicative factor.

Set $W = \mathbb{P}(V)$ of dimension n, endowed with a system of homogeneous coordinates. Let D be a divisor and let $\Lambda \subset \mathbb{P}(V)$ be a linear subspace of dimension $m < n$. Let $f_D = 0$ be a homogeneous equation of D and assume that Λ has a parametric representation given by $\underline{X} = A\underline{\lambda}$, where A a $(m+1) \times (n+1)$-matrix of rank $m+1$, $\underline{X}$ the column-vector of homogeneous coordinates of W and $\underline{\lambda}^t = [\lambda_0, \ldots, \lambda_m] \in \mathbb{P}^m$ homogeneous parameters. The polynomial $f_D(A\underline{\lambda})$ is identically zero if and only if $\Lambda \subseteq \mathrm{supp}(D)$; in such a case, one states that D *contains* Λ and set $\Lambda \subseteq D$. Otherwise, the equation $f_D(A\underline{\lambda}) = 0$ defines a divisor D_Λ in Λ, w.r.t. the homogeneous coordinates $[\lambda_0, \ldots, \lambda_m]$ given by the parametrization of Λ. The divisor D_Λ is called the *intersection* of Λ with D. The previous definition is well-posed and $\deg(D_\Lambda) = \deg(D)$.

When instead $W = \mathbb{A}^n$ one has similar definitions and considerations; the main difference is that in general, for an affine subspace Λ, one has $\deg(D_\Lambda) \leqslant \deg(D)$. One can define in an obvious way the projective closure of a divisor D in $\mathbb{A}^n$. Therefore, one has:

Proposition 6.5.12 (Bezout's theorem for linear sections). *If D is a divisor of either an affine or a projective space and if Z is a subspace, then either Z is contained in D or D intersects Z along a divisor D_Z of degree at most $\deg(D)$; moreover $\deg(D_Z) = \deg(D)$ in the projective case.*

Example 6.5.13 (Intersection multiplicity of a line at a point). If $W = \mathbb{P}^n$ and if $\deg(D) = d$. Take Z a line not contained in D, then $D_Z = m_1 P_1 + \cdots + m_h P_h$, where P_i are distinct points of Z and m_i are positive integers, with $1 \leqslant i \leqslant h$, such that $d = m_1 + \cdots + m_h$. The positive integer m_i is called the *intersection multiplicity* between Z and D at the point P_i and it is denoted by $\mu(D, Z; P_i)$. On the other hand, one poses $\mu(D, Z; P) = 0$ if $P \notin \mathrm{supp}(D_Z)$ and $\mu(D, Z; P) = \infty$ for any $P \in Z$, when $Z \subseteq D$.

For simplicity, from now on let us refer to the case $W = \mathbb{P}(V)$ of dimension n. Fix a positive integer d and consider the set $\mathcal{L}_{\mathbb{P}(V),d}$ of effective divisors of degree d in W. When $W = \mathbb{P}^n$, we will simply write $\mathcal{L}_{n,d}$. One has $\mathcal{L}_{\mathbb{P}(V),d} = \mathbb{P}(\mathrm{Sym}^d(V^*))$.

A r-dimensional subspace of $\mathcal{L}_{\mathbb{P}(V),d}$ is called a *linear system of divisors* of degree d in $\mathbb{P}(V)$. Fixing a homogeneous system of coordinates in $\mathbb{P}(V)$ by means of a projectivity $\phi : \mathbb{P}^n \to \mathbb{P}(V)$, one has a homogeneous isomorphism $\mathcal{S}^{(n)} \cong \mathrm{Sym}(V^*)$ which determines isomorphisms $\mathcal{S}_d^{(n)} \cong \mathrm{Sym}^d(V^*)$, for any $d \geqslant 0$. In this set-up one determines bijective projectivities

$$\phi_d : \mathcal{L}_{n,d} \to \mathcal{L}_{\mathbb{P}(V),d}$$

inducing homogeneous coordinates on $\mathcal{L}_{\mathbb{P}(V),d}$, namely the coordinates $[\cdots, f_I, \cdots]_{|I|=d}$ (where the multi-indexes I are taken, e.g. with the lexicographic order) of a divisor D are simply the coefficients, up to a non-zero constant multiplicative factor, of the polynomial $f_D(\underline{X}) = \Sigma_{|I|=d} f_I \underline{X}^I$ giving an equation of D in the given system of homogeneous coordinates for $\mathbb{P}(V)$.

Let $\mathcal{L} \subseteq \mathcal{L}_{\mathbb{P}(V),d}$ be a linear system of dimension r. If $D_0, \ldots, D_r \in \mathcal{L}$ are linearly independent divisors, whose homogeneous equations in the given system of homogeneous coordinates are $f_{D_i} = 0$, $0 \leqslant i \leqslant r$, then a divisor

D sits in $\mathcal{L}$ if and only if

$$f_D = \lambda_0 f_{D_0} + \cdots + \lambda_r f_{D_r} = 0$$

(up to a non-zero constant multiplicative factor), where $[\lambda_0, \ldots, \lambda_r] \in \mathbb{P}^r$. A linear system of dimension 0 consists of only one effective divisor, if $\dim(\mathcal{L}) = 1$, respectively, 2, then $\mathcal{L}$ is called a *pencil*, respectively, a *web*. If $\mathcal{L} = \emptyset$, one sets $\dim(\mathcal{L}) = -1$. In any case, in this set-up,

$$Z_p(f_{D_0}, \ldots, f_{D_r}) = \bigcap_{D \in \mathcal{L}} D \subset \mathbb{P}(V)$$

is called the *base locus of the linear system* $\mathcal{L}$ and it is denoted by $\mathrm{Bs}(\mathcal{L})$. If $P \in \mathbb{P}(V)$, one denotes by $\mathcal{L}(-P)$ the set of all divisors $D \in \mathcal{L}$ such that $P \in \mathrm{supp}(D)$ (i.e. effective divisors in $\mathcal{L}$ containing the point P). One has $\mathcal{L}(-P) = \mathcal{L}$ if and only if $P \in \mathrm{Bs}(\mathcal{L})$, namely P is a *base point* of $\mathcal{L}$. If $P \notin \mathrm{Bs}(\mathcal{L})$ then $\mathcal{L}(-P)$ is a sub-linear system of $\mathcal{L}$ of dimension $r - 1$.

Example 6.5.14 (Linear series on $\mathbb{P}^1$). A linear system $\mathcal{L}$ of dimension r in $\mathcal{L}_{\mathbb{P}(V),d}$, where $\mathbb{P}(V)$ a projective line, is called *linear series of degree d and dimension r* and will be denoted by $\mathfrak{g}_d^r$. If $P \in \mathbb{P}(V)$ is not a base point for $\mathfrak{g}_d^r$, we know that $\dim(\mathfrak{g}_d^r(-P)) = r - 1$; therefore one has $r \leqslant d$. Note that $\mathcal{L}_{\mathbb{P}(V),d}$ is nothing but the linear series associated to a Veronese morphism $\nu_{1,d}$. This is called *complete linear series* and it is a $\mathfrak{g}_d^d$ which is uniquely determined up to projectivities, i.e. $\mathfrak{g}_d^d$ corresponds to $\nu_{1,d}^{\mathrm{st}}$.

Consider the complete linear series $\mathfrak{g}_2^2$ on $\mathbb{P}(V)$ and take a pencil in it, i.e. a $\mathfrak{g}_2^1$. Choosing a homogeneous system of coordinates on $\mathbb{P}(V)$, there exist two linearly independent homogeneous polynomials f_0, f_1, in the homogeneous indeterminates $[X_0, X_1] \in \mathbb{P}^1 \cong \mathbb{P}(V)$ such that any element of the $\mathfrak{g}_2^1$ is of the form $\lambda_0 f_0 + \lambda_1 f_1 = 0$, where $[\lambda_0, \lambda_1] \in \mathbb{P}^1$. The divisors D_0 and D_1, associated to f_0 and f_1, cannot have the same support otherwise f_0 and f_1 would be proportional, a contradiction. Therefore, we have two possibilities: if P is a common point of D_0 and D_1 then P is a base point of the $\mathfrak{g}_2^1$ so $\mathfrak{g}_2^1(-P) = \mathfrak{g}_2^1$, i.e. any divisor in the $\mathfrak{g}_2^1$ is of the form $P + Q$, where Q varies in the $\mathfrak{g}_1^1$ parametrizing points of $\mathbb{P}(V)$ whereas P is fixed. If otherwise $\mathrm{supp}(D_0) \cap \mathrm{supp}(D_1) = \emptyset$, then the $\mathfrak{g}_2^1$ is *base point free* and any divisor of the $\mathfrak{g}_2^1$ is of the form $P + Q_P$, where $P \in \mathfrak{g}_1^1$ varying and, for any such P there exists one and only one $Q_P \in \mathbb{P}(V)$ such that $P + Q_P \in \mathfrak{g}_2^1$. Any base point free $\mathfrak{g}_2^1$ induces a natural map

$$\sigma : \mathbb{P}(V) \to \mathbb{P}(V), \ P \xrightarrow{\sigma} Q_P,$$

which is a bijection and moreover it is an *involution*, namely $\sigma^2 = \mathrm{Id}_{\mathbb{P}(V)}$.

Exercises

Exercise 6.1. Let $\varphi \;:\; \mathbb{A}^2 \;\rightarrow\; \mathbb{A}^2$ be the map defined by $(x_1, x_2) \xrightarrow{\varphi} (x_1, x_1 x_2)$. Prove that φ is a morphism, find its image and conclude that $\mathrm{Im}(\varphi)$ is neither open nor closed in $\mathbb{A}^2$.

Exercise 6.2. Let $C := Z_a(x_1^3 + x_1^2 - x_2^2) \subset \mathbb{A}^2$. Prove that C is an affine variety and find a parametrization $\varphi \;:\; \mathbb{A}^1 \rightarrow C$ which is a morphism of affine varieties. Establish if φ can be an isomorphism.

Exercise 6.3. Consider the standard Veronese surface $V_{2,2}^{\mathrm{st}} \subset \mathbb{P}^5$. Find explicit parametric equations of the image via $\nu_{2,2}^{\mathrm{st}} : \mathbb{P}^2 \rightarrow \mathbb{P}^5$ of the conic $C := Z_p(X_0 X_2 - X_1^2) \subset \mathbb{P}^2$. Deduce that $\nu_{2,2}^{\mathrm{st}}(C)$ is a rational normal quartic in $\mathbb{P}^4$.

Exercise 6.4. Let $\nu_{2,3}^{\mathrm{st}} : \mathbb{P}^2 \rightarrow \mathbb{P}^9$ be the standard Veronese morphism of indexes $2, 3$. Determine parametric equations of the image of the line $\ell := Z_p(X_1 - X_2) \subset \mathbb{P}^2$ and of the conic $C = Z_p(X_0 X_2 - X_1^2) \subset \mathbb{P}^2$. Determine degrees of $\nu_{2,3}^{\mathrm{st}}(\ell)$ and of $\nu_{2,3}^{\mathrm{st}}(C)$.

Exercise 6.5. Consider any complete linear series $\mathfrak{g}_d^d$ on $\mathbb{P}^1$. Take any point $P \in \mathbb{P}^1$ and consider the sublinear series $\mathfrak{g}_d^d(-P)$ having P as base point. Describe the image of $\mathbb{P}^1$ via $\mathfrak{g}_d^d(-P)$ using suitable projections of $\mathrm{Im}(\mathfrak{g}_d^d)$.

Chapter 7

Products of Algebraic Varieties

In this chapter, we will study products of algebraic varieties. Namely, we want to describe how to give an algebraic variety structure to the set-theoretical cartesian product $V \times W$, when both V and W are algebraic varieties.

7.1 Products of Affine Varieties

Note that Example 2.1.22-(i) already describes the product of two affine spaces $\mathbb{A}^r \times \mathbb{A}^s$, where r and s are non-negative integers. As a set of points, $\mathbb{A}^r \times \mathbb{A}^s$ has been identified with $\mathbb{A}^{r+s}$; in particular, the cartesian product of the two affine spaces has a natural structure of affine variety, the affine space $\mathbb{A}^{r+s}$, where the Zariski topology on $\mathbb{A}^r \times \mathbb{A}^s = \mathbb{A}^{r+s}$ is $\mathfrak{Zar}_a^{r+s}$, which is finer than the product $\mathfrak{Zar}_a^r \times \mathfrak{Zar}_a^s$ of the two Zariski topology (recall Example 2.1.22-(iii)).

From Corollary 5.3.19, we know that any affine variety is completely determined, together with its structural sheaf, by its ring of coordinates. The coordinate ring of $\mathbb{A}^r$, respectively, of $\mathbb{A}^s$, is the polynomial ring $A_{\underline{x}}^{(r)} := \mathbb{K}[x_1, \ldots, x_r]$, respectively, $A_{\underline{y}}^{(s)} := \mathbb{K}[y_1, \ldots, y_s]$. These rings of polynomials are those formed by (evaluating) regular functions which operates on the affine spaces $\mathbb{A}^r$ and $\mathbb{A}^s$, respectively. Similarly, for $\mathbb{A}^r \times \mathbb{A}^s = \mathbb{A}^{r+s}$ the coordinate ring of regular functions operating on this affine space simply is the polynomial ring $A_{\underline{x},\underline{y}}^{(r+s)} := \mathbb{K}[x_1, \ldots, x_r, y_1, \ldots, y_s]$. Note further that $A_{\underline{x},\underline{y}}^{(r+s)} \cong A_{\underline{x}}^{(r)} \otimes_{\mathbb{K}} A_{\underline{y}}^{(s)}$, where the isomorphism is as integral $\mathbb{K}$-algebras of finite type (cf. Section 1.9.2).

From Proposition 6.2.3, the two natural $\mathbb{K}$-algebra monomorphisms $A_{\underline{x}}^{(r)} \hookrightarrow A_{\underline{x},\underline{y}}^{(r+s)} \hookleftarrow A_{\underline{y}}^{(s)}$ correspond to the natural two projections onto the first and the second factor, respectively,

$$\mathbb{A}^r \times \mathbb{A}^s \xrightarrow{\pi_1} \mathbb{A}^r \quad \text{and} \quad \mathbb{A}^r \times \mathbb{A}^s \xrightarrow{\pi_2} \mathbb{A}^s. \tag{7.1}$$

These projections are morphisms since they respectively coincide with the projections π_I and π_J as in (6.11), with multi-indexes $I = (1, 2, \ldots, r)$ and $J = (r + 1, \ldots, r + s)$, respectively (cf. Example 6.2.5-(i)).

Once the product of two affine spaces is clearly described, one would like to extend previous considerations to the more general case of a product of two affine varieties. Concerning the topological structure, we have the following.

Proposition 7.1.1. *Let V and W be affine varieties. Then $V \times W$ is an affine variety.*

Proof. Up to isomorphism, we can assume there exist non-negative integers r and s such that $V \subseteq \mathbb{A}^r$ and $W \subseteq \mathbb{A}^s$, where the inclusions are closed immersions as affine varieties (cf. Definition 6.1.9). We will show that the cartesian product $V \times W$ is endowed with a structure of Zariski closed, irreducible subset of the affine space $\mathbb{A}^{r+s}$. The fact that $V \times W$ is Zariski closed in $\mathbb{A}^{r+s}$ has been already met in Example 2.1.22-(i). Recall indeed that, if $V := Z_a(f_1, \ldots, f_n) = Z_a(I_a(V))$ and $W := Z_a(g_1, \ldots, g_m) = Z_a(I_a(W))$, then (2.8) establish that

$$V \times W = Z_a(f_1, \ldots, f_n) \cap Z_a(g_1, \ldots, g_m) = Z_a(I_a(V)) \cap Z_a(I_a(W)),$$

where $I_a(V), I_a(W)$ are considered as ideals in $A^{(r+s)}$, via the natural $\mathbb{K}$-algebra inclusions $A^{(r)} \hookrightarrow A^{(r+s)}$ and $A^{(s)} \hookrightarrow A^{(r+s)}$ reminded above and where $Z_a(I_a(V)) = V \times \mathbb{A}^s$ is the closed s-dimensional cylinder over V in $\mathbb{A}^{r+s}$ whereas $Z_a(I_a(W)) = \mathbb{A}^r \times W$ is the closed r-dimensional cylinder over W in $\mathbb{A}^{r+s}$. Moreover one has

$$I_a(V \times W) = (f_1, \ldots, f_n, g_1, \ldots, g_m) = I_a(V) + I_a(W), \tag{7.2}$$

where the ideals $I_a(V)$ and $I_a(W)$ are meant as ideals in $\mathbb{A}^{r+s}$ via the previous inclusions. To complete the proof, we are left to showing that $V \times W$ is irreducible as topological space. Assume that $V \times W = Z_1 \cup Z_2$ where Z_i closed subsets of $V \times W$, for $1 \leqslant i \leqslant 2$. For any point $P \in V$, $W_P := \{P\} \times W$ is a subset of $V \times W$ which is homeomorphic to W and so irreducible. Then $W_P \subseteq Z_i$, for $i \in \{1, 2\}$. Consider

$$V_i := \{P \in V : W_P \subseteq Z_i\}, \quad \text{for } 1 \leqslant i \leqslant 2.$$

For any point $Q \in W$, pose $V_i(Q) := \{P \in V \mid (P, Q) \in Z_i\}$, for $1 \leqslant i \leqslant 2$. Then $(V \times \{Q\}) \cap Z_i = V_i(Q) \times \{Q\}$, so $V_i(Q)$ is closed, for any $Q \in W$

and for any $1 \leqslant i \leqslant 2$. Since one has $V_i = \cap_{Q \in W} V_i(Q)$, $1 \leqslant i \leqslant 2$, then V_1, V_2 are closed in V. From $V = V_1 \cup V_2$ and from the irreducibility of V, therefore one has either $V = V_1$ (in which case $V \times W = Z_1$) or $V = V_2$ (consequently, $V \times W = Z_2$). In all cases $V \times W$ is irreducible. $\square$

Since $V \times W$ is an affine variety, from Proposition 6.2.3, we are left with the description of its coordinate ring.

Proposition 7.1.2. *Let* V *and* W *be two affine varieties. Then* $A(V \times W) \cong A(V) \otimes_{\mathbb{K}} A(W)$.

Proof. As in the proof of Proposition 7.1.1, up to isomorphism we can assume there exist non-negative integers r and s such that $V \subseteq \mathbb{A}^r$ and $W \subseteq \mathbb{A}^s$, where the inclusion are closed immersions as affine varieties. From Example 2.1.22-(iii) and the fact that $V \times W = (V \times \mathbb{A}^s) \cap (\mathbb{A}^r \times W)$ in $\mathbb{A}^{r+s}$, the natural $\mathbb{K}$-bilinear map

$$\alpha : A^{(r)} \times A^{(s)} \to A^{(r+s)}, \ (f,g) \overset{\alpha}{\longrightarrow} fg$$

is such that

$$\alpha(I_a(V) \times A^{(s)}) \subseteq I_a(V \times W) \supseteq \alpha(A^{(r)} \times I_a(W)).$$

Therefore, α induces a $\mathbb{K}$-bilinear map

$$\beta : A(V) \times A(W) \to A(V \times W), \ (f,g) \overset{\beta}{\longrightarrow} fg.$$

Since the generators of $A(V)$ and $A(W)$ as $\mathbb{K}$-algebras of finite type are contained in the image of β and these generate $A(V \times W)$, then β is surjective and any element of $A(V \times W)$ is of the form $\Sigma_{i,j} c_{i,j} f_i g_j$, where $f_i \in A(V)$, $g_j \in A(W)$, $c_{i,j} \in \mathbb{K}$. From (7.2), $I_a(V \times W) \subset A^{(r+s)}$ is generated by the elements in $\alpha(I_a(V) \times A^{(s)})$ and in $\alpha(A^{(r)} \times I_a(W))$, in other words

$$A(V \times W) \cong \frac{A^{(r+s)}}{I_a(V) + I_a(W)},$$

where $I_a(V)$ and $I_a(W)$ considered as ideals in $A^{(r+s)}$.

We are left to showing that $A(V \times W)$ is isomorphic to $A(V) \otimes_{\mathbb{K}} A(W)$; to do so, let U be any $\mathbb{K}$-vector space and let $\gamma_U : A(V) \times A(W) \to U$ be any $\mathbb{K}$-bilinear map with target U. Consider the map $\delta_U : A(V \times W) \to U$ defined as follows: $\delta_U (\Sigma_{i,j} c_{i,j} f_i g_j) := \Sigma_{i,j} c_{i,j} \gamma_U(f_i, g_j)$. Note that $\gamma_U = \delta_U \circ \beta$, where $\beta : A(V) \times A(W) \to A(V \times W)$ the $\mathbb{K}$-bilinear map defined above.

Moreover, δ_U is well-posed, $\mathbb{K}$-linear and uniquely determined by γ_U and β. From Proposition 1.9.1, it follows that $A(V \times W) \cong A(V) \otimes_{\mathbb{K}} A(W)$ as $\mathbb{K}$-vector spaces. Since $A(V)$ and $A(W)$ are integral $\mathbb{K}$-algebras of finite type, from Proposition 1.9.7, $A(V \times W)$ is also a $\mathbb{K}$-algebra of finite type. Finally, the integrality of $A(V \times W)$ directly follows from the fact that $V \times W$ is irreducible. $\qquad\square$

Remark 7.1.3. As for the product of two affine spaces, the two natural projections

$$V \times W \xrightarrow{\pi_V} V \quad \text{and} \quad V \times W \xrightarrow{\pi_W} W \tag{7.3}$$

are morphisms of affine varieties since $\pi_V = \pi_1 \circ \iota_{V \times W}$ and $\pi_W = \pi_2 \circ \iota_{V \times W}$ where π_1 and π_2 as in (7.1) whereas $\iota_{V \times W}$ the closed immersion of $V \times W$ in $\mathbb{A}^{r+s}$ (cf. Definition 6.1.9).

7.2 Products of Projective Varieties

Let us consider more generally V and W algebraic varieties. Thus, for some integers n and m, we have $V \subseteq \mathbb{P}^n$, $W \subseteq \mathbb{P}^m$, where the inclusions are locally closed morphisms. To endow the (set theoretical) cartesian product $V \times W$ with a structure of algebraic variety, it will suffice to define a set-theoretical injective map

$$V \times W \overset{\Psi}{\hookrightarrow} \mathbb{P}^N, \tag{7.4}$$

for some positive integer N, such that:

(i) $\psi(V \times W) \subset \mathbb{P}^N$ *is a quasi-projective variety,*
(ii) *for any affine open subsets $U_V \subseteq V$ and $U_W \subseteq W$ (recall Proposition 6.4.2), $\Psi(U_V \times U_W)$ is an affine open subset of $\Psi(V \times W)$; moreover such open sets determine an open covering of $\Psi(V \times W)$,*
(iii) *for any choice of affine open subsets $U_V \subseteq V$ and $U_W \subseteq W$, the map*

$$\Psi|_{U_V \times U_W} : U_V \times U_W \longrightarrow \Psi(U_V \times U_W)$$

is an isomorphism of affine varieties.

Note that (iii) is nothing but a *compatibility condition* between the structure of algebraic variety on $\Psi(V \times W)$ as in (i) and the one given in Section 7.1 on its affine open subsets as in (ii). More precisely, the subset $\Psi(U_V \times U_W)$, which is an affine open subset w.r.t. the structure of algebraic variety of $\psi(V \times W) \subset \mathbb{P}^N$ as in (i), has to be isomorphic (as affine variety) to the affine variety $U_V \times U_W$, whose structure has been defined in Section 7.1.

Lemma 7.2.1. *Given a set-theoretical injective map Ψ as in (7.4), with $\Psi(V \times W) \subset \mathbb{P}^N$ a quasi-projective variety, then Ψ satisfies the compatibility condition (iii) above if and only if, for any points $P \in V$ and $Q \in W$, there exist $U_P \subseteq V$ and $U_Q \subseteq W$ affine open neighborhoods of $P \in V$ and $Q \in W$, respectively, s.t. $\Psi|_{U_P \times U_Q}$ defines an isomorphism of the affine variety $U_P \times U_Q$ onto its image $\Psi(U_P \times U_Q) \subseteq \mathbb{P}^N$ and $\Psi(U_P \times U_Q)$ is an affine open subset of $\psi(V \times W)$.*

Proof. The implication ($\Rightarrow$) is trivial. For the converse, let $U_V \subseteq V$ and $U_W \subseteq W$ be any affine open subsets. Since $U_V = \cup_{P \in U_V} U_P$ and $U_W = \cup_{Q \in U_W} U_Q$, from Section 7.1 we have that $\{U_P \times U_Q\}_{(P,Q) \in U_V \times U_W}$ is an open covering of $U_V \times U_W$, formed by affine open sets by assumptions. The map $\Psi|_{U_V \times U_W}$ is therefore continuous and it is an isomorphism as it follows Proposition 6.4.1. □

A map Ψ for which conditions in Lemma 7.2.1 hold true is said to *satisfy the property of being local*.

Lemma 7.2.2. *If a map Ψ as above exists, it is uniquely determined (up to isomorphism) by the property of being local.*

Proof. Assume there exist a map Ψ as above and another injective map $\Phi : V \times W \hookrightarrow \mathbb{P}^M$, for some integer M, both of them satisfying the property of being local. Thus $\Phi \circ \Psi^{-1} : \Psi(V \times W) \to \Phi(V \times W)$ is bijective. It suffices to show that $\Phi \circ \Psi^{-1}$ is a morphism. The proof of this is identical to that of Lemma 7.2.1, using the fact that both Ψ and Φ satisfy the property of being local. □

The previous result states that, to give a structure of algebraic variety on $V \times W$, one is reduced to showing the existence of a map Ψ satisfying conditions (i), (ii) above and the property of being local and that, once such a map has been constructed, the structure on $V \times W$ is uniquely determined up to isomorphisms. The aim of the next section is to construct such a map Ψ.

7.2.1 Segre morphism and the product of projective spaces

Let us consider first $V = \mathbb{P}^n$ and $W = \mathbb{P}^m$, for some non-negative integers n and m.

Definition 7.2.3. We define the *Segre map of indexes n and m* by

$$\sigma_{n,m} : \qquad \mathbb{P}^n \times \mathbb{P}^m \qquad \longrightarrow \qquad \mathbb{P}^N$$

$$([x_0, \ldots x_n], [y_0, \ldots, y_m]) \longrightarrow [x_0 y_0, x_0 y_1, \ldots, x_i y_j, \ldots, x_n y_m],$$

where $N := (n+1)(m+1) - 1$, with $0 \leqslant i \leqslant n$ and $0 \leqslant j \leqslant m$. This map is well-defined and the image $\Sigma_{n,m} := \mathrm{Im}(\sigma_{n,m})$ is called the *Segre variety of indexes n and m*.

The use of the term "variety" is justified by the following result.

Lemma 7.2.4. *The Segre map $\sigma_{n,m}$ is bijective and its image $\Sigma_{n,m}$ is a projective variety in $\mathbb{P}^N$.*

Proof. Take homogeneous coordinates $[W_{ij}]$ in $\mathbb{P}^N$, with $0 \leqslant i \leqslant n$ and $0 \leqslant j \leqslant m$. These coordinates are lexicographically ordered so that they are compatible with $\sigma_{n,m}$. Thus, points in $\Sigma_{n,m}$ satisfy homogeneous quadratic equations

$$W_{ij} W_{kr} - W_{ir} W_{kj} = 0, \quad 0 \leqslant i, k \leqslant n, \ 0 \leqslant j, r \leqslant m. \tag{7.5}$$

Let Z be the closed subset of $\mathbb{P}^N$ cut out by quadric hypersurfaces as in (7.5); it is clear that $\Sigma_{n,m} \subseteq Z$. We claim that, for any point $R \in Z$, there is a unique pair $(P, Q) \in \mathbb{P}^n \times \mathbb{P}^m$ such that $R = \sigma_{n,m}(P, Q)$; this will imply that $\Sigma_{n,m} = Z$ and that $\sigma_{n,m}$ bijectively maps $\mathbb{P}^n \times \mathbb{P}^m$ onto $\Sigma_{n,m}$. To prove the claim, let $R := [w_{ij}^0] \in Z$ be any point. Without loss of generality, we can assume $w_{00}^0 \neq 0$ and so, by rescaling all the coordinates up to this multiplicative non-zero factor, that $w_{00}^0 = 1$ (the other cases can be handled analogously); since $R \in Z$ then

$$w_{i0}^0 \, w_{0j}^0 = w_{ij}^0 \, w_{00}^0 = w_{ij}^0.$$

Setting

$$P := [1, w_{10}^0, \ldots, w_{n0}^0] \in \mathbb{P}^n \quad \text{and} \quad Q := [1, w_{01}^0, \ldots, w_{0m}^0] \in \mathbb{P}^m,$$

this means that $R = \sigma_{n,m}((P, Q))$ and that P and Q are uniquely determined.

We are left to showing that $\Sigma_{n,m}$ is irreducible. For any $0 \leqslant i \leqslant n$ and $0 \leqslant j \leqslant m$, denote by $U_{ij}^N := \mathbb{P}^N \setminus Z_p(W_{ij})$ the principal affine open set in $\mathbb{P}^N$ isomorphic to $\mathbb{A}^N$ and, similarly, $\mathbb{A}^n \cong Z_p(X_0)^c := U_0^n \subset \mathbb{P}^n$ and $\mathbb{A}^m \cong Z_p(Y_0)^c := U_0^m \subset \mathbb{P}^m$. Previous computations show that $\Sigma_{n,m} \cap U_{00}^N = \sigma_{n,m}(U_0^n \times U_0^m)$ and that $\sigma_{n,m}|_{U_0^n \times U_0^m}$ is an isomorphism from the affine variety $U_0^n \times U_0^m \cong \mathbb{A}^n \times \mathbb{A}^m \cong \mathbb{A}^{n+m}$ onto the affine closed set $\Sigma_{n,m} \cap U_{00}^N$, which is therefore an affine variety. Similar conclusion holds for $\Sigma_{n,m} \cap U_{ij}^N$, for any pair of indexes (i,j) with $0 \leqslant i \leqslant n$ and $0 \leqslant j \leqslant m$.

One has an open covering of $\Sigma_{n,m}$, $\Sigma_{n,m} = \cup_{i,j}(\Sigma_{n,m} \cap U_{ij}^N)$, where each $\Sigma_{n,m} \cap U_{ij}^N$ is an affine open subset of $\Sigma_{n,m}$. Moreover, $W := \cap_{i,j}(\Sigma_{n,m} \cap U_{ij}^N) \neq \emptyset$ is an open subset of $\Sigma_{n,m}$, which is also open in any affine open subset $\Sigma_{n,m} \cap U_{ij}^N$, for any i,j as above; in particular, W is irreducible. Since W is contained in any open set of an affine open covering of $\Sigma_{n,m}$ as a dense open subset, this forces $\Sigma_{n,m}$ to be irreducible. □

As for the Veronese varieties, the fact that $\Sigma_{n,m}$ is closed in $\mathbb{P}^N$ is also a consequence of the *completeness of projective varieties*, which will be discussed later on (cf. Theorem 9.2.1 and Remark 9.2.2-(iii)). At last, we have:

Lemma 7.2.5. *The bijection* $\sigma_{n,m} : \mathbb{P}^n \times \mathbb{P}^m \to \Sigma_{n,m}$ *satisfies condition (ii) as in (7.4) and the property of being local.*

Proof. The proof of Lemma 7.2.4 shows that $\{\sigma_{n,m}(U_i^n \times U_j^m)\}_{i,j}$ is an open covering of the projective variety $\Sigma_{n,m}$, where each $\sigma_{n,m}(U_i^n \times U_j^m)$ is isomorphic to the affine variety $U_i^n \times U_j^m$, i.e. condition (ii) of (7.4) for the map $\sigma_{n,m}$ holds. Moreover, for any pair of indexes (i,j), one has that $\Sigma_{n,m} \cap U_{ij}^N$ is an affine open neighborhood of any of its points over which the Segre map is an isomorphism with $U_i^n \times U_j^m$. From Lemma 7.2.1, $\sigma_{n,m}$ satisfy the property of being local. □

From Lemma 7.2.2, previous results determine (up to isomorphism) a structure of projective variety on $\mathbb{P}^n \times \mathbb{P}^m$, which from now on, will be always identified with $\Sigma_{n,m}$ by means of the Segre morphism $\sigma_{n,m}$.

Remark 7.2.6.

(i) The two natural projections

$$\mathbb{P}^n \times \mathbb{P}^m \xrightarrow{\pi_1} \mathbb{P}^n \text{ and } \mathbb{P}^n \times \mathbb{P}^m \xrightarrow{\pi_2} \mathbb{P}^m \qquad (7.6)$$

are morphisms of projective varieties. This follows from Proposition 6.4.1 and the fact that their restrictions to the affine open subsets $U_0^n \times U_0^m$ are morphisms (cf. Remark 7.1.3).

(ii) Note that $U_0^n \times \mathbb{P}^m = \sigma_{n,m}^{-1} \left(\cup_j \Sigma_{n,m} \cap U_{0j}^N \right)$, therefore it is an open subset of $\mathbb{P}^n \times \mathbb{P}^m$. Namely, identifying $\mathbb{A}^n$ with the affine chart $U_0^n \subset \mathbb{P}^n$, the Segre map $\sigma_{n,m}$ defines a structure of quasi-projective variety on $\mathbb{A}^n \times \mathbb{P}^m$.

(iii) Identify $\mathbb{A}^n$ with the affine open set $U_0^n \subset \mathbb{P}^n$ and similarly $\mathbb{A}^m$ with the affine open set $U_0^m \subset \mathbb{P}^m$. Therefore, $U_0^n \times U_0^m$ is isomorphic to $\mathbb{A}^n \times \mathbb{A}^m = \mathbb{A}^{n+m}$. With explicit computations, if we consider $\mathbb{P}^n \times \mathbb{P}^m$ as the Segre variety $\Sigma_{m,n} \subset \mathbb{P}^N$, where $N = (n+1)(m+1) - 1$, then $U_0^n \times U_0^m$ coincides with $\Sigma_{m,n} \setminus (\Sigma_{m,n} \cap Z_p(W_{00}))$, therefore, for any $[\underline{w}] \in U_0^n \times U_0^m \subset \mathbb{P}^N$, the following map

$$U_0^n \times U_0^m \to \mathbb{A}^{n+m}, \quad [\underline{w}] \to \left(\frac{w_{10}}{w_{00}}, \dots, \frac{w_{n0}}{w_{00}}, \frac{w_{01}}{w_{00}}, \dots, \frac{w_{0m}}{w_{00}} \right)$$

is the required isomorphism.

7.2.2   *Products of projective varieties*

Consider the following preliminary result.

Proposition 7.2.7.

(i) *A subset $T \subseteq \mathbb{P}^n \times \mathbb{P}^m$ is closed if and only if it is defined by polynomials $G_k(X_0, \dots, X_n, Y_0, \dots, Y_m)$, $1 \leqslant k \leqslant s$, where each polynomial G_k is bi-homogeneous, i.e. it is separately homogeneous with respect to the two set of indeterminates $(X_0, \dots, X_n)$ and $(Y_0, \dots, Y_m)$.*

(ii) *A subset $T \subseteq \mathbb{A}^n \times \mathbb{P}^m$ is closed if and only if it is defined by polynomials $F_k(x_1, \dots, x_n, Y_0, \dots, Y_m)$, $1 \leqslant k \leqslant s$, where each polynomial F_k is homogeneous with respect to the set of indeterminates $(Y_0, \dots, Y_m)$.*

Proof. (i) Take any closed subset $Z \subset \mathbb{P}^N$, where $N = (n+1)(m+1) - 1$. Then $Z := Z_p(A_1(W_{00}, \dots, W_{nm}), \dots, A_s(W_{00}, \dots, W_{nm}))$, for some homogeneous polynomials $A_k \in H(S^{(N)})$, $1 \leqslant k \leqslant s$. The closed subset $\sigma_{nm}^{-1}(Z) \subset \mathbb{P}^n \times \mathbb{P}^m$ is defined by the equations

$$G_k(X_0, \dots, X_n, Y_0, \dots, Y_m), \quad 1 \leqslant k \leqslant s,$$

obtained by applying to the polynomials A_k's the indeterminate substitutions $W_{ij} = X_i Y_j$, $0 \leqslant i \leqslant n$, $0 \leqslant j \leqslant m$. Each of the polynomials

$G_1, \ldots, G_s$ are separately homogeneous with respect to the two set of indeterminates $(X_0, \ldots, X_n)$ and $(Y_0, \ldots, Y_m)$.

Conversely, any polynomial $G(X_0, \ldots, X_n, Y_0, \ldots, Y_m)$ which is homogeneous of degree d in the set of indeterminates $(X_0, \ldots, X_n)$ and of degree r in the set of indeterminates $(Y_0, \ldots, Y_m)$ determines the same closed subset $T \subset \mathbb{P}^n \times \mathbb{P}^m$ defined by the vanishing locus of the set of polynomials

$$Y_j^{d-r} G(X_0, \ldots, X_n, Y_0, \ldots, Y_m), \quad 0 \leqslant j \leqslant m, \tag{7.7}$$

where we assumed $d \geqslant r$. All polynomials in (7.7) are now homogeneous of the same degree d with respect to both set of indeterminates $(X_0, \ldots, X_n)$ and $(Y_0, \ldots, Y_m)$. Therefore $T = \sigma_{n,m}^{-1}(Z)$, where Z is the closed subset in $\mathbb{P}^N$ defined by homogeneous polynomials in the indeterminates W_{ij} obtained by polynomials in (7.7) via the indeterminate substitutions $W_{ij} = X_i Y_j$, $0 \leqslant i \leqslant n$, $0 \leqslant j \leqslant m$. (Note that the indeterminate substitutions are not uniquely determined, so Z is not uniquely determined; on the other hand, two different choices differ by elements of the ideal generated by polynomials as in (7.5), so $Z \cap \Sigma_{n,m}$ is independent from the choices considered.)

(ii) $\mathbb{A}^n \times \mathbb{P}^m$ is open in $\mathbb{P}^n \times \mathbb{P}^m$ since it is $Z_p(X_0)^c$; therefore, the assertion follows by considering suitable dehomogenization of a system of equations for $\overline{T} \subset \mathbb{P}^n \times \mathbb{P}^m$, which is a closed subset as proved in (i). □

Let now $V \subset \mathbb{P}^n$ and $W \subset \mathbb{P}^m$ be projective varieties. We will show that $V \times W$ is a closed subvariety of $\mathbb{P}^n \times \mathbb{P}^m$; from Section 7.2.1, this will endow $V \times W$ with a structure of projective variety. Suppose that $I_p(V)$ is generated by homogeneous polynomials

$$P_h(X_0, \ldots, X_n) = 0, \quad h = 1, \ldots, s$$

and that $I_p(W)$ is generated by homogeneous polynomials

$$Q_k(Y_0, \ldots, Y_m) = 0, \quad k = 1, \ldots, t.$$

Then $V \times W$ is defined by the system of homogeneous polynomials

$$P_h(X_0, \ldots, X_n) = 0 = Q_k(Y_0, \ldots, Y_m), \quad h = 1, \ldots, s, \quad k = 1, \ldots, t.$$

From Proposition 7.2.7 (i), $V \times W$ is closed in $\mathbb{P}^n \times \mathbb{P}^m$. We are therefore left to showing that $V \times W$ is irreducible. If either V or W is a point, then $V \times W$ is irreducible since homeomorphic to an irreducible topological space. Assume therefore that neither V nor W are points; without loss of generality, we can assume that $V_0 := V \cap U_0^n \neq \emptyset$ and $W_0 := V \cap U_0^m \neq \emptyset$. Now $V_0 \times W_0$ is an affine variety so it is irreducible. Any polynomial $G(X_0, \ldots, X_n, Y_0, \ldots, Y_m)$ which is separately homogeneous with respect to the two sets of indeterminates $(X_0, \ldots, X_n)$ and $(Y_0, \ldots, Y_m)$ and which vanishes along $V_0 \times W_0$ is such that, for any $[q_0, \ldots, q_m] \in W_0$,

$$G(X_0, \ldots, X_n, q_0, \ldots, q_m) \in I_p(V_0) = I_p(V)$$

(recall (3.21)), i.e. $G(X_0, \ldots, X_n, Y_0, \ldots, Y_m)$ vanishes also along $V \times W_0$. Similarly, it vanishes along $V_0 \times W$, so it also vanishes along $V \times W$. In other words, $\overline{V_0 \times W_0}^p = V \times W$, which implies that $V \times W$ is irreducible.

Remark 7.2.8. The two natural projections

$$\pi_V : V \times W \longrightarrow V \quad \text{and} \quad \pi_W : V \times W \longrightarrow W \tag{7.8}$$

are morphisms of projective varieties since they are compositions of morphisms, i.e. $\pi_V = \pi_1 \circ \iota_{V \times W}$ and $\pi_W = \pi_2 \circ \iota_{V \times W}$ where π_1 and π_2 projections as in (7.6) whereas $\iota_{V \times W}$ the closed immersion of $V \times W$ into $\mathbb{P}^n \times \mathbb{P}^m \cong \Sigma_{n,m} \subset \mathbb{P}^N$, $N = (n+1)(m+1) - 1$ (cf. Definition 6.1.9).

7.3 Products of Algebraic Varieties

Previous considerations allow to easily extend similar reasoning to the more general case of algebraic varieties, i.e. quasi-projective varieties.

Let $V \subset \mathbb{P}^n$ and $W \subset \mathbb{P}^m$ be any quasi-projective varieties. Denote by $\overline{V}$ and $\overline{W}$ their projective closures, respectively. Then $C := \overline{V} \setminus V$ is a closed subset of $\overline{V}$ as well as $D := \overline{W} \setminus W$ is a closed subset of $\overline{W}$. Therefore, one has

$$V \times W = (\overline{V} \setminus C) \times (\overline{W} \setminus D) = (\overline{V} \times \overline{W}) \setminus \{(\overline{V} \times D) \cup (C \times \overline{W})\}.$$

The previous equality shows that $V \times W$ is an open set of the projective variety $\overline{V} \times \overline{W}$; thus $V \times W$ has a natural structure of a quasi-projective variety. The fact that the natural projections

$$\pi_V : V \times W \to V \quad \text{and} \quad \pi_W : V \times W \to W$$

are morphisms of quasi-projective varieties follows from the same reasoning as in Remark 7.2.8, where $\iota_{V \times W}$ is a locally closed immersion of $V \times W$ into $\mathbb{P}^n \times \mathbb{P}^m \cong \Sigma_{n,m} \subset \mathbb{P}^N$, $N = (n+1)(m+1) - 1$ (cf. Definition 6.1.9).

7.4 Products of Morphisms

Let V, W, Z be algebraic varieties and $\varphi \in \text{Morph}(Z, V)$ and $\psi \in \text{Morph}(Z, W)$. Define set-theoretically the map

$$\varphi \times \psi : Z \to V \times W, \quad P \xrightarrow{\varphi \times \psi} (\varphi(P), \psi(P)), \quad \forall\ P \in Z. \quad (7.9)$$

This is the unique map which let the diagrams

$$
\begin{array}{ccc}
Z \xrightarrow{\varphi \times \psi} V \times W & & Z \xrightarrow{\varphi \times \psi} V \times W \\
\searrow^{\varphi} \quad \swarrow^{\pi_V} & \text{and} & \searrow^{\psi} \quad \swarrow^{\pi_W} \\
V & & W
\end{array}
$$

to be commutative, where π_V, respectively, π_W, the projection onto the first, respectively, the second, factor of the product $V \times W$. In this set-up, one has the following.

Proposition 7.4.1. *Let V, W, Z be algebraic varieties and $\varphi \in \text{Morph}(Z, V)$ and $\psi \in \text{Morph}(Z, W)$. Then $\varphi \times \psi \in \text{Morph}(Z, V \times W)$.*

Proof. Up to isomorphisms, we may assume that $Z \subseteq \mathbb{P}^r$, $V \subseteq \mathbb{P}^n$ and $W \subseteq \mathbb{P}^m$ are quasi-projective varieties, for some non-negative integers r, n, m. By Proposition 6.3.1, for any point $P \in Z$ there exist an open neighborhood U_P of P, positive integers h and k and homogeneous polynomials $F_0, \ldots, F_n \in \mathcal{S}_k^{(r)}$, $G_0, \ldots, G_m \in \mathcal{S}_h^{(r)}$ such that, for any $P' \in U_P$ there exist $i \in \{0, \ldots, n\}$ and $j \in \{0, \ldots, m\}$, for which $F_i(P') \neq 0$ and $G_j(P') \neq 0$, such that

$$\varphi(P') = [F_0(P'), \ldots, F_n(P')] \in V \subseteq \mathbb{P}^n$$

$$\psi(P') = [G_0(P'), \ldots, G_m(P')] \in W \subseteq \mathbb{P}^m, \ \forall\ P' \in U_P.$$

Therefore, for any $P' \in U_P$, one has

$$(\varphi \times \psi)(P') = [\ldots, F_i(P') \cdot G_j(P'), \ldots] \in V \times W \subseteq \Sigma_{m,n} \subset \mathbb{P}^N,$$

where $N = (n+1)(m+1) - 1$ and where $F_i \cdot G_j \in \mathcal{S}_{k+h}^{(r)}$, $0 \leqslant i \leqslant n$, $0 \leqslant j \leqslant m$. Once again from Proposition 6.3.1, it follows that $\varphi \times \psi$ is a morphism. $\qquad\square$

As a consequence we have the following useful.

Theorem 7.4.2. *Let V, W be algebraic varieties. If X is an algebraic variety for which there exist $\rho_1 \in \text{Morph}(X, V)$, $\rho_2 \in \text{Morph}(X, W)$ such* *that, for any algebraic variety Z and for any pair $(\varphi, \psi) \in \operatorname{Morph}(Z, V) \times$ $\operatorname{Morph}(Z, W)$, there exists a unique $\eta \in \operatorname{Morph}(Z, X)$ which let the diagrams*

$$
\begin{array}{ccc}
Z & \xrightarrow{\eta} & X \\
& \searrow{\scriptstyle\varphi} \quad \swarrow{\scriptstyle\rho_1} & \\
& V &
\end{array}
\quad \text{and} \quad
\begin{array}{ccc}
Z & \xrightarrow{\eta} & X \\
& \searrow{\scriptstyle\psi} \quad \swarrow{\scriptstyle\rho_2} & \\
& W &
\end{array}
$$

to be commutative, then there exists a unique isomorphism $\alpha : X \to V \times W$ for which the diagrams

$$
\begin{array}{ccc}
X & \xrightarrow{\alpha} & V \times W \\
& \searrow{\scriptstyle\rho_1} \quad \swarrow{\scriptstyle\pi_V} & \\
& V &
\end{array}
\quad \text{and} \quad
\begin{array}{ccc}
X & \xrightarrow{\alpha} & V \times W \\
& \searrow{\scriptstyle\rho_2} \quad \swarrow{\scriptstyle\pi_W} & \\
& W &
\end{array}
\qquad (7.10)
$$

are commutative, where π_V and π_W the projections onto the factors of $V \times W$.

Proof. From Proposition 7.4.1, $\alpha := \rho_1 \times \rho_2 \in \operatorname{Morph}(X, V \times W)$ is uniquely defined by ρ_1 and ρ_2 and it is a morphism which let the diagrams (7.10) to be commutative. We need to show that α is an isomorphism. From the assumptions, there exists a unique $\beta \in \operatorname{Morph}(V \times W, X)$ for which the diagrams

$$
\begin{array}{ccc}
V \times W & \xrightarrow{\beta} & X \\
& \searrow{\scriptstyle\pi_V} \quad \swarrow{\scriptstyle\rho_1} & \\
& V &
\end{array}
\quad \text{and} \quad
\begin{array}{ccc}
V \times W & \xrightarrow{\beta} & X \\
& \searrow{\scriptstyle\pi_W} \quad \swarrow{\scriptstyle\rho_2} & \\
& W &
\end{array}
$$

are commutative. Then $\beta \circ \alpha \in \operatorname{Morph}(X, X)$ let the diagrams

$$
\begin{array}{ccc}
X & \xrightarrow{\beta \circ \alpha} & X \\
& \searrow{\scriptstyle\rho_1} \quad \swarrow{\scriptstyle\rho_1} & \\
& V &
\end{array}
\quad \text{and} \quad
\begin{array}{ccc}
X & \xrightarrow{\beta \circ \alpha} & X \\
& \searrow{\scriptstyle\rho_2} \quad \swarrow{\scriptstyle\rho_2} & \\
& V &
\end{array}
$$

to be commutative. By uniqueness, $\beta \circ \alpha = \operatorname{Id}_X$; similarly one shows that $\alpha \circ \beta = \operatorname{Id}_{V \times W}$, so α is an isomorphism whose inverse is β. $\qquad \square$

Corollary 7.4.3.

(i) *If $V \cong V'$ and $W \cong W'$ are algebraic varieties, then $V \times W \cong V' \times W'$ as algebraic varieties.*

(ii) *If V and W are algebraic varieties, then $V \times W \cong W \times V$ as algebraic varieties.*

(iii) *If V, W and Z are algebraic varieties, then $V \times (W \times Z) \cong (V \times W) \times Z$; in particular one can simply writes $V \times W \times Z$. More generally, if V_1, $V_2, \ldots, V_n$ are algebraic varieties, then $V_1 \times V_2 \times \cdots V_n$, defined as $((((V_1 \times V_2) \times V_3) \times \cdots) \times V_n)$, is isomorphic to $V_{i_1} \times \cdots \times V_{i_n}$, where $(i_1, \ldots, i_n)$ is any permutation of $\{1, \ldots, n\}$.*

(iv) *For any point $P \in V$, the subset of $V \times W$ given by*

$$\pi_V^{-1}(P) := \{(P, Q) \in V \times W \mid \forall\, Q \in W\}$$

is a closed subvariety of $V \times W$ which is isomorphic to W. Similarly, for any point $Q \in W$,

$$\pi_W^{-1}(Q) := \{(P, Q) \in V \times W \mid \forall\, P \in V\}$$

is a closed subvariety of $V \times W$ which is isomorphic to V.

(v) *If V, W and Z are algebraic varieties, a map $\varphi : Z \to V \times W$ is a morphism if and only if $\pi_V \circ \varphi$ and $\pi_W \circ \varphi$ are morphisms, where π_V and π_W the projections onto the factors of $V \times W$.*

Proof. Left as an exercise to the reader (cf. Exercise 7.2). □

7.5 Diagonals, Graph of a Morphism and Fiber-Products

From the previous analysis, one can deduce further interesting properties of algebraic varieties. Let V be an algebraic variety; the subset

$$\Delta(V) := \{(P, P) \mid P \in V\} \subset V \times V \tag{7.11}$$

is called the *diagonal of $V \times V$*.

Proposition 7.5.1. *For any algebraic variety V, $\Delta(V)$ is a closed subvariety of $V \times V$ which is isomorphic to V.*

Proof. Up to isomorphism, we may assume that $V \subseteq \mathbb{P}^n$, for some integer n, as a quasi-projective variety. Thus $\Delta(V) = \Delta(\mathbb{P}^n) \cap (V \times V)$. To show that $\Delta(V)$ is closed in $V \times V$ it is therefore enough to prove that $\Delta(\mathbb{P}^n)$ is closed in $\mathbb{P}^n \times \mathbb{P}^n$. To show this, it suffices to noticing that $\Delta(\mathbb{P}^n) = Z_p(X_1 Y_0 - X_0 Y_1, \ldots, X_n Y_0 - X_0 Y_n)$ and then one concludes by using Proposition 7.2.7-(i).

To show that $\Delta(V)$ is irreducible and isomorphic to V observe that we have a morphism of algebraic varieties

$$\mathrm{Id}_V \times \mathrm{Id}_V : V \longrightarrow V \times V, \quad P \overset{\mathrm{Id}_V \times \mathrm{Id}_V}{\longrightarrow} (P, P),$$

whose image is Δ_V. Therefore, $\Delta(V)$ is irreducible by Corollary 4.1.6 and so it is a closed subvariety of $V \times V$. Let now $\iota_{\Delta(V)}$ be the closed immersion of $\Delta(V)$ into $V \times V$ and let $\pi_1 : V \times V \to V$ be the projection onto, e.g. the first factor. Then $\alpha := \pi_1 \circ \iota_{\Delta(V)} \in \mathrm{Morph}(\Delta(V), V)$ which is the inverse of $\mathrm{Id}_V \times \mathrm{Id}_V$, showing that $\Delta(V) \cong V$. $\qquad\square$

Recalling Proposition 6.4.2 and using a similar strategy as in Proposition 7.5.1, one can also prove the following.

Lemma 7.5.2. *Let V be an algebraic variety and U_1 and U_2 be non-empty affine open subsets of V. Then $U_1 \cap U_2$ is an affine open subset of V.*

Proof. One needs only to show that $U_1 \cap U_2$ is isomorphic to a closed subset in some affine space. By assumption U_i is isomorphic to an affine variety $V_i \subseteq \mathbb{A}^{n_i}$, for some non-negative integer n_i, $1 \leqslant i \leqslant 2$. Thus, $U_1 \times U_2 \subset \mathbb{A}^{n_1+n_2}$ is an affine variety. One concludes by observing that

$$U_1 \cap U_2 = (U_1 \times U_2) \cap \Delta(\mathbb{A}^{n_1+n_2})$$

and that $\Delta(\mathbb{A}^{n_1+n_2})$ is closed in $\mathbb{A}^{n_1+n_2}$. $\qquad\square$

Let now $\varphi : V \to W$ be a morphism of algebraic varieties; the set

$$\Gamma_\varphi := \{(P, \varphi(P)) \mid P \in V\} \subseteq V \times W, \qquad (7.12)$$

is called the *graph of the morphism φ.*

Proposition 7.5.3. *For any algebraic varieties V and W and for any $\varphi \in \mathrm{Morph}(V, W)$, Γ_φ is a closed subvariety of $V \times W$ which is isomorphic to V.*

Proof. From Proposition 7.5.1, $\Delta(W)$ is closed in $W \times W$. Now

$$\varphi \times \mathrm{Id}_W : V \times W \to W \times W$$

is a morphism since it is the product of the two morphisms φ and Id_W (cf. Proposition 7.4.1). Since $\Gamma_\varphi = (\varphi \times \mathrm{Id}_W)^{-1}(\Delta(W))$, the continuity of $(\varphi \times \mathrm{Id}_W)$ ensures that Γ_φ is closed in $V \times W$. Let ι_{Γ_φ} be the closed immersion of Γ_φ into $V \times W$, then the morphism $\pi_V \circ \iota_{\Gamma_\varphi} : \Gamma_\varphi \to V$ is actually an isomorphism, whose inverse is given by the morphism $\mathrm{Id}_V \times \varphi : V \to V \times W$. This at once shows that Γ_φ is irreducible, so it is a subvariety of $V \times W$, which is isomorphic to V. $\qquad\square$

Let now V, W and Z be algebraic varieties and let $\varphi \in \mathrm{Morph}(V, Z)$ and $\psi \in \mathrm{Morph}(W, Z)$. Consider the subset of $V \times W$ defined as follows:

$$V \times_Z W := \{(P, Q) \in V \times W \mid \varphi(P) = \psi(Q)\} \subseteq V \times W; \qquad (7.13)$$

this is called the *fiber-product of V and W over Z* (w.r.t. the morphisms φ and ψ).

Proposition 7.5.4. *Let V, W and Z be algebraic varieties let $\varphi \in \mathrm{Morph}(V, Z)$ and $\psi \in \mathrm{Morph}(W, Z)$. Then the fiber-product $V \times_Z W$ is a closed subset of $V \times W$.*

Proof. Since φ and ψ are morphisms, from Proposition 7.4.1 it follows that $\varphi \times \psi$ is a morphism. Moreover, from Proposition 7.5.1, $\Delta(Z)$ is closed in $Z \times Z$. Thus, by the continuity of $(\varphi \times \psi)$, one concludes since $V \times_Z W = (\varphi \times \psi)^{-1}(\Delta(Z))$. $\qquad \square$

Remark 7.5.5. In general $V \times_Z W$ is not irreducible; for example if V and W are subvarieties of Z and if $\varphi = \iota_V : V \hookrightarrow Z$ and $\psi = \iota_W : W \hookrightarrow Z$ are respective immersions as subvarieties, then $V \times_Z W = V \cap W$ and we know that in general intersections of subvarieties may be reducible (cf. Remark 4.2.10). On the other hand, in some cases, it can happen that $V \times_Z W$ is irreducible and so a closed subvariety of $V \times W$; e.g. it occurs when $W = Z$ and $\psi = \mathrm{Id}_W$ indeed in such a case $V \times_Z W = \Gamma_\varphi$ the graph of the morphism $\varphi : V \to W$, which is irreducible by Proposition 7.5.3.

Exercises

Exercise 7.1. Describe the Segre variety $\Sigma_{1,1}$ and show that it is *doubly ruled*, namely it contains two distinct one-dimensional families of lines where lines belonging to the same family are skew whereas any line of the first family intersects any line of the second family.

Exercise 7.2. Prove Corollary 7.4.3.

Exercise 7.3. Let $\mathbb{K}$ be an algebraically closed field and let A and B be integral $\mathbb{K}$-algebras of finite type. Show that $A \otimes_{\mathbb{K}} B$ is an integral $\mathbb{K}$-algebra of finite type. Deduce in an algebraic way that if V and W are affine varieties defined over an algebraically closed field, then $V \times W$ is an affine variety.

Exercise 7.4. Find a counterexample to Exercise 7.3 when $\mathbb{K}$ is not algebraically closed.

Exercise 7.5. Find examples, different from those in Remark 7.5.5, of algebraic varieties V, W and Z and of morphisms $\varphi \in \mathrm{Morph}(V, Z)$ and $\psi \in \mathrm{Morph}(W, Z)$ such that $V \times_Z W$ turns out to be irreducible (respectively, reducible).

Chapter 8

Rational Maps of Algebraic Varieties

In this chapter, we introduce the notion of *rational maps, birational maps* and *birational equivalence* of algebraic varieties, which are mile-stone concepts for the classification of algebraic varieties. As we will see, a rational map is a morphism which is defined only on some non-empty, open subset of an algebraic variety. On the other hand, since any non-empty, open set in an algebraic variety is dense, this already carries a lot of information about the map.

8.1 Rational and Birational Maps

We start with the following preliminary result.

Lemma 8.1.1. *Let* V *and* W *be algebraic varieties, let* $\varphi, \psi \in \mathrm{Morph}(V, W)$ *and assume there exists a non-empty, open subset* $U \subseteq V$ *s.t.* $\varphi|_U = \psi|_U$. *Then* $\varphi = \psi$ *on* V.

Proof. φ and ψ determine the morphism $\varphi \times \psi : V \to W \times W$ (cf. Proposition 7.4.1). From the assumption, one has $(\varphi \times \psi)(U) \subseteq \Delta(W)$, where $\Delta(W)$ the diagonal in $W \times W$ (cf. (7.11)). Since U is dense in V and $\varphi \times \psi$ is continuous, then $(\varphi \times \psi)(\overline{U}) = (\varphi \times \psi)(V) \subseteq \overline{(\varphi \times \psi)(U)} \subseteq \Delta(W)$, the latter inclusion following from the fact that $\Delta(W)$ is closed in $W \times W$ (cf. Proposition 7.5.1). This implies $\varphi = \psi$. $\square$

Definition 8.1.2. Let V and W be algebraic varieties. A *rational map* Φ between V and W, denoted by $\Phi : V \dashrightarrow W$, is an equivalence class of pairs (U, φ_U), where $U \subseteq V$ is a non-empty, open subset of V, $\varphi_U \in \mathrm{Morph}(U, W)$ and where (U_1, φ_{U_1}) and (U_2, φ_{U_2}) are said to be *equivalent* if φ_{U_1} and φ_{U_2} agree on $U_1 \cap U_2$. Thus, $\Phi = [U, \varphi_U]$, where $[U, \varphi_U]$ the equivalence class of the pair (U, φ_U). In such a case, φ_U is said to be a *representative morphism* of Φ over the open set U.

If Φ is a rational map and if (U_1, φ_{U_1}), (U_2, φ_{U_2}) are representative morphism of Φ, then $(U_1 \cup U_2, \varphi_{U_1 \cup U_2})$ is another representative of Φ, where $\varphi_{U_1 \cup U_2}|_{U_i} := \varphi_{U_i}$ for $i = 1, 2$. The previous observation shows that there exists a non-empty, open subset of V, say U_Φ, which is the biggest open subset of V where $\Phi|_{U_\Phi}$ is a morphism. This open set is called the *open set of definition* (or even the *domain*) of the rational map Φ; sometimes U_Φ is also denoted by $\mathrm{Dom}(\Phi)$.

From Examples 5.3.23-(i), (iii), 6.2.5-(iv) and Exercise 5.1, the maps ϕ_1 and ϕ_3 are rational maps s.t. $U_{\phi_1} = U_{\phi_3} = W = \mathbb{A}^1 \setminus \{0\}$. Similarly, from Example 6.3.2-(i), when $B \neq \emptyset$, the map ν_L as in (6.16) defines a rational map $\nu_L : \mathbb{P}^n \dashrightarrow \mathbb{P}^r$ such that $U_{\nu_L} = \mathbb{P}^n \setminus B$.

In general the composition of rational maps is not well-defined. For example, consider the map $\varphi : \mathbb{A}^1 \to \mathbb{A}^2$, $x_1 \to (x_1, 0)$, which is a morphism, and

$$\Phi : \mathbb{A}^2 \dashrightarrow \mathbb{A}^1, \quad (x_1, x_2) \dashrightarrow \frac{x_1}{x_2},$$

which is a rational map whose domain is $U_\Phi = \mathbb{A}^2 \setminus Z_a(x_2)$. From the fact that $\varphi(\mathbb{A}^1) \cap U_\Phi = \emptyset$, the composition $\Phi \circ \varphi$ is not defined. To avoid this kind of phenomena, one gives the following definition.

Definition 8.1.3. A rational map $\Phi : V \dashrightarrow W$ is said to be *dominant* if for some (and so every) representative (U, φ_U) is a dominant morphism in the sense of Definition 6.1.5.

To have some examples, note that on any algebraic variety V, dominant rational maps $\Phi : V \dashrightarrow \mathbb{A}^1$ are nothing but rational functions $\Phi \in K(V) \setminus \mathbb{K}$. Furthermore, if V is a $(n+1)$-dimensional $\mathbb{K}$-vector space and if we identify it with the affine space $\mathbb{A}^{n+1}$, the map π in (3.1) is a morphism (as it follows from Proposition 6.3.1) so it defines a rational surjective (so dominant) map $\pi : \mathbb{A}^{n+1} \dashrightarrow \mathbb{P}^n$ such that $\mathrm{Dom}(\pi) = \mathbb{A}^{n+1} \setminus \{\underline{0}\}$.

Lemma 8.1.4. *Let V, W, Z be algebraic varieties and let $\Phi : V \dashrightarrow W$ and $\Psi : W \dashrightarrow Z$ be dominant rational maps. Then $\Psi \circ \Phi : V \dashrightarrow Z$ is a dominant rational map.*

Proof. By assumptions on Φ and Ψ, the representatives $(U_\Phi, \varphi_{U_\Phi})$ and (U_Ψ, ψ_{U_Ψ}) are dominant morphisms (which for simplicity will be simply denoted in the sequel by φ and ψ, respectively). Then $\mathrm{Im}(\varphi) \cap U_\Psi \neq \emptyset$ and it contains a dense open subset U_0 of U_Ψ, which is therefore also an open subset of W. Then $\varphi^{-1}(U_0) \subseteq V$ is an open subset where the composition $\psi \circ \varphi$ is defined. On this open set $\psi \circ \varphi$ is a dominant morphism. Indeed, by the dominance of Ψ, for any proper closed subset $K \subsetneq Z$ one has that $\mathrm{Im}(\psi)$ is not contained in K and $\psi^{-1}(K)$ is a proper closed subset of W. For the same reason, $\mathrm{Im}(\varphi)$ is not contained in $\psi^{-1}(K)$, so $\mathrm{Im}(\psi \circ \varphi)$ is not contained in $K \subsetneq Z$; since this holds for any proper closed subset $K \subsetneq Z$, this implies that $\Psi \circ \Phi$ is a dominant rational map. $\qquad\square$

Definition 8.1.5. A *birational map* $\Phi : V \dashrightarrow W$ is a dominant rational map which admits an inverse rational map, namely a rational map $\Psi : W \dashrightarrow V$ s.t. $\Psi \circ \Phi = \mathrm{Id}_V$ and $\Phi \circ \Psi = \mathrm{Id}_W$, where the previous equalities are intended as rational maps. If there exists a birational map from V to W, then V and W are said to be *birationally equivalent* (or simply *birational*).

By the very definition of birational maps, V and W are birational varieties if and only if there exist non-empty, open subsets $U_V \subseteq V$ and $U_W \subseteq W$ which are isomorphic. For this reason, birational maps are sometimes called also *birational isomorphisms*.

Theorem 8.1.6. *Let V and W be algebraic varieties. There is a bijective correspondence between:*

(i) *the set of dominant rational maps $V \dashrightarrow W$, and*
(ii) *the set of $\mathbb{K}$-algebra monomorphisms $K(W) \hookrightarrow K(V)$.*

In this bijective correspondence, birational maps correspond to field isomorphisms.

Proof. Let $\Phi : V \dashrightarrow W$ be any dominant rational map; since (U_Φ, φ) is a dominant morphism, from (6.5) we get a $\mathbb{K}$-algebra monomorphism $\varphi^* : K(W) \hookrightarrow K(U_\Phi)$. We then conclude by $K(U_\Phi) \cong K(V)$. Conversely, let $K(W) \overset{\theta}{\hookrightarrow} K(V)$ be any K-algebra monomorphism. From Proposition 6.4.2,

W is covered by affine open subsets. Thus, by Lemma 5.3.13-(ii) we can assume W to be affine. Let therefore $y_1, \ldots, y_n$ be generators of the affine coordinate ring $A(W)$ as a $\mathbb{K}$-algebra of finite type. Since $A(W) \subseteq K(W)$, then $\theta(y_1), \ldots, \theta(y_n) \in K(V)$. There exists a non-empty, open subset $U \subseteq V$ s.t. $\theta(y_1), \ldots, \theta(y_n) \in \mathcal{O}_V(U) \subset K(V)$ (it suffices to take the intersection of all the domains of definition of the rational functions $\theta(y_i)$, $1 \leqslant i \leqslant n$).

This defines a $\mathbb{K}$-algebra homomorphism $A(W) \xrightarrow{\theta|} \mathcal{O}_V(U)$, $y_i \to \theta(y_i)$ which is injective, since θ is. Since W is affine, from Proposition 6.2.3, the homomoprhism $\theta|$ corresponds to $\varphi^\theta \in \mathrm{Morph}(U, W)$. By the injectivity of $\theta|$ and by Corollary 6.2.8, we get that φ^θ is dominant. This correspondence defines therefore a dominant rational map $\Phi : V \dashrightarrow W$, with representative φ^θ, which is such that $(\varphi^\theta)^* = \theta$, i.e. this correspondence is the inverse of that reminded at the beginning of the proof. $\qquad \square$

Corollary 8.1.7.

(i) V and W are birationally equivalent if and only if $K(V) \cong K(W)$.

(ii) Any algebraic variety V is birationally equivalent to any of its non-empty, open subset.

(iii) Any algebraic variety V is birationally equivalent to an affine variety and to a projective variety.

Proof.

(i) This directly follows from Theorem 8.1.6.

(ii) Since for any non-empty, open subset $U \subseteq V$ one has $K(U) \cong K(V)$, the statement follows from (i).

(iii) From Proposition 6.4.2, any algebraic variety has an affine open set U. Moreover, since U is affine, its projective closure $\overline{U}$ is a projective variety for which U is an open dense subset. Then one concludes by (ii). $\qquad \square$

Remark 8.1.8.

(i) Birationality is an equivalence relation among algebraic varieties. For any algebraic variety V, the symbol $[V]_{\mathrm{bir}}$ will denote the equivalence class consisting of all algebraic varieties birationally equivalent to V. Any representative of $[V]_{\mathrm{bir}}$ will be called a *model* of the birational class; it is clear that $[V]_{\mathrm{bir}}$ contains all algebraic varieties which are isomorphic to V on the other hand it contains also several other models as the following examples show.

(ii) The rational map $\phi_1 : \mathbb{A}^1 \dashrightarrow Y := Z_a(x_1 x_2 - 1)$ in Examples 5.3.23-(i), 6.2.5-(iv) and in Exercise 5.1 is birational, since Y is isomorphic to $\mathbb{A}^1 \setminus \{0\}$ via ϕ_1. On the other hand, we already observed that Y is not isomorphic to $\mathbb{A}^1$. Same occurs for, e.g. the ellipse in Examples 5.3.23-(iii), 6.2.5-(iv) and for the semi-cubic parabola in Examples 5.3.24, 6.2.5-(v), since it has been showed therein that both $Z_a(x_1^2 + x_2^2 - 1)$ and $Z_a(x_1^3 - x_2^2) \setminus \{(0,0)\}$ are isomorphic to $\mathbb{A}^1 \setminus \{0\}$.

(iii) Recalling examples discussed in the previous chapters, we also get that $\mathbb{A}^1, \mathbb{P}^1, \mathbb{A}^1 \setminus \{0\}$, the ellipse, the hyperbola, the parabola, the semi-cubic parabola, the cubic $Z_a(x_2^2 - x_1^2(x_1 - 1))$ (recall Example 5.3.25), any affine and projective rational normal curve of degree d, etc. these are all models of $[\mathbb{P}^1]_{\text{bir}}$: indeed, in all these cases, we proved that the field of rational functions is isomorphic to $\mathbb{K}(x_1)$ (these are all examples of *rational curves*, cf. Section 8.2).

(iv) On the contrary the affine curve $Y_a : Z_a(x_2^2 - x_1(x_1 - 1)(x_1 - a))$, with $1, 0 \neq a \in \mathbb{K}$, cannot be a model of $[\mathbb{P}^1]_{\text{bir}}$: in example Example 5.3.25 we showed that $K(Y_a)$ is an algebraic extension of degree 2 of the field $\mathbb{K}(x_1)$ (Y_a is a *smooth plane cubic*, as such it is called *elliptic curve*).

For V any algebraic variety, $\text{Bir}(V)$ will denote the group of birational transformations of V, which contains $Aut(V)$ as a subgroup. When $V = \mathbb{P}^n$, for some $n \geq 1$, $\text{Bir}(\mathbb{P}^n)$ is also called *Cremona group* in honor of Luigi Cremona, who was a pioneer in the study of birational maps of projective spaces. As a consequence of Example 8.1.12, we will deduce that $\text{Bir}(\mathbb{P}^1) = \text{Aut}(\mathbb{P}^1) = \text{PGL}(2, \mathbb{K})$. On the other hand, for $n \geq 2$, $\text{Aut}(\mathbb{P}^n) = \text{PGL}(n + 1, \mathbb{K})$ is a proper subgroup of the Cremona group of $\mathbb{P}^n$, whose structure turns out to be in general quite intricate (cf., e.g. Verra, 2005).

8.1.1 *Some properties and some examples of (bi)rational maps*

In this section, we shall discuss interesting examples of rational and birational maps. Let V be any algebraic variety; if one has a rational map $\Psi : V \dashrightarrow \mathbb{A}^n$, for some positive integer n, composition with the isomorphism $\phi_0 : \mathbb{A}^n \to U_0$ and with the open inclusion $\iota_{U_0} : U_0 \hookrightarrow \mathbb{P}^n$ determines a rational map $\widetilde{\Psi} := \iota_{U_0} \circ \phi_0 \circ \Psi : V \dashrightarrow \mathbb{P}^n$.

Conversely, take a rational map $\widetilde{\Psi} : V \dashrightarrow \mathbb{P}^n$. Let $U_{\widetilde{\Psi}}$ be its domain and let $\widetilde{\psi}$ be its representative morphism over the open set $U_{\widetilde{\Psi}}$. If $\widetilde{\psi}(U_{\widetilde{\Psi}}) \cap H_0 = \emptyset$, then $\widetilde{\Psi}$ corresponds to a rational map $\Psi : V \dashrightarrow \mathbb{A}^n$.

Corollary 8.1.9. *Let V be any algebraic variety. Any non-constant morphism $\phi : V \to \mathbb{P}^1$ determines a rational map $\Phi : V \dashrightarrow \mathbb{A}^1$, i.e. an element $\Phi \in K(V)$. If moreover ϕ is not surjective, the map Φ is a morphism, i.e. $\Phi \in \mathcal{O}_V(V)$.*

Proof. Only the last assertion needs some comment: if $\phi(V) \subsetneq \mathbb{P}^1$, for any choice of $P \in \mathbb{P}^1 \setminus \phi(V)$ one has that $\mathbb{P}^1 \setminus \{P\} \cong \mathbb{A}^1$, from which one concludes.   □

One basic question is: what about the converse of Corollary 8.1.9? In other words, since any $\Phi \in K(V)$ induces a morphism $\phi : U_\Phi \to \mathbb{A}^1$ and so a rational map $\Phi : V \dashrightarrow \mathbb{A}^1$, is it true that Φ always extends to a morphism to $\mathbb{P}^1$? The answer in general is negative, as the following example shows.

Example 8.1.10. Let $\Phi := \frac{X_1}{X_0} \in K(\mathbb{P}^2)$. This rational function is regular on the principal affine open set $U_0^2 \subset \mathbb{P}^2$, so it defines a rational map $\Phi : \mathbb{P}^2 \dashrightarrow \mathbb{A}^1$ with a representative morphism $\phi : \mathbb{A}^2 \cong U_0^2 \to \mathbb{A}^1$, which is the projection

$$\pi_1 : \mathbb{A}^2 \to \mathbb{A}^1, \quad [1, x_1, x_2] = (x_1, x_2) \xrightarrow{\pi_1} x_1 = [1, x_1]$$

onto the first coordinate (when one identifies $\mathbb{A}^1$ with $U_0^1 \subset \mathbb{P}^1$). Note that π_1 is the restriction to the open set U_0^2 of the projection

$$\pi_\Lambda : \mathbb{P}^2 \dashrightarrow \mathbb{P}^1, \quad [X_0, X_1, X_2] \xdashrightarrow{\pi_\Lambda} [X_0, X_1]$$

from the linear subspace $\Lambda := \{[0, 0, 1]\} \subset \mathbb{P}^2$. Thus, π_Λ is a rational map whose domain is $\mathbb{P}^2 \setminus \Lambda$. In other words, Φ extends to a rational map from $\mathbb{P}^2$ to $\mathbb{P}^1$ but not to a morphism defined on the whole $\mathbb{P}^2$ with target $\mathbb{P}^1$.

On the other hand, in some cases we have affirmative answer to the previous question, as the following examples show.

Example 8.1.11. *Any rational map $\Psi : \mathbb{A}^1 \dashrightarrow \mathbb{A}^n$ extends to a unique morphism $\widetilde{\psi} : \mathbb{P}^1 \to \mathbb{P}^n$.*

Proof. It is enough to show it for $n = 1$, the general case being similar (using Proposition 6.2.2). The rational map $\Psi : \mathbb{A}^1 \dashrightarrow \mathbb{A}^1$ corresponds to a rational function $\Psi \in K(\mathbb{A}^1) \cong \mathcal{Q}^{(1)} \cong \mathbb{K}(t)$, where t is an indeterminate over $\mathbb{K}$. Thus, there exist $f, g \in A^{(1)} \cong \mathbb{K}[t]$ s.t. $\Psi = \frac{f}{g}$, where

$$U_\Psi = U_a(g) = Z_a(g)^c, \quad \text{i.e } \Psi \in \mathcal{O}_{\mathbb{A}^1}(U_a(g)).$$

We can assume that g.c.d.$(f, g) = 1$. The map Ψ defines a morphism $\psi : U_a(g) \to \mathbb{A}^1$. We first show that the morphism ψ uniquely extends to a morphism $\mathbb{A}^1 \to \mathbb{P}^1$. To do this, identify the target $\mathbb{A}^1$ of ψ with $U_0 \subset \mathbb{P}^1$ and consider the map

$$\psi' : \mathbb{A}^1 \to \mathbb{P}^1, \quad P \to [g(P), f(P)].$$

This map is well-defined as, for points $P \in \mathbb{A}^1 \setminus U_a(g)$, one has

$$[g(P), f(P)] = [0, f(P)] = [0, 1]$$

since $f(P) \neq 0$ (otherwise $(t - P)$ would be a common divisor of f and g against the assumption on g.c.d.(f, g)). Observe moreover that ψ' is a morphism; indeed let

$$f(t) = f_0 + f_1 t + \ldots f_n t^n \quad \text{and} \quad g(t) = g_0 + g_1 t + \ldots g_m t^m,$$

with $f_i, g_j \in \mathbb{K}$, $f_n, g_m \neq 0$, $0 \leqslant i \leqslant n$, $0 \leqslant j \leqslant m$, i.e. $\deg(f) = n$ and $\deg(g) = m$. Assume, e.g. $n \geqslant m$ (the other case being similar). Identify once again $\mathbb{A}^1$ (the domain of ψ') with $U_0 \subset \mathbb{P}^1$; thus $t = \frac{X_1}{X_0}$ and the homogeneous polynomials

$$F(X_0, X_1) = X_0^n f\left(\frac{X_1}{X_0}\right) = \mathfrak{h}_0(f) \quad \text{and}$$

$$G(X_0, X_1) := X_0^n g\left(\frac{X_1}{X_0}\right) = X_0^{n-m} \mathfrak{h}_0(g)$$

are such that

$$[g(P), f(P)] = [G(1, P), F(1, P)]$$

for any $P \in \mathbb{A}^1$. This implies that ψ' is a morphism (cf. Proposition 6.3.1); moreover its uniqueness follows from Lemma 8.1.1. The next step is to show that ψ' uniquely extends to a morphism $\mathbb{P}^1 \to \mathbb{P}^1$. Since $\mathbb{P}^1 = U_0 \cup \{[0, 1]\}$, from the expression of F and G above, one has $[G(0, 1), F(0, 1)] = [0, f_n] = [0, 1]$, since $f_n \neq 0$. In particular, ψ' is the restriction to U_0 of the map

$$\widetilde{\psi} : \mathbb{P}^1 \to \mathbb{P}^1, \quad [X_0, X_1] \to [G(X_0, X_1), F(X_0, X_1)],$$

which is a morphism from Proposition 6.3.1 (its uniqueness once again follows from Lemma 8.1.1).

Similarly, as in the previous example, one has the following.

Example 8.1.12. *Any rational map* $\Phi : \mathbb{P}^1 \dashrightarrow \mathbb{P}^n$ *is always a morphism.*

Proof. We can assume Φ to be non-constant, otherwise there is nothing to prove. Let U_Φ be the domain of Φ and let $\phi : U_\Phi \to \mathbb{P}^n$ be the representative morphism. Since being a morphism is a local property (cf. Proposition 6.4.1), from Proposition 6.3.1 we can directly assume that ϕ is globally defined on U_Φ by a collection of homogeneous polynomials $F_0, \ldots, F_n \in \mathcal{S}_k^{(1)}$, for some integer $k \geqslant 1$, with $Z_p(F_0, \ldots, F_n) = \emptyset$. Then

$$\phi(P) = [F_0(P), \ldots, F_n(P)], \quad \forall \; P \in U_\Phi,$$

where, with no loss of generality, we can assume g.c.d$(F_0, \ldots, F_n) = 1$. Thus, for any point $P = [p_0, p_1] \in \mathbb{P}^1 \setminus U_\Phi$, $(F_0(P), \ldots, F_n(P)) \neq (0, \ldots 0)$, otherwise the polynomials $F_0, \ldots, F_n$ would have as common divisor the homogeneous polynomial $p_1 X_0 - p_0 X_1$, against the assumption. This means that Φ is wherever defined, so it is a morphism.

To have concrete applications of previous statements, consider:

Example 8.1.13. Let $\mathbb{K} = \mathbb{C}$ and let $\Psi : \mathbb{A}^1 \dashrightarrow \mathbb{A}^2$ be the rational map defined by

$$t \dashrightarrow \left(\frac{t^2 - 1}{t^2 + 1}, \frac{-2t}{t^2 + 1} \right).$$

Its domain is $U_\Psi = \mathbb{A}^1 \setminus \{i, -i\}$ and its representative morphism ψ over U_Ψ is such that $\psi(U_\Psi) \subset C = Z_a(x_1^2 + x_2^2 - 1) \subset \mathbb{A}^2$, the first inclusion being strict since $(1, 0) \in C$ is not contained in Im(ψ) as one can easily check. The map Ψ is more precisely birational; indeed, consider the pencil of lines through $(1, 0) \in C$, i.e. $x_2 = t(x_1 - 1)$, where $t \in \mathbb{K}$. This pencil defines a morphism

$$C \setminus \{(1, 0)\} \to \mathbb{A}^1, \quad (p_1, p_2) \to \frac{p_2}{p_1 - 1} = t$$

which is the inverse of the morphism ψ over the open set $C \setminus \{(1, 0)\}$, as for any $t \in U_\Psi$ the intersection $Z_a(x_2 - t(x_1 - 1), x_1^2 + x_2^2 - 1)$ (off the point of coordinates $(1, 0)$, which is the fixed point of the pencil of lines) determines the point of coordinates $\left(\frac{t^2 - 1}{t^2 + 1}, \frac{-2t}{t^2 + 1} \right) \in C \setminus \{(1, 0)\}$. In particular Ψ is another rational parametrization of the ellipse, which differs from the parametrization considered in Examples 5.3.23-(iii) and 6.2.5-(iv).

As in Example 8.1.11, ψ uniquely extends to a morphism

$$\psi' : \mathbb{A}^1 \to \mathbb{P}^1, \quad t \to [t^2 + 1, t^2 - 1, -2t].$$

Note indeed that ψ' is well-defined since $\psi'(\pm i) = [0, 1, \pm i] \in \overline{C} \cap H_0$, where $\overline{C}$ denotes the projective closure of C in $\mathbb{P}^2$ and where we identified $\mathbb{A}^2$ with the affine chart $U_0^2 \subset \mathbb{P}^2$. Moreover, identifying $\mathbb{A}^1$ with $U_0^1 \subset \mathbb{P}^1$, one has $t = \frac{X_1}{X_0}$ so

$$[X_0, X_1] \xrightarrow{\psi'} [X_0^2 + X_1^2, X_1^2 - X_0^2, -2X_0X_1],$$

which shows that the previous map is a morphism since

$$Z_p(X_0^2 + X_1^2, X_1^2 - X_0^2, -2X_0X_1) = Z_p(X_0^2, X_1^2, X_0X_1) = Z_p((\mathcal{S}_+^{(1)})^2) = \emptyset.$$

Note that $\mathrm{Im}(\psi') = \overline{C} \setminus \{[1,1,0]\}$ nonetheless, as in Example 8.1.11, ψ' extends to a morphism $\tilde{\psi} : \mathbb{P}^1 \to \overline{C}$ which maps $[0,1]$ to $[1,1,0] \in \overline{C}$ (this is the point where originally the inverse of ψ was not defined).

Example 8.1.10 shows that the projection of $\mathbb{P}^2$ from one of its point gives rise to a rational map which cannot be extended to a morphism. Below, we consider in more generalities these examples of rational maps, namely the *projections* of a projective space from a linear subspace.

Example 8.1.14 (Projections). Consider the projective space $\mathbb{P}^n$ with homogeneous coordinates $[X_0, \ldots, X_n]$. Let us denote with $\underline{X}$ the column-vector whose entries are these homogeneous coordinates. Let A be a $(n+1) \times (m+1)$-matrix with entries in $\mathbb{K}$ and such that $\mathrm{rank}(A) := r \leqslant n+1$. Then $\mathbb{P}(\mathrm{Ker}(A))$ is either empty, when $r = n + 1$, or it is a linear subspace $\Lambda := \Lambda_A \subset \mathbb{P}^n$ of dimension $n - r$. In the empty case, i.e. $r = n + 1$, the map $\mathbb{P}^n \to \mathbb{P}^m$, given by $[X_0, \ldots, X_n] \xrightarrow{\pi_\Lambda} [A\underline{X}]$ is a homography as in Section 3.3.8, in particular $n \leqslant m$, the map is injective and everywhere defined.

From now on we therefore shall focus on the case when $\Lambda \neq \emptyset$; in such a case the map

$$\pi_\Lambda : \mathbb{P}^n \setminus \Lambda \to \mathbb{P}^m, \quad [X_0, \ldots, X_n] \xrightarrow{\pi_\Lambda} [A\underline{X}] \tag{8.1}$$

is well-defined and it is a morphism on the open set $U := \mathbb{P}^n \setminus \Lambda$ of $\mathbb{P}^n$. So, π_Λ defines a rational map $\pi_\Lambda : \mathbb{P}^n \dashrightarrow \mathbb{P}^m$ which is also called *degenerate homography* or even *projection* of $\mathbb{P}^n$ from the linear subspace Λ to the projective space $\mathbb{P}^m$. The subspace Λ is also called the *center of the projection* π_Λ. The domain of π_Λ is the open set U, where it is defined as a morphism. If, e.g. $r = 1$, then Λ is a hyperplane in $\mathbb{P}^n$ and π_Λ is a constant map.

A typical example of degenerate homography is the following: let $\Lambda \subset \mathbb{P}^n$ be a linear subspace of dimension $n - r$, with $r \geqslant 1$ and let $\Sigma \subset \mathbb{P}^n$ be another linear subspace of dimension $r-1$ which is skew to Λ, i.e. $\Lambda \cap \Sigma = \emptyset$. From the projective Grassmann formula (3.23) one has that $\Lambda \vee \Sigma = \mathbb{P}^n$. Therefore, for any point $P \in \mathbb{P}^n \setminus \Lambda := U$, the linear space $P \vee \Sigma$ is of dimension $n-r+1$ which, once again by the projective Grassmann formula, intersects Σ at a unique point P'. The map

$$\tau : \mathbb{P}^n \setminus \Lambda \to \Sigma, \quad P \xrightarrow{\tau} P' := (P \vee \Lambda) \cap \Sigma$$

is a degenerate homography; indeed up to projectivities, we may fix homogeneous coordinates on $\mathbb{P}^n$ in such a way that

$$\Lambda = Z_p(X_0, \ldots, X_{r-1}) \quad \text{and} \quad \Sigma = Z_p(X_r, \ldots, X_n).$$

In this set-up the map τ is simply given by

$$[X_0, \ldots, X_n] \xrightarrow{\tau} [X_0, \ldots, X_{r-1}], \tag{8.2}$$

i.e. $\tau = \pi_\Lambda$. From the previous expression for τ we note that the map is surjective and, moreover, that $\tau(P) = \tau(Q)$ if and only if $P \vee \Lambda = Q \vee \Lambda$ as linear subspaces of $\mathbb{P}^n$. Any projection of $\mathbb{P}^n$ from a linear subspace Λ of dimension $n - r$ onto a skew subspace Σ of dimension $r - 1$ can be obtained as the map τ above after a projective transformation, i.e. any such projection is the composition of τ and of a projectivity of $\mathbb{P}^n$.

Identify $\mathbb{A}^n$ with the affine chart $U_0 \subset \mathbb{P}^n$ and consider the degenerate homography in (8.2). Since Σ is not contained in the hyperplane $H_0 = Z_p(X_0)$, then $\Sigma \setminus \Sigma_\infty \cong \mathbb{A}^{r-1}$ is an affine subspace of $\mathbb{A}^n = U_0$. The restriction of τ to $U_0 \setminus (U_0 \cap \Lambda) \cong \mathbb{A}^n \setminus \Lambda_0$, where Λ_0 the affine subspace cut-out by Λ in U_0, is a morphism. If $\Lambda_0 \neq \emptyset$, then $\tau_0 := \tau_{|\mathbb{A}^n \setminus \Lambda_0}$ is called *projection of $\mathbb{A}^n$ with center Λ_0 onto $\mathbb{A}^{r-1}$*. If otherwise $\Lambda_0 = \emptyset$, then τ_0 is called *projection of $\mathbb{A}^n$ onto $\mathbb{A}^{r-1}$ with direction parallel to Λ*. In the case of parallel projections, τ_0 is a morphism. When otherwise the center is not empty, τ_0 is a rational map from $\mathbb{A}^n$ to $\mathbb{A}^{r-1}$ but not a morphism.

Consider now $\Lambda \subset \mathbb{P}^n$ a linear subspace, with $\Lambda \cong \mathbb{P}^{n-k-1}$, and let $\pi_\Lambda : \mathbb{P}^n \dashrightarrow \mathbb{P}^k$ be the projection with center Λ. If V is any algebraic variety not contained in Λ, the restriction of π_Λ to V defines a rational map

$$\pi_V : V \dashrightarrow \mathbb{P}^k, \tag{8.3}$$

which is called *projection of V on the subspace $\mathbb{P}^k$*; π_V is a morphism on $V \setminus (V \cap \Lambda)$; in particular, in $V \cap \Lambda = \emptyset$, then $\pi_V \in \mathrm{Morph}(V, \mathbb{P}^k)$. On the other hand, in some cases the set $V \setminus (V \cap \Lambda)$ is strictly contained in the whole domain of the map π_V, as the following example shows.

Example 8.1.15. Let $\overline{C} \subset \mathbb{P}^2$ be the conic as in Example 8.1.13. The pencil of lines therein defines a projection of $\mathbb{A}^2$ onto the x_2-axis and so a linear projection $\pi_P : \mathbb{P}^2 \dashrightarrow \mathbb{P}^1$, where $\mathbb{P}^1$ identified with the line $H_1 = Z_p(X_1) \subset \mathbb{P}^2$ and where the center of π_P is the base point of the pencil of lines $P = [1, 1, 0]$. Since P is on $\overline{C}$, π_P defines a rational map $\pi_{\overline{C}}$ which is a projection of $\overline{C}$ to the linear subspace $\mathbb{P}^1$. This is a morphism certainly over $\overline{C} \setminus \{P\}$ and it is an example of *stereographic projection of an irreducible conic onto* $\mathbb{P}^1$. For any point $Q = (0, y_0)$ on the x_2-axis in $\mathbb{A}^2 = U_0$, with $y_0 \neq \pm i$, the unique line of the pencil $P \vee Q$ intersects C outside P at a unique further point, say $P_Q \in C$, which is $P_Q = \left(\frac{y_0^2 - 1}{y_0^2 + 1}, \frac{2y_0}{y_0^2 + 1} \right)$. From Example 8.1.13, this is a morphism. Identifying the line H_1 with $\mathbb{P}^1$, $\pi_{\overline{C}}$ is therefore a birational map $\overline{C} \dashrightarrow \mathbb{P}^1$ whose domain certainly contains $\overline{C} \setminus \{P\}$. On the other hand, $\overline{C} \cong \mathbb{P}^1$ via a Veronese morphism $\nu_{1,2}$. Up to the composition with $\nu_{1,2}$, from Example 8.1.12 $\pi_{\overline{C}}$ extends to a unique morphism defined on the whole $\overline{C}$. The geometric interpretation of such an extension is given by considering the tangent line $X_0 - X_1 = 0$ to $\overline{C}$ at P and then intersecting it with the line H_1, which gives the point $[0, 0, 1]$.

8.2   Unirational and Rational Varieties

We start with some important definitions.

Definition 8.2.1. An algebraic variety V is said to be *unirational* if there exist a positive integer r, a dominant rational map $\Phi : \mathbb{P}^n \dashrightarrow V$ and a dense open subset $U \subseteq \text{Im}(\Phi)$ s.t., for any $P \in U$, $\Phi^{-1}(P)$ is a finite set. Such a map Φ is said to be a *generically finite* dominant rational map.

From Theorem 8.1.6, the map Φ is associated with a field extension $K(V) \subseteq \mathbb{K}(x_1, \ldots, x_r) = \mathcal{Q}^{(n)}$.

Definition 8.2.2. An algebraic variety V is said to be *rational* if it is birational to $\mathbb{P}^r$ (or $\mathbb{A}^r$), for some non-negative integer r.

In particular, if V is rational then it is unirational. Moreover, from Corollary 8.1.7-(i), we get:

Corollary 8.2.3. *V is a rational algebraic variety if and only if $K(V) \cong \mathcal{Q}^{(r)}$, i.e. if and only if the field $K(V)$ is obtained as a purely transcendental* *extension of the ground field* $\mathbb{K}$ *of some transcendental degree* r, *for some non-negative integer* r.

All the curves listed in Remark 8.1.8-(ii) and (iii) are rational curves, whereas the curve Y_a in (iv) therein is not rational (recall indeed it has been called *elliptic curve*). The so-called *Lüroth problem* is a fundamental problem in Algebraic Geometry:

Lüroth problem *Does unirationality imply rationality?*

In 1876, Lüroth proved that in the curve case the answer to the previous question is yes, i.e. any unirational curve is also rational (cf., e.g. Cohn, 1991, cf. p. 148). In 1894, Castelnuovo showed that also unirational surfaces are rational (cf., e.g. Barth *et al.*, 2004; Beauville, 1996). Only after almost one century, e.g. Clemens and Griffiths (1972) showed that a (smooth) cubic *threefold* (i.e. an hypersurface of degree 3 in $\mathbb{P}^4$ with no singular points) is unirational but in general not a rational variety, providing an example for three dimensions that unirationality does not imply rationality.

In what follows we shall give several examples of rational varieties which are birational but not isomorphic to $\mathbb{P}^n$, for $n \geqslant 2$.

8.2.1 Stereographic projection of a rank-four quadric surface

A nice example of a surface which is birational (but not isomorphic) to $\mathbb{P}^2$ is given by an irreducible, doubly ruled quadric $\mathfrak{Q} \subset \mathbb{P}^3$. This is a *rank-four quadric* in $\mathbb{P}^3$, i.e. the homogeneous quadratic polynomial defining $\mathfrak{Q}$ is a quadratic form in the indeterminates $X_0, \ldots, X_3$ whose associated symmetric matrix has non-zero determinant. Since $\mathbb{K}$ is algebraically closed, all these quadrics are projectively equivalent. Therefore, with no loss of generality, we may assume that $\mathfrak{Q} = Z_p(X_0 X_3 - X_1 X_2)$, which is the isomorphic image of $\mathbb{P}^1 \times \mathbb{P}^1$ embedded, via the Segre morphism $\sigma_{2,2}$, as an irreducible doubly ruled quadric in $\mathbb{P}^3$ (cf. Exercise 7.1). Take $P_0 = [1, 0, 0, 0] \in \mathfrak{Q}$. From (8.3), the rational map

$$\pi_{P_0} : \mathbb{P}^3 \dashrightarrow \mathbb{P}^2, \quad [X_0, X_1, X_2, X_3] \dashrightarrow [X_1, X_2, X_3]$$

induces the projection $\pi_{\mathfrak{Q}} : \mathfrak{Q} \dashrightarrow \mathbb{P}^2$ which is called *stereographic projection of the rank-four quadric surface* to a plane. This rational map is a morphism over the open subset $\mathfrak{Q} \setminus \{P_0\}$.

Differently from the stereographic projection in Example 8.1.15, the expression of $\pi_{\mathfrak{Q}}$ shows that the map cannot be extended (as a morphism) at the point P_0, moreover that its target space is the plane $H_0 = Z_p(X_0) \subset \mathbb{P}^3$. Our aim is to show that $\pi_{\mathfrak{Q}}$ is a birational map to such a plane. To do this consider $H_3 = Z_p(X_3) \subset \mathbb{P}^3$ and the open set $U := \mathfrak{Q} \setminus (\mathfrak{Q} \cap H_3) \subset \mathfrak{Q}$. Note that

$$\mathfrak{Q} \cap H_3 = Z_p(X_1, X_3) \cup Z_p(X_2, X_3) = \ell_{13} \cup \ell_{23},$$

where ℓ_{13} and ℓ_{23} are the two lines passing through P_0 of the two different rulings of $\mathfrak{Q}$. Thus, $\varphi := \pi_{\mathfrak{Q}}|_U : U \to U' \subset H_0 \cong \mathbb{P}^2$ is a morphism (since $P_0 \in \ell_{13} \cup \ell_{23}$) whose image is the open set $U' := H_0 \setminus (H_0 \cap H_3)$, which is isomorphic to $\mathbb{P}^2$ minus a line (such a line is given by $\ell_{03} = Z_p(X_0, X_3) \subset H_0$).

Claim 8.2.4. $\varphi : U \to U'$ *is an isomorphism.*

Proof. For any point $Q = [0, q_1, q_2, q_3] \in U'$, consider the line $P_0 \vee Q$, which has parametric equations $X_0 = s$, $X_1 = q_1 t$, $X_2 = q_2 t$, $X_3 = q_3 t$, $[s, t] \in \mathbb{P}^1$. Such a line intersects U at the point $[\frac{q_1 q_2}{q_3} t, q_1 t, q_2 t, q_3 t]$, which is $[\frac{q_1 q_2}{q_3}, q_1, q_2, q_3]$, as $t \neq 0$ since $P_0 \notin U$. In other words, $\varphi^{-1}(Q) = [q_1 q_2, q_1 q_3, q_2 q_3, q_3^2]$ which is a morphism, as it follows from its expression and the fact that $q_3 \neq 0$ since $Q \in U'$.

In particular, the map $\pi_{\mathfrak{Q}}$ is birational (but not a morphism) so $\mathfrak{Q}$ is a *rational surface*. Identifying the plane H_0 with $\mathbb{P}^2$, with homogeneous coordinates $[X_1, X_2, X_3]$, from Claim 8.2.4, the inverse of φ is induced by the rational map

$$\nu : \mathbb{P}^2 \dashrightarrow \mathbb{P}^3, \ [X_1, X_2, X_3] \dashrightarrow [X_1 X_2, X_1 X_3, X_2 X_3, X_3^2],$$

where $\nu = \nu_L$ as in (6.16) with $L = \mathrm{Span}\{X_1 X_2, X_1 X_3, X_2 X_3, X_3^2\} \subset \mathcal{S}_2^{(2)}$. In particular ν_L is not defined at $B := \{P_1 = [1, 0, 0], P_2 = [0, 1, 0]\} \subset \mathbb{P}^2$, which is the base locus of the linear system of plane conics $\mathbb{P}(L)$, and $\mathrm{Im}(\nu_L) \subseteq \mathfrak{Q}$.

To sum up, $\pi_{\mathfrak{Q}}$ and ν_L are two birational maps, one is the inverse of the other (as rational maps), none of them is a morphism. Note, moreover, that $\ell_{03} \subset H_0$ is the line joining P_1, P_2, the two points of indeterminacy of ν_L, that ν_L contracts $\ell_{03} \setminus \{P_1, P_2\}$ to $P_0 \in \mathfrak{Q}$ and, conversely, that $\pi_{\mathfrak{Q}}$ contracts $\ell_{13} \setminus P_0 := \{[s, 0, t, 0], \ s, t \in \mathbb{K}, \ t \neq 0\}$ to $P_1 \in H_0$ and $\ell_{23} \setminus P_0 := \{[s, t, 0, 0], \ s, t \in \mathbb{K}, \ t \neq 0\}$ to $P_2 \in H_0$. This also explains the choice of the open set U and U' to construct the isomorphism φ. $\quad\square$

The previous construction shows that $\mathfrak{Q}$ is birational to $\mathbb{P}^2$; to conclude that they cannot be isomorphic it suffices to observing that $\mathfrak{Q}$ contains pairs of skew lines, which cannot occur in $\mathbb{P}^2$.

8.2.2   Monoids

Other examples of rational varieties in any *dimension* $n \geqslant 1$ can be easily constructed as follows. Let $F_d(x_1, \ldots, x_n), F_{d-1}(x_1, \ldots, x_n) \in \mathbb{K}[x_1, \ldots, x_n]$ be reduced homogeneous polynomials of degree d and $d-1$, respectively. Consider the hypersurface

$$Z := Z_a(F_d(x_1, \ldots, x_n) + F_{d-1}(x_1, \ldots, x_n)) \subset \mathbb{A}^n, \qquad (8.4)$$

which is called (*affine*) *monoid of degree* d *with vertex the origin* $O \in \mathbb{A}^n$. The projective closure of Z is the hypersurface $\overline{Z} \subset \mathbb{P}^n$ given by

$$\overline{Z} = Z_p(F_d(X_1, \ldots, X_n) + X_0 F_{d-1}(X_1, \ldots, X_n)), \qquad (8.5)$$

which is called (*projective*) *monoid of degree* d *with vertex* $P_0 = [1, 0, 0, \ldots, 0] \in \mathbb{P}^n$. More generally, any hypersurface $V \subset \mathbb{A}^n$ which is the transform of Z via a linear transformation of $\mathbb{A}^n$ as well as its projective closure $\overline{V}$ will be called *monoid of degree* d.

Example 8.2.5.

(i) The parabola $Z = Z_a(x_2 - x_1^2) \subset \mathbb{A}^2$ is a (affine) monoid of degree two with vertex $O = (0, 0)$.

(ii) The semi-cubic parabola $Z = Z_a(x_2^2 - x_1^3) \subset \mathbb{A}^2$ is a (affine) monoid of degree three with vertex O. Identifying $\mathbb{A}^2$ with the affine chart U_0 of $\mathbb{P}^2$, then $\overline{Z}$ is given by $Z_p(X_0 X_2^2 - X_1^3)$ which is a (projective) monoid with vertex the fundamental point $P_0 = [1, 0, 0] \in \mathbb{P}^2$.

(iii) Similarly, the plane *nodal cubic* $Z := Z_a(x_1^3 + x_1^2 - x_2^2) \subset \mathbb{A}^2$ is a (affine) monoidal curve, whose vertex is $O = (0, 0)$ and whose projective closure is the (projective) monoid $Z_p(X_1^3 + X_1^2 X_0 - X_2^2 X_0) \subset \mathbb{P}^2$ with vertex the fundamental point $P_0 = [1, 0, 0]$.

(iv) The rank-four quadric surface $\mathfrak{Q} = Z_p(X_0 X_3 - X_1 X_2) \subset \mathbb{P}^3$ is a (projective monoid) of degree two with vertex $P_0 = [1, 0, 0, 0]$, being $\mathfrak{Q}_0 = \mathfrak{Q} \cap U_0 = Z_a(x_3 - x_1 x_2)$ an affine monoid of vertex O. Similarly, the quadric cone $Z_p(X_0 X_3 - X_1^2) \subset \mathbb{P}^3$ (i.e. a rank-three quadric) is a monoidal surface too, always with vertex the point P_0.

As in Section 8.2.1, we can prove the following result.

Proposition 8.2.6. *For any* $n \geqslant 2$, *any monoid of degree* $d \geqslant 2$ *is rational.*

Proof. With no loss of generality, we can focus on the case of $\overline{Z} \subset \mathbb{P}^n$ a monoid with vertex $P_0 = [1, 0, \ldots, 0]$, i.e. whose reduced equation is given by (8.5). In such a case, we will more precisely show that the projection $\pi_{\overline{Z}} : \overline{Z} \dashrightarrow H_0 = Z_p(X_0) \subset \mathbb{P}^n$ from the point $P_0 \in \overline{Z}$ is birational onto $H_0 \cong \mathbb{P}^{n-1}$. The restriction of $\pi_{\overline{Z}}$ at $\overline{Z} \setminus \{P_0\}$ is a surjective morphism

$$\varphi : \overline{Z} \setminus \{P_0\} \to H_0 \cong \mathbb{P}^{n-1}, \quad [X_0, \ldots, X_n] \to [X_1, \ldots, X_n].$$

Similarly to the quadric case, the dominant rational map $\pi_{\overline{Z}}$ is called *stereographic projection of the monoid from its vertex P_0*. It suffices to find a dominant rational map inverting $\pi_{\overline{Z}}$. To do this observe that for any $[a_1, \ldots, a_n] \in H_0 \setminus (Z_p(F_{d-1}) \cap H_0)$, i.e. s.t. $F_{d-1}(a_1, \ldots, a_n) \neq 0$, one has

$$\varphi^{-1}([a_1, \ldots, a_n]) = \left[\frac{F_d(a_1, \ldots, a_n)}{F_{d-1}(a_1, \ldots, a_n)}, a_1, \ldots, a_n \right],$$

i.e. the rational map $\Psi : \mathbb{P}^{n-1} \cong H_0 \dashrightarrow \mathbb{P}^n$ defined by

$$[X_1, \ldots, X_n] \overset{\Psi}{\dashrightarrow} [F_d(X_1, \ldots, X_n), X_1 F_{d-1}(X_1, \ldots, X_n), \ldots, X_n F_{d-1}(X_1, \ldots, X_n)]$$

is the desired rational inverse. $\qquad\square$

8.2.3 *Blow-up of $\mathbb{P}^n$ at a point*

Here, we use products of projective varieties to introduce an important birational transformation. Consider $\mathbb{P}^n$ with homogeneous coordinates $[\underline{X}] := [X_0, \ldots, X_n]$ and $\mathbb{P}^{n-1}$ with homogeneous coordinates $[\underline{Y}] := [Y_1, \ldots, Y_n]$. In $\mathbb{P}^n \times \mathbb{P}^{n-1}$ consider the closed subset (cf. Proposition 7.2.7-(i)):

$$\widetilde{\mathbb{P}^n} := Z_p (X_i Y_j - X_j Y_i), \quad \text{for } 1 \leqslant i, j \leqslant n. \tag{8.6}$$

Let $\sigma : \widetilde{\mathbb{P}^n} \to \mathbb{P}^n$ denote the morphism defined by the restriction to $\widetilde{\mathbb{P}^n}$ of the first projection $\pi_1 : \mathbb{P}^n \times \mathbb{P}^{n-1} \to \mathbb{P}^n$. The map σ is called the *blow-up of $\mathbb{P}^n$ at the point P_0*, where $P_0 = [1, 0, \ldots, 0]$ the fundamental point of $\mathbb{P}^n$. Similar approach holds for an arbitrary point $P \in \mathbb{P}^n$ (cf., e.g. Hartshorne, 1977); for simplicity in what follows we will focus only on the case of P_0.

Proposition 8.2.7. $\widetilde{\mathbb{P}^n} \setminus \sigma^{-1}(P_0)$ *is a quasi-projective variety isomorphic (via σ) to $\mathbb{P}^n \setminus \{P_0\}$. In particular, $\widetilde{\mathbb{P}^n}$ is rational, being birational to $\mathbb{P}^n$.*

Proof. For any $Q = [q_0, \ldots, q_n] \in \mathbb{P}^n \setminus \{P_0\}$ there exists $i \in \{1, \ldots, n\}$ for which $q_i \neq 0$. Thus, $([q_0, \ldots, q_n], [y_1, \ldots, y_n]) \in \sigma^{-1}(Q)$ if and only if

$y_j = y_i \frac{q_j}{q_i}$, for any $1 \leqslant j \leqslant n$; in other words

$$\sigma^{-1}(Q) = ([q_0, \dots, q_n], [q_1, \dots, q_n]) \in \widetilde{\mathbb{P}^n} \subset \mathbb{P}^n \times \mathbb{P}^{n-1}$$

is a unique point. Consider the map

$$
\begin{array}{ccc}
\mathbb{P}^n \setminus \{P_0\} & \overset{\tau}{\longrightarrow} & \mathbb{P}^n \times \mathbb{P}^{n-1} \\
Q = [q_0, \dots, q_n] & \longrightarrow & \sigma^{-1}(Q) = ([q_0, \dots, q_n], [q_1, \dots, q_n]).
\end{array}
\tag{8.7}
$$

Composing τ with the Segre isomorphism $\sigma_{n,n-1}$ gives a map $\gamma = \sigma_{n,n-1} \circ \tau$ defined by

$$\gamma([q_0, \dots, q_n]) = [q_0 q_1, q_0 q_2, \dots, q_n^2],$$

which is therefore a morphism by Proposition 6.3.1. Since $\sigma_{n,n-1}$ is an isomorphism, one deduces that τ is a morphism. In particular, one gets that $\widetilde{\mathbb{P}^n} \setminus \sigma^{-1}(P_0)$ is a quasi-projective variety. Indeed $\tau(\mathbb{P}^n \setminus \{P_0\}) = \widetilde{\mathbb{P}^n} \setminus \sigma^{-1}(P_0)$. The latter is locally closed in $\mathbb{P}^n \times \mathbb{P}^{n-1}$: indeed $\widetilde{\mathbb{P}^n}$ is closed and also $\sigma^{-1}(P_0)$ is closed, being σ a morphism. Moreover, $\widetilde{\mathbb{P}^n} \setminus \sigma^{-1}(P_0)$ is also irreducible, since $\mathbb{P}^n \setminus \{P_0\}$ is irreducible and τ is a morphism (cf. Corollary 4.1.6). At last, by its expression, $\tau = (\sigma|_{\mathbb{P}^n \setminus \{P_0\}})^{-1}$, i.e. τ is an isomorphism. $\qquad \square$

From (8.6), one gets

$$\sigma^{-1}(P_0) = \pi_1^{-1}(P_0) \cong \mathbb{P}^{n-1},\tag{8.8}$$

which is called the *exceptional divisor of σ* in $\widetilde{\mathbb{P}^n}$.

Remark 8.2.8. A geometric interpretation of the isomorphism in (8.8) can be given as follows. Let H be any hyperplane in $\mathbb{P}^n$ not passing through P_0; up to a projectivity of $\mathbb{P}^n$ fixing P_0, we can assume with no loss of generality that this hyperplane coincides with $H_0 = Z_p(X_0)$. Any point $Q \in H$ uniquely determines the line $r_Q := P_0 \vee Q$ in $\mathbb{P}^n$. Conversely, any line $r \subset \mathbb{P}^n$ passing through P_0 intersects H at a unique point $Q_r = r \cap H$ such that $P_0 \vee Q_r = r$.

In other words, the set $\mathcal{L}_{P_0} := \{\text{lines in } \mathbb{P}^n \text{ through } P_0\}$ bijectively corresponds to the hyperplane $H \cong \mathbb{P}^{n-1}$. For any point $Q = [0, q_1, \dots, q_n] \in H$, the line $r_Q = P_0 \vee Q$ has parametric equations

$$X_0 = \lambda, \; X_i = \mu \, q_i, \quad 1 \leqslant i \leqslant n, \quad \text{with } [\lambda, \mu] \in \mathbb{P}^1.\tag{8.9}$$

By (8.7), at the points of $r_Q \setminus \{P_0\}$ the map τ restricts to the morphism:

$$\tau|_{r_Q \setminus \{P_0\}} : \quad r_Q \setminus \{P_0\} \quad \longrightarrow \quad \widetilde{\mathbb{P}^n} \setminus \sigma^{-1}(P_0) \subset \mathbb{P}^n \times \mathbb{P}^{n-1}$$
$$[\lambda, \mu q_1, \ldots, \mu q_n] \longrightarrow ([\lambda, \mu q_1, \ldots, \mu q_n], [q_1, \ldots, q_n]), \tag{8.10}$$

where the previous expression follows from the fact that

$$([\lambda, \mu q_1, \ldots, \mu q_n], [\mu q_1, \ldots, \mu q_n]) = ([\lambda, \mu q_1, \ldots, \mu q_n], [q_1, \ldots, q_n]),$$

as $\mu \neq 0$ on $r_Q \setminus \{P_0\}$. Composing $\tau|_{r_Q \setminus \{P_0\}}$ with the Segre morphism $\sigma_{n,n-1}$, one gets a rational map $\tau|_{r_Q} : r_Q \dashrightarrow \mathbb{P}^{n(n+1)-1}$. Since $r_Q \cong \mathbb{P}^1$, from Example 8.1.12, $\tau|_{r_Q}$ is a morphism extending $\tau|_{r_Q \setminus \{P_0\}}$ at the point P_0; by (8.9), this extension is given by

$$\tau|_{r_Q}(P_0) := ([1, 0, \ldots, 0], [q_1, \ldots, q_n]). \tag{8.11}$$

Consider $\widetilde{r_Q} := \tau(r_Q) \subset \widetilde{\mathbb{P}^n}$. From Proposition 8.2.7 and (8.11), $\widetilde{r_Q}$ is isomorphic to $\mathbb{P}^1$ in particular it is an irreducible rational curve. Moreover, by its definition

$$\widetilde{r_Q} \cap \sigma^{-1}(P_0) = \tau|_{r_Q}(P_0) \cap \sigma^{-1}(P_0) = [q_1, \ldots, q_n] \in \sigma^{-1}(P_0).$$

In particular, the map

$$\begin{aligned}
H &\cong \mathcal{L}_{P_0} \longrightarrow \sigma^{-1}(P_0) \\
Q &\cong r_Q \longrightarrow \widetilde{r_Q} \cap \sigma^{-1}(P_0) \\
[0, q_1, \ldots, q_n] &\longleftrightarrow [q_1, \ldots, q_n]
\end{aligned}$$

can be interpreted as the isomorphism (8.8).

Proposition 8.2.9. $\widetilde{\mathbb{P}^n}$ *is irreducible.*

Proof. Note that $\widetilde{\mathbb{P}^n} = (\widetilde{\mathbb{P}^n} \setminus \sigma^{-1}(P_0)) \cup \sigma^{-1}(P_0)$. From Proposition 8.2.7, $(\widetilde{\mathbb{P}^n} \setminus \sigma^{-1}(P_0))$ is irreducible and $\sigma^{-1}(P_0)$ is closed and irreducible in $\widetilde{\mathbb{P}^n}$; it therefore suffices to show that $\widetilde{\mathbb{P}^n} \setminus \sigma^{-1}(P_0)$ is dense in $\widetilde{\mathbb{P}^n}$. To prove this, it is enough to show that any point of $\sigma^{-1}(P_0)$ is in the closure in $\widetilde{\mathbb{P}^n}$ of some algebraic subset contained in $\widetilde{\mathbb{P}^n} \setminus \sigma^{-1}(P_0)$. In Remark 8.2.8, it has been proved that any point of $\sigma^{-1}(P_0)$ is given by $\tau|_{r_Q}(P_0) = \widetilde{r_Q} \cap \sigma^{-1}(P_0)$, for some point $Q \in H$, and moreover for any $Q \in H$ the map $\tau|_{r_Q}$ is an isomorphism. Thus, $\widetilde{r_Q} \cong \mathbb{P}^1$ is the closure in $\widetilde{\mathbb{P}^n}$ of $\tau(r_Q \setminus \{P_0\}) \subset \widetilde{\mathbb{P}^n} \setminus \sigma^{-1}(P_0)$, where $\tau(r_Q \setminus \{P_0\}) \cong \mathbb{A}^1$. $\square$

Example 8.2.10.

(i) By definition of blow-up, it is clear that $\widetilde{\mathbb{P}^1} \cong \mathbb{P}^1$.

(ii) For $n = 2$, $\widetilde{\mathbb{P}^2}$ is the rational surface in $\mathbb{P}^2 \times \mathbb{P}^1$ given by the equation $Z_p(X_1 Y_2 - X_2 Y_1)$. Repeating verbatim the proof of Proposition 8.2.7, one easily realizes that, e.g. $\widetilde{\mathbb{P}^2} \cap (U_0^2 \times U_0^1)$ is isomorphic to the hyperbolic paraboloid $\Sigma := Z_a(x_2 - yx_1) \subset \mathbb{A}^3$, where $\mathbb{A}^3$ is $U_0^2 \times U_0^1 \cong \mathbb{A}^2 \times \mathbb{A}^1$ and where affine coordinates (x_1, x_2, y) are given by $x_1 = \frac{X_1}{X_0}$, $x_2 = \frac{X_2}{X_0}$, $y = \frac{Y_2}{Y_1}$. The exceptional divisor $\sigma^{-1}(P_0) \cap (U_0^2 \times U_0^1)$ is the line $Z_a(x_1, x_2)$ contained in Σ. Its points $(0, 0, y)$, with y varying in $\mathbb{A}^1$, bijectively correspond to direction coefficients y of lines through the origin of $\mathbb{A}^2$ (i.e. P_0 in the chart U_0^2) in the pencil $\{x_2 - yx_1 = 0\} \subset \mathbb{A}^2$.

8.2.4 *Blow-ups and resolution of singularities*

Here, we discuss some examples which show how blow-ups can be used to resolve *singularities* of algebraic varieties (see also Mumford, 1995, Example 2.21, p. 32). As in the previous section, things more generally hold for a blow-up at any arbitrary point P of $\mathbb{P}^n$. On the other hand, up to projectivities, one can always reduce to the basic case of the blow-up at the fundamental point $P_0 = [1, 0, \ldots, 0]$.

Let $V \subseteq \mathbb{P}^n$ be any algebraic variety passing through P_0. Consider the blow-up $\sigma : \widetilde{\mathbb{P}^n} \to \mathbb{P}^n$ as above and let

$$W_{P_0} := \overline{\sigma^{-1}(V \setminus \{P_0\})}, \tag{8.12}$$

where the closure is taken in $\widetilde{\mathbb{P}^n}$.

Claim 8.2.11. W_{P_0} is a projective variety which is birational to V.

Proof. $V \setminus \{P_0\}$ is an algebraic variety, being an open dense subset of V; furthermore, since $\sigma|_{V \setminus \{P_0\}}$ is an isomorphism between $V \setminus \{P_0\}$ and its image, then $\sigma(V \setminus \{P_0\})$ is irreducible and locally closed in $\widetilde{\mathbb{P}^n}$, i.e. it is a quasi-projective variety. This implies that W_{P_0} is irreducible and birational to V. Moreover, since $\widetilde{\mathbb{P}^n}$ is closed in $\mathbb{P}^n \times \mathbb{P}^{n-1}$, W_{P_0} is also projective. $\square$

$\sigma^{-1}(V)$ is called the *total transform* of V *via* σ . We let

$$\widetilde{V_{P_0}} := W_{P_0} \cap \sigma^{-1}(V); \tag{8.13}$$

since V is quasi-projective and σ is a morphism, $\widetilde{V_{P_0}}$ is locally closed in W_{P_0} (so irreducible) and birational to V. $\widetilde{V_{P_0}}$ is called the *proper transform of V via* σ. The restriction of σ at $\widetilde{V_{P_0}}$ is a birational morphism which will be denoted by $\sigma_V : \widetilde{V_{P_0}} \to V$ and called *blow-up of V at* P_0.

Example 8.2.12. (i) Let $V := Z_p(X_0 X_1^2 - X_2^3) \subset \mathbb{P}^2$; note that V is the projective closure of the semi-cubic parabola $V_0 = V \cap U_0 = Z_a(x_1^2 - x_2^3) \subset \mathbb{A}^2 \cong U_0$, so V is a monoid in $\mathbb{P}^2$ with vertex the fundamental point $P_0 = [1, 0, 0]$ (cf. Example 8.2.5-(ii)). Since V is projective, then $T := V \times \mathbb{P}^1$ is a projective subvariety of $\mathbb{P}^2 \times \mathbb{P}^1$, whose defining equation is simply $Z_p(X_0 X_1^2 - X_2^3)$. Let now $\sigma : \widetilde{\mathbb{P}^2} \to \mathbb{P}^2$ be the blow-up of $\mathbb{P}^2$ at P_0. As in Example 8.2.12-(ii), $\widetilde{\mathbb{P}^2}$ is defined by $Z_p(X_1 Y_2 - X_2 Y_1)$. The total transform of V is therefore:

$$\sigma^{-1}(V) = T \cap \widetilde{\mathbb{P}^2} = Z_p(X_0 X_1^2 - X_2^3, X_1 Y_2 - X_2 Y_1). \tag{8.14}$$

Take affine charts $U_0^2 \subset \mathbb{P}^2$, with affine coordinates $x_1 = \frac{X_1}{X_0}$, $x_2 = \frac{X_2}{X_0}$, and $U_2^1 \subset \mathbb{P}^1$, with affine coordinate $t = \frac{Y_1}{Y_2}$. From (8.14), the corresponding open subset of the total transform is:

$$\sigma^{-1}(V) \cap (U_0^2 \times U_2^1) = Z_a(x_1^2 - x_2^3, x_1 - x_2 t) = Z_a(x_2^2(x_2 - t^2), x_1 - x_2 t),$$

which reads also $Z_a(x_1 - t^3, x_2 - t^2) \cup Z_a(x_2, x_1)$. In the affine chart $U_0^2 \times U_2^1 \cong \mathbb{A}^2 \times \mathbb{A}^1 \cong \mathbb{A}^3$, with affine coordinates (x_1, x_2, t), the set $Z_a(x_1 - t^3, x_2 - t^2)$ is an affine twisted cubic in $\mathbb{A}^3$, precisely the image of the morphism $\mathbb{A}^1 \longrightarrow \mathbb{A}^3$ given by $t \longrightarrow (t^3, t^2, t)$, whereas $Z_a(x_2, x_1)$ is simply the t-axis in $\mathbb{A}^3$, which coincides with the affine part of the exceptional divisor $\sigma^{-1}(P_0) \cap (U_0^2 \times U_2^1)$ (cf. Example 8.2.10-(ii)).

To sum up, in the affine chart $\widetilde{\mathbb{P}^2} \cap (U_0^2 \times U_2^1)$, the proper transform $(\widetilde{V_0})_{P_0}$ of the semi-cubic parabola $V_0 \subset \mathbb{A}^2$ is isomorphic to an affine twisted cubic $C \subset \mathbb{A}^3$. The projective closure in $\widetilde{\mathbb{P}^2}$ of $(\widetilde{V_0})_{P_0}$ is the proper transform $\widetilde{V_{P_0}}$ (which in this case coincides with W_{P_0}, since V is already projective). From Claim 3.3.18, $\widetilde{V_{P_0}}$ is therefore isomorphic to the standard projective twisted

$$\nu_{1,2} : \mathbb{P}^1 \longrightarrow \mathbb{P}^3, \quad [Y_1, Y_2] \longrightarrow [Y_1^3, Y_1^2 Y_2, Y_1 Y_2^2, Y_2^3].$$

The total transform $\sigma^{-1}(V)$ coincides with $\widetilde{V_{P_0}} \cup \sigma^{-1}(P_0)$, where $\sigma^{-1}(P_0) \cong \mathbb{P}^1$. As for the blow-up morphism $\sigma_V : \widetilde{V_{P_0}} \to V$, previous computations show that $\sigma_V|_{\widetilde{V_{P_0}} \cap (U_0^2 \times U_2^1)}$ is nothing but the restriction to C of the projection $\pi_I : \mathbb{A}^3 \to \mathbb{A}^2$ onto the first two coordinates, i.e. with multi-index $I = (1, 2)$.

With a more geometrical perspective, π_I is the projection onto the plane $Z_a(t) \subset \mathbb{A}^3$ from the point at infinity of the t-axis; under this projection one therefore has

$$\pi_I|_C : C \longrightarrow V_0, \quad (t^3, t^2, t) \longrightarrow (t^3, t^2),$$

i.e. $C \subset \mathbb{A}^3$ maps onto the semi-cubic parabola $V_0 \subset \mathbb{A}^2$. We will see that V_0 is *singular* at the origin (cf. Example 12.1.3-(ii)), whereas the affine twisted cubic $C \subset \mathbb{A}^3$ is a *smooth* curve, i.e. it is non-singular at any of its points (cf. Example 12.1.3). In other words, the semi-cubic parabola acquires its *cuspidal singularity* at O because of the projection $\pi_I|_C$; indeed, the intersection between $C \subset \mathbb{A}^3$ (i.e. the proper transform of V_0) and the t-axis (i.e. the affine part of the exceptional divisor) is given by $Z_a(x_1 - t^3, x_2 - t^2, x_1, x_2)$. This intersection is set-theoretically supported at the unique point $O = \{(0,0,0)\} \in C$, which corresponds to the coefficient direction $m = 0$ of the line $Z_a(x_2) \subset \mathbb{A}^2$. On the other hand, the *intersection multiplicity* at O of the system of equations $Z_a(x_1 - t^3, x_2 - t^2, x_1, x_2)$ is 2 (cf. Definition 12.1.4) as the t-axis is the *tangent line* to $C \subset \mathbb{A}^3$ at O. Thus, the cuspidal singularity of V_0 at O deals with the fact that $\pi_I|_C$ is a *tangential projection* and the singular point is where the parametrization is *not regular* (i.e. where the Jacobian matrix vanishes). Conversely, the affine twisted cubic $C \subset \mathbb{A}^3$ can be viewed as a *non-singular* birational model of V_0, since C is the proper transform of V_0 under the blow-up $\pi_I|_C$ and one says that the morphism $\pi_I|_C$ *resolves* the cuspidal singularity of V_0 at O (cf. also Harris, 1995, Example 1.26, pp. 14–15).

(ii) Let $V := Z_p(X_1^3 + X_0(X_1^2 - X_2^2)) \subset \mathbb{P}^2$. This is a (projective) monoid of vertex $P_0 = [1, 0, 0]$ being the projective closure of the (affine) monoid $V_0 = V \cap U_0^2 = Z_a(x_1^3 + x_1^2 - x_2^2) \subset \mathbb{A}^2$ (cf. Example 8.2.5-(iii)). Computations as in (i) show that the total transform of V is given by

$$\sigma^{-1}(V) = Z_p(X_1^3 + X_0(X_1^2 - X_2^2), \ X_1 Y_2 - X_2 Y_1). \tag{8.15}$$

Take affine charts $U_0^2 \subset \mathbb{P}^2$, with affine coordinates $x_1 = \frac{X_1}{X_0}$, $x_2 = \frac{X_2}{X_0}$, and $U_1^1 \subset \mathbb{P}^1$, with affine coordinate $s = \frac{Y_2}{Y_1}$. From (8.15), we get that $\sigma^{-1}(V) \cap (U_0^2 \times U_1^1)$ is given by

$$Z_a(x_1^3 + x_1^2 - x_2^2, \ x_2 - x_1 s) = Z_a(x_1^2(x_1 + 1 - s^2), \ x_2 - x_1 s),$$

which is $Z_a(x_1 - s^2 + 1, \ x_2 - s^3 + s) \cup Z_a(x_1, x_2)$. The first algebraic subset $Z_a(x_1 - s^2 + 1, \ x_2 - s^3 + s)$ is the image of the morphism $\mathbb{A}^1 \longrightarrow \mathbb{A}^3$, defined by $s \rightarrow (s^2 - 1, s^3 - s, s)$, i.e. it is an affine twisted cubic in $\mathbb{A}^3$ so the proper transform $\widetilde{V_{0 P_0}}$ is irreducible. The second algebraic subset $Z_a(x_1, x_2)$

is the s-axis, i.e. the affine part of the exceptional divisor. The proper transform $\widetilde{V_{P_0}}$ of the monoid V_0 coincides with W_{P_0} and it is isomorphic to the projective closure of C.

The blow-up morphism $\sigma_V : \widetilde{V_{P_0}} \to V$ in the above affine charts reads as the restriction to C of the projection onto the plane $Z_a(s) \subset \mathbb{A}^3$ from the point at infinity of the s-axis. In this case, the intersection between C and the s-axis is given by $Z_a(x_1 - s^2 + 1, x_2 - s^3 + s, x_1, x_2)$; this system of equations gives $\{Q_1 = (0, 0, 1),\ Q_{-1} = (0, 0, -1)\} \in C$, where the points of intersection correspond to the coefficient directions $m = \pm 1$ of the two lines $Z_a(x_2 - x_1)$ and $Z_a(x_2 + x_1)$ in the plane $Z_a(s)$. The *nodal singularity* of the curve V_0 at O appears since σ_V is a *secant projection* of C to the s-plane. This projection $C \longrightarrow V_0$ is given by $(s^2 - 1, s^3 - s, s) \to (s^2 - 1, s^3 - s)$; the s-axis is a secant line to C at the points Q_1 and Q_2 which are identified under the projection, i.e. they both map to $O \in V_0$. This creates the nodal singularity at O (i.e. a point where the parametrization is not injective). To sum-up, the affine twisted cubic $C \subset \mathbb{A}^3$ is a non-singular birational model of the plane nodal cubic V_0 and σ_V resolves its (nodal) singularity at O (see also Harris, 1995, Example 1.26, pp. 14–15).

Exercises

Exercise 8.1. Let $X := Z_a(x_1 x_2 \cdots x_n - 1) \subset \mathbb{A}^n$, for any integer $n \geqslant 2$. Show that X is an affine variety which is birational to $\mathbb{A}^{n-1}$.

Exercise 8.2. For any integer $n \geqslant 1$, give an example of an affine variety which is birational but not isomorphic to $\mathbb{A}^n$.

Exercise 8.3. Consider the affine irreducible curve $C := Z_a(x_1^4 + x_1^3 - x_2^3) \subset \mathbb{A}^2$. Show that C is a rational curve determining an explicit birational parametrization $\mathbb{A}^1 \overset{\Phi}{\dashrightarrow} C$.

Exercise 8.4. With the use of blow-ups, determine a birational model of the curve $C \subset \mathbb{A}^2$ as in Exercise 8.3 which is isomorphic to $\mathbb{A}^1$.

Exercise 8.5. Consider the map $q : \mathbb{P}^2 \dashrightarrow \mathbb{P}^2$ defined by $[X_0, X_1, X_2] \overset{q}{\dashrightarrow} [\frac{1}{X_0}, \frac{1}{X_1}, \frac{1}{X_2}]$ which is called *elementary quadratic* (*or Cremona*) *transformation of* $\mathbb{P}^2$. Show that q is a birational automorphism of $\mathbb{P}^2$, explicitly finding the open set $U \subset \mathbb{P}^2$ where $q_{|U}$ is an isomorphism. Describe the indeterminacy locus of q and of its rational inverse. Finally, describe the image via q of any line of $\mathbb{P}^2$ (cf. Mumford, 1995, Example 2.20, p. 31; Ciliberto, 2021, § 18.2.2, pp. 270–271).

Chapter 9

Completeness of Projective Varieties

Here, we discuss an important topological property of projective varieties.

9.1 Complete Algebraic Varieties

Complete varieties are, in the category of algebraic varieties, the analogues of compact topological spaces in the category of Hausdorff topological spaces. Recall that the image of a compact topological space under a continuous map is compact and, hence, it is closed if the target space is Hausdorff. Moreover, a Hausdorff topological space X is compact if and only if, for all topological spaces Y, the projection $\pi_Y : X \times Y \to Y$ (with $X \times Y$ endowed with the product topology) is a *closed map*, i.e. it maps closed subsets of $X \times Y$ to closed subsets of Y (cf. Bourbaki, 1966, 10.2).

For algebraic varieties, some of the previous requests are empty some others make no sense. Indeed:

- compactness (in the Zariski topology) is a property which is satisfied by all algebraic varieties (cf. Section 4.2);
- any algebraic variety V, which is not reduced to a single point, is never a Hausdorff space;
- the Zariski topology $\mathfrak{Z}_{V \times W}$ of the product $V \times W$ of two algebraic varieties V and W never coincides with the product topology $\mathfrak{Z}\mathfrak{ar}_V \times \mathfrak{Z}\mathfrak{ar}_W$, unless at least one of them is reduced to a single point.

Thus, in the category of algebraic varieties we will consider the following.

Definition 9.1.1. An algebraic variety V is said to be *complete* if, for all algebraic varieties W, the projection morphism $\pi_W : V \times W \to W$ onto the second factor is a closed map.

Note that, in the previous definition, $V \times W$ has a structure of algebraic variety as discussed in Chapter 7, i.e. it is endowed with the Zariski topology $\mathfrak{Zar}_{V \times W}$.

Corollary 9.1.2. *Let V be a complete algebraic variety. Let W be an algebraic variety and let $\varphi : V \to W$ be a morphism. Then $\varphi(V) \subseteq W$ is a closed subvariety. In particular, if φ is dominant then it must be surjective, i.e. $\varphi(V) = W$.*

Proof. For the first part of the statement, note first that $\varphi(V)$ is irreducible since V is irreducible and φ is a morphism (cf. Corollary 4.1.6). Thus, we are left to prove that $\varphi(V)$ is closed in W. From Proposition 7.5.3, the graph Γ_φ is closed in $V \times W$. Then one concludes by using that V is complete and that $\pi_W(\Gamma_\varphi) = \varphi(V)$. The second assertion follows from the fact that $\varphi(V)$ is a closed subset of W containing an open (so dense) subset of W. □

The previous corollary gives motivation for the terminology *complete*: if V and W are algebraic varieties, with V a subvariety of W, and V is complete then, if one has $\dim(V) = \dim(W)$, one necessarily must have $V = W$. This behavior is in contrast with, e.g. the open immersion $\mathbb{A}^n \cong U_0 \overset{\iota_{U_0}}{\hookrightarrow} \mathbb{P}^n$, for $n \geqslant 1$. In other words, for any $n \geqslant 1$, $\mathbb{A}^n$ cannot be a complete variety. The next example more generally show other affine varieties which are not complete.

Example 9.1.3. Let $V := Z_a(x_1 x_2 - x_3)$ be the hyperbolic paraboloid in $\mathbb{A}^3$. From Exercise 7.1, V is the image of $\mathbb{A}^1 \times \mathbb{A}^1 \cong U_0^1 \times U_0^1$ in $\mathbb{A}^3$ via the restriction to $U_0^1 \times U_0^1$ of the Segre morphism $\sigma_{1,1} : \mathbb{P}^1 \times \mathbb{P}^1 \to \mathbb{P}^3$. Take the multi-index $I = (1, 2)$ and consider the projection morphism $\pi_I : \mathbb{A}^3 \to \mathbb{A}^2$. The map $\phi := (\pi_I)|_V : V \to \mathbb{A}^2$ is the morphism ϕ in Remark 6.1.10, where we showed that $\phi(V)$ is neither closed nor open in $\mathbb{A}^2$ (it is a *constructible set*). In particular, Corollary 9.1.2 does not hold so V cannot be complete.

9.2 The Main Theorem of Elimination Theory

The core of this section is to show that, among algebraic varieties, projective varieties play a special role namely they are complete varieties.

Theorem 9.2.1. *If V is a projective variety, then V is complete.*

Proof. We have to show that, for any algebraic variety W, the projection $\pi_W : V \times W \to W$ is a closed morphism. Since V is projective, up to an isomorphism, we can assume V to be a closed, irreducible subset of $\mathbb{P}^r$, for some integer r. From the fact that $V \times W$ is therefore closed in $\mathbb{P}^r \times W$, it suffices to consider $V = \mathbb{P}^r$. From Proposition 6.4.2, $\mathfrak{Z}\mathfrak{ar}_W$ has a basis consisting of affine open sets. Since the property of being closed is local (recall, e.g. the proof of Proposition 5.2.4-(i)), we can verify the property of being closed in any open set of an affine open covering of W; with no loss of generality, we can therefore reduce to the case W to be affine. In such a case, W can be considered as an irreducible, closed subset of $\mathbb{A}^n$, for some integer n. By the induced topology $\mathfrak{Z}\mathfrak{ar}_W$, we can reduce to the case $W = \mathbb{A}^n$.

To sum up, we need to show that the second projection

$$\pi_2 : \mathbb{P}^r \times \mathbb{A}^n \to \mathbb{A}^n$$

is closed. To prove this, let $Z \subseteq \mathbb{P}^r \times \mathbb{A}^n$ be any closed subset. From Proposition 7.2.7, Z is defined by polynomial equations

$$g_j(\underline{X}, \underline{y}) = 0, \quad 1 \leqslant j \leqslant s,$$

for some integer s, where $\underline{X} = (X_0, \ldots, X_r)$, $\underline{y} = (y_1, \ldots, y_n)$ and where the polynomials g_j are homogeneous of degree d_j with respect to the set of indeterminates $\underline{X}$, $1 \leqslant j \leqslant s$.

A point $P = (p_1, \ldots p_n) \in \mathbb{A}^n$ is such that $P \in \pi_2(Z)$ if and only if $\pi_2^{-1}(P) \cap Z \neq \emptyset$, i.e. if and only if

$$Z_p(g_1(\underline{X}, P), \ldots, g_s(\underline{X}, P)) \neq \emptyset, \tag{9.1}$$

where $g_j(\underline{X}, P) := g_j(X_0, \ldots, X_r, p_1, \ldots, p_n) \in \mathcal{S}_{d_j}^{(r)}$, $1 \leqslant j \leqslant s$. Since $\mathbb{K}$ is algebraically closed, by the Homogeneous Hilbert "Nullstellensatz"-weak form (cf. Theorem 3.2.4), (9.1) is equivalent to

$$(\mathcal{S}_+^{(r)})^d \not\subseteq (g_1(\underline{X}, P), \ldots, g_s(\underline{X}, P)), \quad \forall \, d \geqslant 1.$$

Let $Z_d := \{P \in \mathbb{A}^n \mid (g_1(\underline{X}, P), \ldots, g_s(\underline{X}, P)) \not\supseteq (\mathcal{S}_+^{(r)})^d\}$. Since $\pi_2(Z) = \cap_{d \geqslant 1} Z_d$, it suffices to show that Z_d is closed for any $d \geqslant 1$.

For any point $P \in \mathbb{A}^n$ and for any $d \geqslant 1$ consider the vector space homomorphism

$$\rho_d(P) : \mathcal{S}_{d-d_1}^{(r)} \oplus \cdots \mathcal{S}_{d-d_s}^{(r)} \longrightarrow \mathcal{S}_d^{(r)}$$

defined by $\rho_d(P)\,(F_1(\underline{X}), \ldots, F_s(\underline{X})) := \sum_{j=1}^{s} F_j(\underline{X})\,g_j(\underline{X}, P)$, where $\mathcal{S}_{d-d_t}^{(r)} = (0)$ for those $d_t > d$. With this set-up, $P \in Z_d$ if and only if

$$\mathrm{rk}(\rho_d(P)) < \binom{r+d}{d} = \dim(\mathcal{S}_d^{(r)}). \tag{9.2}$$

Condition (9.2) is equivalent to the vanishing of all the minors of order $\binom{r+d}{d}$ of the matrix associated to the linear map $\rho_d(P)$ in the canonical bases of these vector spaces. These minors are polynomial expressions in the p_i's, $1 \leqslant i \leqslant n$. Replacing the coordinate p_i with the indeterminate y_i, $1 \leqslant i \leqslant n$, one gets that Z_d is the closed subset of $\mathbb{A}^n$ defined as the vanishing locus of such polynomial minors, as desired. □

The previous result is called the *Main Theorem of Elimination Theory*. Motivation for the terminology is clearly described by the strategy of the proof: we started with some equations $g_j(\underline{X}, \underline{y}) = 0, 1 \leqslant j \leqslant s$, and we asked for the image of the projection map $(\underline{X}, \underline{y}) \to \underline{y}$, which can be written as $\{\underline{y} \in \mathbb{A}^n \mid \exists\,\underline{X} \in \mathbb{P}^r \text{ s.t. } g_j(\underline{X}, \underline{y}) = 0, \ \forall\ 1 \leqslant j \leqslant s\}$. In other words, the indeterminates $\underline{X}$ have been "*eliminated*" from the problem. Theorem 9.2.1 states that the set of all such $\underline{y} \in \mathbb{A}^n$ can itself be written as the zero-set of suitable polynomial equations. Elimination Theory is concerned with providing algorithms for passing from the equations defining $Z \subseteq \mathbb{P}^r \times \mathbb{A}^n$ to equations defining $\pi_2(Z) \subset \mathbb{A}^n$.

For example, let $Z \subseteq \mathbb{P}^1 \times \mathbb{A}^n$ be the closed subset defined by two polynomials $g_1(X_0, X_1, \underline{y}) := s_0(\underline{y})X_0^m + s_1(\underline{y})X_0^{m-1}X_1 + \cdots + s_m X_1^m$ and $g_2(X_0, X_1, \underline{y}) := t_0(\underline{y})X_0^n + t_1(\underline{y})X_0^{n-1}X_1 + \cdots + t_n X_1^n$, with $s_i(\underline{y}), t_k(\underline{y}) \in \mathbb{K}[\underline{y}] = A^{(n)}, 0 \leqslant i \leqslant m, 0 \leqslant k \leqslant n$. For any $P \in \mathbb{A}^n$, the two polynomials $g_1^P(X_0, X_1) := g_1(X_0, X_1, P) \in \mathcal{S}_m^{(1)}$, $g_2^P(X_0, X_1) := g_1(X_0, X_1, P) \in \mathcal{S}_n^{(1)}$ have a common zero in $\mathbb{P}^1$ if and only if the resultant is such that $R(\delta_0(g_1^P), \delta_0(g_2^P)) = 0$. Indeed, if this common zero is $[0, 1] \in \mathbb{P}^1$ then $g_1^P(0, 1) = g_2^P(0, 1) = 0$ gives $s_m(P) = t_n(P) = 0$, i.e. the last column of the Sylvester matrix (1.7) of $\delta_0(g_1^P), \delta_0(g_2^P)$ is zero. If otherwise the common zero is of the form $[1, a] \in \mathbb{P}^1$, for some $a \in \mathbb{K}$, the assertion follows from Theorem 1.3.16 (cf. also the proof of Proposition 1.10.19)). From this observation, since $\delta_0(g_1(X_0, X_1, \underline{y})), \delta_0(g_2(X_0, X_1, \underline{y})) \in (\mathbb{K}[\underline{y}]) [X_1]$, then

$\pi_2(Z) \subset \mathbb{A}^n$ is given by $Z_a(R_{X_1}(\delta_0(g_2(X_0, X_1, \underline{y})), \delta_0(g_2(X_0, X_1, \underline{y}))) \subseteq \mathbb{A}^n$, as $R_{X_1}(\delta_0(g_2(X_0, X_1, \underline{y})), \delta_0(g_2(X_0, X_1, \underline{y}))) \in \mathbb{K}[\underline{y}]$. Elimination Theory runs this procedure in general, with the use of *elimination ideals* (for details, see, e.g. Cohn *et al.*, 2015, Chapter 8, § 5).

9.2.1 Consequences of the main theorem of elimination theory

In the previous section, we have just seen that every projective variety is complete; the converse is not true. On the other hand, it is quite hard to exhibit explicit examples of complete varieties which are not projective; there are examples in dimension 2 due to Nagata and in dimension 3 to Hironaka (cf., e.g. Hartshorne, 1977, Exercise 7.13, p. 171, and Example 3.4.1, p. 443). We will certainly do not treat such examples, so for practical purposes the terms "projective variety" and "complete variety" can be considered almost synonymous.

Remark 9.2.2. Theorem 9.2.1 gives different (and shorter) proofs of some results already encountered in the previous chapters, as well as some new consequences. For example:

(i) If V is a projective variety and W is an affine variety, any morphism $\varphi : V \to W$ is constant; in particular $\mathcal{O}_V(V) \cong \mathbb{K}$.

Proof. This has been already proved in Theorem 5.3.14-(e) and in Corollary 6.2.7. We can give an alternative proof, using the completeness of V. Indeed, since W is affine, we can assume that W is a closed subvariety of $\mathbb{A}^n$, for some $n \geqslant 0$. If we identify $\mathbb{A}^n$ with the affine chart $U_0 \subset \mathbb{P}^n$ and if we consider the morphism $\varphi' := \iota_{U_0} \circ \iota_W \circ \varphi : V \to \mathbb{P}^n$, by the completeness of V we get that $\varphi'(V)$ is closed in $\mathbb{P}^n$. On the other hand, $\varphi'(V) = \varphi(V) \subseteq W \subseteq \mathbb{A}^n$ is closed in W which is closed in $\mathbb{A}^n$; then one concludes by Corollary 6.2.6. $\square$

(ii) If V is a projective variety and $\varphi : V \to \mathbb{P}^n$ is a morphism, then $\varphi(V) \subset \mathbb{P}^n$ is closed; in particular it is a projective variety.

Recall that this does not occur for arbitrary algebraic varieties (cf., Example 9.1.3).

(iii) From (ii) one easily deduces that any Veronese variety $V_{n,d}$, any Segre variety $\Sigma_{n,m}$, etc., are always projective varieties.

(iv) Let $\Lambda \subset \mathbb{P}^n$ be a linear space inducing a projection $\pi_\Lambda : \mathbb{P}^n \dashrightarrow \mathbb{P}^k$, for suitable $k < n$ (cf. Example 8.1.14). We know that if $V \subset \mathbb{P}^n$ is a projective variety s.t. $V \cap \Lambda = \emptyset$, then the projection $\pi_V : V \to \mathbb{P}^k$ is a morphism. Thus, $\pi_V(V) \subset \mathbb{P}^k$ is again a projective variety which is called the *projection of V* on the given $\mathbb{P}^k$.

(v) Let $V \subset \mathbb{P}^n$ be a projective variety which contains more than one point and let $F \in H(\mathcal{S}^{(n)})$ be a non-constant homogeneous polynomial. Then $V \cap Z_p(F) \neq \emptyset$.

Proof. This has already been proved in Corollary 6.5.7. An alternative proof which uses completeness of V runs as follows. Assume by contradiction there exists a homogeneous polynomial $F \in \mathcal{S}_d^{(n)}$, for some integer $d \geqslant 1$, such that $F(P) \neq 0$ for any $P \in V$. Let $P, Q \in V$ be two distinct points and let $G \in I_p(P)$ be homogeneous, of degree d, such that $G(Q) \neq 0$. Consider the morphism

$$\varphi : V \to \mathbb{P}^1, \quad R \to [F(R), G(R)], \quad \forall \ R \in V$$

(this is well-defined by the assumption on F). Then, $\varphi(V)$ is closed and irreducible in $\mathbb{P}^1$; therefore φ is either constant or it is surjective. Since $[0, 1] \notin \varphi(V)$, this would imply φ is constant. This is a contradiction as $\varphi(P) = [F(P), G(P)] = [1, 0]$ and $\varphi(Q) = [F(Q), G(Q)] \neq [1, 0]$ by the choice of G. $\qquad \square$

Exercises

Exercise 9.1. Let $V = W = \mathbb{A}^1$ with $A(V) = \mathbb{K}[t]$ and $A(W) = \mathbb{K}[s]$. Let $Z := Z_a(ts - 1) \subset \mathbb{A}^1 \times \mathbb{A}^1$.

(i) Show that Z is a closed variety in $\mathbb{A}^1 \times \mathbb{A}^1$.
(ii) Let $\pi_2 : \mathbb{A}^1 \times \mathbb{A}^1 \to \mathbb{A}^1$ be the projection onto the second factor. Using (i), deduce in an alternative way that $\mathbb{A}^1$ cannot be a complete variety.
(iii) If otherwise $V = \mathbb{P}^1$ and $W = \mathbb{A}^1$, set $\widehat{Z}$ to be the closure of Z in $\mathbb{P}^1 \times \mathbb{A}^1$. Show that $\pi_{\mathbb{A}^1}(\widehat{Z})$ is closed in $\mathbb{A}^1$.

Exercise 9.2. Consider in $\mathbb{P}_{\mathbb{R}}^3$ the line $\bar{\ell} := Z_p(X_1 + X_2 - X_0, \ X_3 - X_0)$. Find $F \in \mathcal{S}^{(3)}$ such that $Z_p(F) \cap \bar{\ell} = \emptyset$ deducing that, in Remark 9.2.2-(v), the assumption for the field $\mathbb{K}$ to be algebraically closed is essential.

Exercise 9.3. For any integers $n \geqslant 1$ and $d \geqslant 2$, consider the projective space $\mathbb{P}(\mathcal{S}_d^{(n)})$. Let

$$\mathcal{R} := \left\{ [F] \in \mathbb{P}(\mathcal{S}_d^{(n)}) \mid F = F' \cdot F'', \ F' \in \mathcal{S}_h^{(n)}, \ F'' \in \mathcal{S}_{d-h}^{(n)}, \ 1 \leqslant h \leqslant d-1 \right\}.$$

Show that $\mathcal{R}$ is a proper closed subset of $\mathbb{P}(\mathcal{S}_d^{(n)})$ (cf. Shafarevich, 1994, Proposition p. 60).

Exercise 9.4.

(i) Let V and W be algebraic varieties and $\varphi \in \mathrm{Morph}(V, W)$. The morphism φ is said to be a *finite morphism* if φ is dominant and, for any point $P \in W$, there exists an affine open neighborhood U_P of P s.t. $\varphi^{-1}(U_P)$ is an affine open set of V and moreover $\varphi^{U_P} : \mathcal{O}_W(U_P) \hookrightarrow \mathcal{O}_V(\varphi^{-1}(U_P))$ is an integral extension of $\mathbb{K}$–algebras. Show that if φ is a finite morphism then, for any point $P \in \mathrm{Im}(\varphi)$, one has $\varphi^{-1}(P)$ is a finite set of V.

(ii) Consider the hypersurface $Z = Z_a(x_3^2 + x_3(x_1 - x_2) + x_1^2 - 1) \subset \mathbb{A}^3$. Show that Z is irreducible and that the restriction to Z of the projection $\pi_I : \mathbb{A}^3 \to \mathbb{A}^2$, where $I = (1, 2)$, defines a morphism $\varphi \in \mathrm{Morph}(Z, \mathbb{A}^2)$ which is finite and such that $\varphi(Z)$ is closed.

Exercise 9.5. Let $P \in \mathbb{P}^n$ be a point and let $V \subset \mathbb{P}^n$ be an irreducible hypersurface not containing P. Let $\varphi : V \to \mathbb{P}^{n-1}$ be the restriction to V of the projection from P onto $\mathbb{P}^{n-1}$. Show that φ is a finite morphism.

Chapter 10

Dimension of Algebraic Varieties

Here, we first define the dimension of an algebraic variety V with the use of its field of rational functions $K(V)$. Then, we will compare the given general definition with, e.g. the *combinatorial dimension* introduced in Section 4.3 for topological spaces and with some other notions of "dimension" (cf. Section 10.2 and, e.g. Theorem 12.3.9).

10.1 Dimension of an Algebraic Variety

Let V be any algebraic variety. In Corollary 5.3.21, we proved that $K(V)$ is a finitely generated field extension of the base field $\mathbb{K}$. In particular, $\text{trdeg}_{\mathbb{K}}(K(V)) < +\infty$, as this number equals the maximal number of algebraically independent elements among the generators of $K(V)$ over $\mathbb{K}$ (recall the proof of Corollary 5.3.21). In this set-up, one sets:

Definition 10.1.1. Let V be any algebraic variety. One poses $\dim(V) := \text{trdeg}_{\mathbb{K}}(K(V))$.

Immediate consequences of the previous definition are the following.

Proposition 10.1.2.

(i) $\dim(\mathbb{A}^n) = \dim(\mathbb{P}^n) = n$.

(ii) *If V is any algebraic variety and if $U \subseteq V$ is any non-empty open subset, then $\dim(U) = \dim(V)$.*

(iii) *Let $n \geqslant 1$ be an integer and let Λ be a projective (respectively, affine) subspace of $\mathbb{P}^n$ (respectively, of $\mathbb{A}^n$) s.t. $\Lambda \cong \mathbb{P}^h$ (respectively, $\Lambda \cong \mathbb{A}^h$), with $h \geqslant 0$. Then $\dim(\Lambda) = h$.*

(iv) *If V is any algebraic variety, then $\dim(V) = 0$ if and only if V is a point.*

(v) *If V and W are algebraic varieties and if $\Phi : V \dashrightarrow W$ is a dominant rational map, then $\dim(W) \leqslant \dim(V)$.*

(vi) *If W is a closed, proper subvariety of an algebraic variety V, then $\dim(W) < \dim(V)$.*

(vii) *If $V \subset \mathbb{P}^n$ is a projective variety and if $C_a(V) \subset \mathbb{A}^{n+1}$ is the affine cone over V, then $\dim(C_a(V)) = \dim(V) + 1$.*

Proof.

(i) It follows from Corollary 5.3.18 and from Remark 1.8.4-(iii).

(ii) It follows from Lemma 5.3.13-(ii).

(iii) If $\Lambda \subset \mathbb{A}^n$ is an affine subspace s.t. $\Lambda \cong \mathbb{A}^h$, identifying $\mathbb{A}^n$ with the affine chart U_0, the projective closure $\overline{\Lambda}$ of Λ in $\mathbb{P}^n$ is a linear subspace isomorphic to $\mathbb{P}^h$ and Λ is a dense open subset of $\overline{\Lambda}$ (cf. Section 3.3.5). From (ii) above, we can therefore focus on the case $\Lambda \subset \mathbb{A}^n$ an affine subspace. Since $\Lambda \cong \mathbb{A}^h$, then $A(\Lambda) \cong A^{(h)} = \mathbb{K}[x_1, \ldots, x_h]$, where the x_i's are indeterminates over $\mathbb{K}$. Since Λ is affine, from Theorem 5.3.14-(d), $K(\Lambda) = \mathcal{Q}(A(\Lambda)) \cong \mathcal{Q}(A^{(h)}) = \mathcal{Q}^{(h)}$ and we are done.

(iv) If V is a point, then $V \cong \mathbb{A}^0$ and we conclude using (iii). Conversely, assume $\dim(V) = 0$. From Proposition 6.4.2 and (ii) above, with no loss of generality, we can assume V to be affine so V can be considered as an irreducible, closed subset of $\mathbb{A}^n$, for some positive integer n. From Theorem 5.3.14, we have $\mathbb{K} \subseteq A(V) \subseteq K(V) = \mathcal{Q}(A(V))$. Since $\dim(V) = 0$, it follows that $\mathbb{K} \subseteq K(V)$ is an algebraic extension. On the other hand, since $\mathbb{K}$ is algebraically closed, we must have $\mathbb{K} = K(V)$. This forces $A(V) = \mathbb{K}$ and so $I_a(V) \subset A^{(n)}$ to be a maximal ideal. From the Hilbert "Nullstellensatz"-weak form (cf. Theorem 2.2.1), V is a point.

(v) If $\Phi : V \dashrightarrow W$ is dominant, from Theorem 8.1.6, we have a field extension $K(W) \subseteq K(V)$. Since $\mathrm{trdeg}_{\mathbb{K}}(K(V)) = \dim(V) < +\infty$, from the field extensions $\mathbb{K} \subseteq K(W) \subseteq K(V)$ we get also $\mathrm{trdeg}_{K(W)}(K(V)) < +\infty$. We conclude by Proposition 1.8.5.

(vi) Using (ii) and Proposition 6.4.2, up to replace V with an affine open subset U intersecting W, with no loss of generality we may assume V to be affine, with coordinate ring $A(V)$, and W to be an affine subvariety of V, with coordinate ring $A(W)$. Since $W \subset V$ is a proper subvariety, from the bijective correspondence between closed subvarieties of V and prime ideals in $A(V)$, one deduces that there exists a prime ideal $(0) \neq \mathfrak{p} \subset A(V)$ such that $A(W) \cong \frac{A(V)}{\mathfrak{p}}$. Since both $A(V)$ and $A(W)$ are integral $\mathbb{K}$-algebras of finite type, from the proof of Theorem 1.12.3, we get that $\mathrm{trdeg}_{\mathbb{K}}(A(W)) < \mathrm{trdeg}_{\mathbb{K}}(A(V))$, which therefore concludes the proof of (vi).

(vii) As in the proof of Proposition 4.3.9, one has a $\mathbb{K}$-algebra isomorphism $\mathcal{S}(V) \cong A(C_a(V))$. Moreover, since V is a projective variety, then $C_a(V)$ is an affine variety in $\mathbb{A}^{n+1}$. From Theorem 5.3.14-(d) one has $K(C_a(V)) \cong \mathcal{Q}(A(C_a(V))) \cong \mathcal{Q}(\mathcal{S}(V))$.

Since $A(C_a(V))$ is an integral $\mathbb{K}$-algebra of finite type, from Definition 1.8.7, one has also $\dim(C_a(V)) = \mathrm{trdeg}_{\mathbb{K}}(A(C_a(V))) = \mathrm{trdeg}_{\mathbb{K}}(\mathcal{S}(V))$. On the other hand, from Corollary 5.3.17 one has that $\mathcal{Q}(\mathcal{S}(V)) \cong K(V)(t)$ where t is an indeterminate over $K(V)$, i.e. $K(V) \subset \mathbb{K}(V)(t)$ is a simple, purely transcendental field extension. Thus, from the definition of $\dim(V) := \mathrm{trdeg}_{\mathbb{K}}(K(V))$ and from the transitivity of transcendence degree as in Proposition 1.8.5, the proof is complete. $\qquad\square$

Similarly as for the combinatorial dimension (cf. Section 4.3), when $\dim(V) = 1$ then V is called an *irreducible curve*, if otherwise $\dim(V) = 2$ then V is called an *irreducible surface*, if $\dim(V) = 3$ then V is an *irreducible threefold* and so on. More generally, any algebraic set Y is said to be *of pure dimension* (or simply, *pure*, cf. also Definition 4.3.3) if all its irreducible components have the same dimension; in particular, a *curve* is an algebraic set of pure dimension 1, a *surface* is an algebraic set of pure dimension 2 and so on. When otherwise Y is *not pure*, since Y has finitely many irreducible components then $\dim(Y)$ is defined to be the *maximum* among all the dimensions of its irreducible components (cf. (4.5)).

Important consequences are the following.

Remark 10.1.3.

(i) From Proposition 10.1.2-(v), the dimension is a *birational invariant* of algebraic varieties.

(ii) In particular *if V is a rational variety of* $\dim(V) = n$, *then it is birational to* $\mathbb{P}^n$ *(and even to* $\mathbb{A}^n$*). Moreover,* $\mathbb{K}(V)$ *is a purely transcendental extension of* $\mathbb{K}$. Indeed, in such a case, $K(V)$ is isomorphic to $\mathcal{Q}^n$ so $\mathrm{trdeg}_{\mathbb{K}}(K(V)) = n$.

(iii) Revisiting Examples 5.3.23, 5.3.24, 6.2.5 (ii)–(v), 8.1.13 and 8.2.12 (i)–(ii) and taking into account, e.g. any affine (respectively, projective) rational normal curve of degree $d \geqslant 1$, these are all irreducible rational curves since they are all birational, sometimes isomorphic, to $\mathbb{P}^1$.

(iv) The elliptic curve Y_a in Example 5.3.25, with $a \neq 0, 1$, is an example of an irreducible curve which is not rational. Indeed, therein we proved that $K(Y_a)$ is an algebraic extension of degree 2 of $\mathbb{K}(x_1) \cong \mathcal{Q}^{(1)}$, as $x_2 \in K(Y_a)$ has minimal polynomial $P_{x_1}(t) = t^2 - x_1(x_1 - 1)(x_1 - a)$ in $\mathcal{Q}^{(1)}[t]$, where t an indeterminate over $\mathbb{K}$. Thus, the field extensions $\mathbb{K} \subset \mathbb{K}(x_1) \subset K(Y_a)$ are such that the first extension is purely transcendental (of transcendence degree 1) whereas the second one is algebraic of degree two, so $\mathrm{trdeg}_{\mathbb{K}}(K(Y_a)) = \mathrm{trdeg}_{\mathbb{K}}(\mathcal{Q}^{(1)}) = 1 = \dim(Y_a)$. Instead, for $a = 0, 1$ we obtain plane (nodal) irreducible cubic curves which are monoids, so they are irreducible rational curves.

(v) For any subvariety W of an algebraic variety V, one can define its *codimension* in V, which is denoted by

$$\mathrm{codim}_V(W) := \dim(V) - \dim(W). \tag{10.1}$$

From Proposition 10.1.2-(ii) and (vi), this integer is always non-negative; it is zero when W is a non-empty, open subset of V whereas it is strictly positive when W is a closed and proper subvariety of V (or more generally a locally closed subset of a proper, closed subvariety of V).

Proposition 10.1.4. *For any algebraic varieties V and W, one has* $\dim(V \times W) = \dim(V) + \dim(W)$.

Proof. From Propositions 6.4.2 and 10.1.2-(ii), we can reduce to the case that V and W are both affine varieties. In particular, we can consider them as irreducible, closed subsets $V \subset \mathbb{A}^n$ and $W \subset \mathbb{A}^m$, for some integers n and m. By Proposition 7.1.1, $V \times W \subset \mathbb{A}^n \times \mathbb{A}^m \cong \mathbb{A}^{n+m}$ is an affine variety. Let $v := \dim(V)$ and $w := \dim(W)$. By assumptions on V and W, we have $A(V) = \frac{\mathbb{K}[x_1,\ldots,x_n]}{I_a(V)}$ and $A(W) = \frac{\mathbb{K}[y_1,\ldots,y_m]}{I_a(W)}$. From (1.16), we have $\mathrm{trdeg}_{\mathbb{K}}(A(V)) = v$ and $\mathrm{trdeg}_{\mathbb{K}}(A(W)) = w$. Since $A(V)$ is generated by (the images of) $x_1,\ldots x_n$ as an integral $\mathbb{K}$-algebra of finite type, the previous equality implies that $\{x_1,\ldots,x_n\}$ contains a transcendence basis over $\mathbb{K}$ determined by $v \leqslant n$ elements. For simplicity, assume this basis to be $\{x_1,\ldots,x_v\}$, i.e. $\mathcal{Q}^{(v)} \cong \mathbb{K}(x_1,\ldots,x_v) \subset \mathcal{Q}(A(V))$, where the field extension is algebraic. As for $A(W)$, assume that $\{y_1,\ldots,y_w\}$ is a transcendence basis over $\mathbb{K}$.

From Corollary 6.2.8, the two projection morphisms $V \xleftarrow{\pi_V} V \times W \xrightarrow{\pi_W} W$ give rise to injective $\mathbb{K}$-algebra homomorphisms $A(V) \xhookrightarrow{\pi_V^\#} A(V \times W) \xhookleftarrow{\pi_W^\#} A(W)$ (cf. (6.2) and Theorem 5.3.14-(a)). Since $\{x_1, \ldots, x_v\}$ is a set of elements of $A(V)$ which are algebraically independent over $\mathbb{K}$, these elements remain algebraically independent over $\mathbb{K}$ when viewed as elements of $A(V \times W)$. Same occurs for the set $\{y_1, \ldots, y_w\}$.

We want to show that $\{x_1, \ldots, x_v, y_1, \ldots, y_w\}$ is a set of elements in $A(V \times W)$ which are algebraically independent over $\mathbb{K}$. Reasoning recursively, it suffices to show that $\{x_1, \ldots, x_v, y_1\}$ is formed by algebraically independent elements over $\mathbb{K}$. Assume there exists a polynomial $f(t_1, \ldots, t_{v+1}) \in \mathbb{K}[t_1, \ldots, t_{v+1}]$ such that $f(x_1, \ldots, x_v, y_1) = 0$. Since we have $\mathbb{K}[t_1, \ldots, t_{v+1}] = (\mathbb{K}[t_1, \ldots, t_v])[t_{v+1}]$, the polynomial $f(t_1, \ldots, t_{v+1})$ can be viewed as a polynomial in the indeterminate t_{v+1} and with coefficients from $\mathbb{K}[t_1, \ldots, t_v]$, namely

$$f(t_1, \ldots, t_{v+1}) = c_0(t_1, \ldots, t_v) + c_1(t_1, \ldots, t_v)t_{v+1} + \cdots + c_d(t_1, \ldots, t_v)t_{v+1}^d,$$

where d the degree w.r.t. the indeterminates t_{v+1} of the polynomial $f(t_1, \ldots, t_{v+1})$ and where $c_j(t_1, \ldots, t_v) \in \mathbb{K}[t_1, \ldots, t_v]$, $0 \leqslant j \leqslant d$. Since V is an affine variety, then $A(V) \cong \mathcal{O}_V(V)$ (cf. Theorem 5.3.14-(a)). Thus, since $x_1, \ldots, x_v \in A(V)$, in particular any polynomial expression evaluated at $x_1, \ldots x_v$ gives an element in the ground field $\mathbb{K}$. Thus, $c_j(x_1, \ldots, x_v) \in \mathbb{K}$, for any $0 \leqslant j \leqslant d$, so $g(t_{v+1}) := f(x_1, \ldots, x_v, t_{v+1}) \in \mathbb{K}[t_{v+1}]$. From the hypothesis one has $g(y_1) = f(x_1, \ldots, x_v, y_1) = 0$. Thus, since $y_1 \in A(W)$ is by assumption algebraically independent over $\mathbb{K}$, the polynomial $g(t_{v+1}) \in \mathbb{K}[t_{v+1}]$ has to be identically zero. This means that any coefficient $c_j(x_1, \ldots, x_v)$ has to be zero, $0 \leqslant j \leqslant d$. On the other hand, the elements $x_1, \ldots, x_v \in A(V)$ are by assumption algebraically independent over $\mathbb{K}$, therefore all the polynomials $c_j(t_1, \ldots, t_v) \in \mathbb{K}[t_1, \ldots, t_v]$, $0 \leqslant j \leqslant d$, have to be identically zero (otherwise any non-zero polynomial among them would give a non-trivial algebraic relation among $x_1, \ldots, x_v$, which cannot occur from their algebraic independence over $\mathbb{K}$). This shows that $f(t_1, \ldots, t_{v+1})$ is identically zero, i.e. it is the zero-polynomial and so that $x_1, \ldots, x_v, y_1$ are algebraically independent elements over $\mathbb{K}$.

Finally, from (2.8) one has $A(V \times W) \cong \frac{\mathbb{K}[x_1, \ldots, x_n, y_1, \ldots, y_m]}{I_a(V) + I_a(W)}$ so $A(V \times W)$ is generated as an integral $\mathbb{K}$-algebra of finite type by (the images of) $x_1, \ldots, x_n, y_1, \ldots, y_m$. Thus, $\mathbb{K}(x_1, \ldots, x_v, y_1, \ldots, y_w) \subset \mathcal{Q}(A(V \times W))$ is an algebraic extension and so one concludes. $\qquad\square$

Proposition 10.1.5. *Let V be any closed subvariety of $\mathbb{A}^n$ (equivalently, $\mathbb{P}^n$). Then $\dim(V) = n - 1$ if and only if V is an irreducible hypersurface. In particular any hypersurface of $\mathbb{A}^n$ (equivalently, $\mathbb{P}^n$) is of pure dimension $n - 1$.*

Proof. As usual, by Propositions 6.4.2 and 10.1.2-(ii), it suffices to consider V to be affine.

($\Rightarrow$) If $\dim(V) = n - 1 < n = \dim(\mathbb{A}^n)$, V is a proper, closed subvariety of $\mathbb{A}^n$ so $I_a(V) \neq (0)$. Take any non-zero $f \in I_a(V)$; since $I_a(V)$ is prime, we can moreover assume f to be irreducible. Then $V \subseteq Z_a(f) \subset \mathbb{A}^n$, where the second inclusion is strict. Therefore, $\dim(Z_a(f)) < n$ so necessarily $n - 1 = \dim(V) = \dim(Z_a(f))$. Since V is closed in $\mathbb{A}^n$, then V is also closed in $Z_a(f)$. By Proposition 10.1.2-(vi), we must have $V = Z_a(f)$.

($\Leftarrow$) Let $V = Z_a(f)$, for some non-zero irreducible polynomial $f \in \mathbb{K}[x_1, \ldots, x_n]$. Up to re-labeling the indeterminates, we can assume f to be non-constant with respect to the indeterminate x_1; then any $g \in (f)$ is non-constant with respect to x_1, i.e. (f) does not contain any non-zero polynomial $h(x_2, \ldots, x_n)$. Since $A(V) = \frac{\mathbb{K}[x_1, \ldots, x_n]}{(f)}$ is generated by (the images of) $x_1, \ldots, x_n$, in particular (the images of) $x_2, \ldots, x_n \in A(V)$ are algebraically independent over $\mathbb{K}$; thus $\dim(V) \geqslant n - 1$. By Proposition 10.1.2-(vi) equality must hold.

For the last part of the statement, any (possibly reducible) affine hypersurface is given by $Z_a(f) = \cup_{i=1}^{\ell} Z_a(f_i)$, where $f = f_1 \cdots f_\ell$ is its reduced equation (recall (2.6) and (2.7)). Any $Z_a(f_i)$ is an irreducible component of $Z_a(f)$ to which one applies the first part of the proof.   □

10.2   Comparison on Various Definitions of "Dimension"

Up to now, we have introduced several concepts related to the word *"dimension"*, e.g. the Krull-dimension for (commutative and with identity) rings, the combinatorial dimension for topological spaces and the dimension of algebraic varieties. We want to compare all these notions. We start with the following easy result.

Proposition 10.2.1. *Let V be any affine variety and let $A(V)$ denote its coordinate ring. Then $\dim(V) = \dim_c(V) = K\dim(A(V))$.*

Proof. The second equality has been proved in Proposition 4.3.7. Using Definition 1.8.7, we have that $\dim(V) = \mathrm{trdeg}_{\mathbb{K}}(A(V))$ so, since $A(V)$

is an integral $\mathbb{K}$-algebra of finite type, by Theorem 1.12.3, we have that $\operatorname{trdeg}_{\mathbb{K}}(A(V)) = K \dim(A(V))$, and the proof is complete. $\square$

Corollary 10.2.2.

(i) *Let V be any algebraic variety. Then $\dim(V) = \dim_c(V)$. In particular, if $\dim(V) = d$, then:*

    (a) *all maximal chains of closed subvarieties of V have length d;*

    (b) *for any closed subvariety $\emptyset \neq W \subset V$, there exists a maximal chain of closed subvarieties of V containing W;*

    (c) *for any integer $0 \leqslant m \leqslant d$, there exists a non-empty, closed subvariety $W_m \subset V$ such that $\dim(W_m) = m$.*

(ii) *If $V \subset \mathbb{P}^n$ is in particular a projective variety and if $\mathcal{S}(V)$ denotes its homogeneous coordinate ring, then (4.7) holds true.*

Proof.

(i) From Proposition 10.1.2-(ii), for any non-empty, open subset $U \subseteq V$ one has $\dim(U) = \dim(V)$. Since, by Proposition 6.4.2, any algebraic variety V has an affine open covering, with no loss of generality, one can reduce to the case of V an affine variety and conclude that $\dim(V) = \dim_c(V)$ by using Proposition 10.2.1.

Since V can be assumed to be affine, by Theorem 1.12.3 applied to the coordinate ring $A(V)$, (a) directly follows from Proposition 4.3.7. To prove (b), it suffices to observe that $W \subset V$ is a chain of length 1 and that any chain of closed subvarieties of V can be always extended to a maximal chain of length d, by definition of $\dim_c(V) = d$. As for (c), if

$$\emptyset \neq W_0 \subset W_1 \subset \cdots W_d = V$$

is a maximal chain of closed subvarieties then, by Proposition 10.1.2-(vi), for any $0 \leqslant m \leqslant d - 1$ one has $\dim(W_m) < \dim(W_{m+1})$, therefore one more precisely has $\dim(W_m) = m$. Since, by (a), all maximal chains of closed subvarieties of V have precisely length d, one concludes.

(ii) Since $V \subset \mathbb{P}^n$ is a projective variety, then $C_a(V) \subset \mathbb{A}^{n+1}$ is an affine variety. From Proposition 10.2.1 we have that $\dim(C_a(V)) = \dim_c(C_a(V))$. Proposition 4.3.9 shows that $K \dim(\mathcal{S}(V)) = \dim_c(C_a(V))$. Moreover, by Proposition 10.1.2-(vi), we also have $\dim(C_a(V)) = \dim(V) + 1$. Since, by (i), we have $\dim(V) = \dim_c(V)$, in particular (4.7) holds true. $\square$

Another important identification, which also provides a *local interpretation* of dimension, is given by the following.

Proposition 10.2.3. *Let V be any algebraic variety and let $P \in V$ be any point. Then $\dim(V) = K \dim(\mathcal{O}_{V,p})$, where $\mathcal{O}_{V,P}$ the local ring of V at P.*

Proof. From Corollary 6.4.3-(i), there is a bijective correspondence between prime ideals of $\mathcal{O}_{V,p}$ and closed subvarieties of V containing the point P. Therefore, prime ideal chains in $\mathcal{O}_{V,p}$ bijectively correspond to chains of closed subvarieties of V containing the point P. From Corollary 10.2.2-(i), (a), any such maximal chain has length $\dim(V)$.   □

Remark 10.2.4. Recall that, for any point $P \in V$, $\mathcal{O}_{V,p}$ is an integral $\mathbb{K}$-algebra such that $\mathcal{O}_{V,p} \subseteq K(V)$. From Corollary 6.4.3-(ii) one has $Q(\mathcal{O}_{V,p}) \cong K(V)$ whereas, from Definition 1.8.7, $\mathrm{trdeg}_{\mathbb{K}}(\mathcal{O}_{V,p}) = \mathrm{trdeg}_{\mathbb{K}}(K(V))$, the latter being equal to $\dim(V)$. On the other hand, Proposition 10.2.3 is not a consequence of Theorem 1.12.3 since, in general, $\mathcal{O}_{V,p}$ is not a $\mathbb{K}$-algebra of finite type (recall Remark 1.12.5).

10.3   Dimension and Intersections

Here, we discuss some results related to the dimension of the locus obtained by intersecting an algebraic variety V with a certain finite number of irreducible projective hypersuperfaces. These results will have significant consequences on the structure of algebraic varieties.

Theorem 10.3.1. *Let $V \subset \mathbb{P}^n$ be an algebraic variety and let $W_1, \ldots, W_s$, $s \geqslant 1$, be irreducible hypersurfaces of $\mathbb{P}^n$. Then:*

(i) *if $V \cap W_1 \cap \cdots \cap W_s \neq \emptyset$, any irreducible component X of $V \cap W_1 \cap \cdots \cap W_s$ is such that $\dim(X) \geqslant \dim(V) - s$;*

(ii) *if V is projective with $\dim(V) \geqslant s$, then $V \cap W_1 \cap \cdots \cap W_s \neq \emptyset$.*

Proof.

(i) Using induction on s, we can reduce to the case $s = 1$. Since X is closed in V, up to replacing V with an affine open set intersecting X, we may assume that V is an affine variety and that X is an irreducible component of $Z_V(f)$, for some $f \in A(V)$. If $f = 0$ in $A(V)$ this means that W_1 is defined by a polynomial lying in $I_a(V)$, i.e. $V \subseteq W_1$ so that $V \cap W_1 = V$ and $X = V$ since V is irreducible; in this case

$\dim(X) = \dim(V) > \dim(V) - 1$. If otherwise $f \in A(V) \setminus \{0\}$, then f is not invertible in $A(V)$ since $\emptyset \neq X \subseteq V \cap W_1$. Let $\mathfrak{p} \subset A(V)$ be the prime ideal defining X. Since $X \subseteq Z_V(f)$, by reversing inclusion we have $\mathfrak{p} \supset (f)$. Moreover, since X is an irreducible component of $V \cap W_1$, then $\mathfrak{p}$ is a *minimal prime ideal* containing (f) (in general, the *minimal prime ideals* of an ideal I are defined to be the minimal elements of the set of all prime ideals containing I). Since $f \neq 0$ and not invertible in $A(V)$, from *Krull's principal ideal theorem* (also *Hauptidealsatz*, cf., e.g. Milne, 2017, Theorem 15.3), it follows that $\mathrm{ht}(\mathfrak{p}) = 1$. From the bijective correspondence prime ideals-closed subvarieties, this implies that it cannot exist any closed subvariety $Z \subset V$ satisfying $X \subset Z \subset V$. Thus $X \subset V$ is a maximal chain of closed subvarieties of V containing Z so $\dim(X) = \dim(V) - 1$.

(ii) We use induction on s. Consider first $s = 1$ and let $F_1 \in \mathcal{S}_d^{(n)}$ be the homogeneous polynomial such that $W_1 = Z_p(F_1)$, where $d = \deg(W_1) \geqslant 1$. Set $\mathcal{S}(V)$ the homogeneous coordinate ring of V and let $\overline{F}_1 \in \mathcal{S}(V)$ be the image of F_1 under the canonical epimorphism $\pi : \mathcal{S}^{(n)} \twoheadrightarrow \mathcal{S}(V)$; thus $W_1 \cap V = Z_V(\overline{F}_1)$. If $\overline{F}_1 = 0$ in $\mathcal{S}(V)$, as in (i), this implies that $V \subseteq W_1$ so $V \cap W_1 = V \neq \emptyset$ since $\dim(V) \geqslant 1$. If otherwise $\overline{F}_1 \neq 0$ in $\mathcal{S}(V)$, the $\overline{F}_1$ is a homogeneous element of positive degree in $\mathcal{S}(V)$. Let $C_a(V) \subset \mathbb{A}^{n+1}$ be the affine cone over V; since $A(C_a(V)) \cong \mathcal{S}(V)$ as integral $\mathbb{K}$-algebras, then $\overline{F}_1 \in A(C_a(V))$ defines a closed subset $Y \subset C_a(V)$ which certainly is not empty since, being $\overline{F}_1$ homogeneous of positive degree, its vanishing locus Y contains the vertex O of the affine cone $C_a(V)$. Applying (i), we get that any irreducible component of Y has dimension at least $\dim(C_a(V)) - 1 \geqslant 1$, the latter inequality following from Proposition 10.1.2-(vii) and from the assumption $\dim(V) \geqslant 1$. Thus, $V \cap W_1 \neq \emptyset$.

Assume now $s \geqslant 2$. By inductive hypothesis, $V \cap W_1 \cap \cdots \cap W_{s-1} \neq \emptyset$; thus, from (i), $\dim(X) \geqslant \dim(V) - (s - 1) \geqslant 1$ for any irreducible component X of $V \cap W_1 \cap \cdots \cap W_{s-1}$, the second inequality following from $\dim(V) \geqslant s$, as in the assumptions. Applying reasoning for the case $s = 1$ as above to X and W_{s-1}, one concludes. $\qquad\square$

As a consequence of the previous result, we also find a different proof of Corollary 6.2.6.

Corollary 10.3.2. *Let $V \subset \mathbb{A}^n$ be an affine variety which coincides with its projective closure in $\mathbb{P}^n$. Then V is a point.*

Proof. Identifying $\mathbb{A}^n$ with the affine chart U_0 of $\mathbb{P}^n$, we have inclusions $V \subset U_0 \subset \mathbb{P}^n$. From the assumption, $\overline{V}^p \cap H_0 = \emptyset$, where $H_0 = Z_p(X_0)$ the fundamental hyperplane of $\mathbb{P}^n$. From Theorem 10.3.1-(ii), it follows that $\dim(\overline{V}^p) < 1$ which implies $\dim(V) = 0$. $\qquad\square$

Another fundamental property, which will also be used later on in the proof of Theorem 11.1.1, is the following.

Proposition 10.3.3. *Let $V \subseteq \mathbb{P}^r$ be an algebraic variety and let $Z \subset V$ be a closed subvariety such that $\operatorname{codim}_V(Z) = s \geqslant 1$. Then, there exist hypersurfaces $W_1, \ldots, W_s \subset \mathbb{P}^r$ such that any irreducible component of $V \cap W_1 \cap \cdots \cap W_s$ has codimension s in V and moreover Z is one of such irreducible components.*

Proof. We argue by induction on the integer s. Let $s = 1$; since Z is a proper, closed subvariety of V, there exists $F_1 \in I_p(Z) \setminus I_p(V)$ and, by Theorem 10.3.1-(i), any irreducible component Y of $Z_p(F_1) \cap V$ satisfies $\operatorname{codim}_V(Y) = 1$. Since $Z \subseteq Z_p(F_1) \cap V$ and Z is irreducible, then Z must be one of such irreducible components.

Assume now $s \geqslant 2$. From Corollary 10.2.2-(i), (c), there exists a closed subvariety $Z' \subset V$ such that $Z \subset Z' \subset V$ and that $\dim(Z') = \dim(Z) + 1$. From the inductive hypothesis, there exist $W_1, \ldots, W_{s-1}$ hypersurfaces of $\mathbb{P}^r$ such that

$$V \cap W_1 \cap \cdots \cap W_{s-1} = Z'_1 \cup \cdots \cup Z'_k,$$

where Z'_j is irreducible with $\dim(Z'_j) = \dim(V) - (s-1)$, for any $1 \leqslant j \leqslant k$, and moreover there exists $j_0 \in \{1, \ldots, k\}$ such that $Z' = Z'_{j_0}$. Note that since Z is a proper, closed irreducible subset of Z', in particular $I_p(Z)$ is not contained in $I_p(Z'_j)$, for any $1 \leqslant j \leqslant k$. From the fact that any $I_p(Z'_j)$ is a prime ideal, it follows that $I_p(Z)$ is not contained in the union $\cup_{j=1}^k I_p(Z'_j)$. Therefore, there exists a homogeneous polynomial $F_s \in I_p(Z) \setminus (\cup_{j=1}^k I_p(Z'_j))$ which gives rise to the hypersurface $W_s := Z_p(F_s) \subset \mathbb{P}^r$ such that W_s contains Z but it does not contain any of the Z'_j, $1 \leqslant j \leqslant k$. Therefore,

$$V \cap W_1 \cap \cdots \cap W_s = (Z'_1 \cup \cdots \cup Z'_k) \cap W_s,$$

so any irreducible component of this intersection has exactly codimension s in V and Z, being irreducible and closed, is one of such irreducible components. $\qquad\square$

10.4 Complete Intersections

We conclude this chapter by discussing special cases of intersections of an algebraic variety with finitely many hypersurfaces of an affine, respectively, projective space. We start with the following.

Definition 10.4.1. Let R be an integral $\mathbb{K}$-algebra of finite type and let $f_1, \ldots, f_s \in R$. The elements $f_1, \ldots, f_s$ form a *regular sequence* in R if, for any $1 \leqslant i \leqslant s$, the multiplication map

$$\frac{R}{(f_1, \ldots, f_{i-1})} \xrightarrow{\cdot f_i} \frac{R}{(f_1, \ldots, f_{i-1})}$$

is injective.

In other words, $f_1, \ldots, f_s$ form a regular sequence if and only if, for any $1 \leqslant i \leqslant s$, the image of f_i via the quotient epimorphism $\pi_{i-1} : R \twoheadrightarrow \frac{R}{(f_1,\ldots,f_{i-1})}$ is not a zero-divisor in $\frac{R}{(f_1,\ldots,f_{i-1})}$ (equivalently, $f_i \in R$ is not a zero-divisor *modulo the ideal* $(f_1, \ldots, f_{i-1})$, for any $1 \leqslant i \leqslant s$).

We have the following.

Proposition 10.4.2.

(i) *Let V be an affine variety and let $A(V)$ be its coordinate ring. If $f_1, \ldots, f_s \in A(V)$ form a regular sequence then, for any $1 \leqslant i \leqslant s$, any irreducible component of $Z_V(f_1, \ldots, f_i)$ has dimension exactly $\dim(V) - i$.*

(ii) *Let $V \subset \mathbb{P}^n$ be a projective variety and let $\mathcal{S}(V)$ be its homogeneous coordinate ring. If $F_1, \ldots, F_s \in \mathcal{S}(V)$ are homogeneous elements forming a regular sequence then, for any $1 \leqslant i \leqslant s$, any irreducible component of $Z_V(F_1, \ldots, F_i)$ has dimension exactly $\dim(V) - i$.*

Proof. We focus on (i), since the proof of (ii) is similar. Note that the statement is equivalent to saying that, for any $1 \leqslant i \leqslant n$, $Z_V(f_i)$ does not contain any irreducible component of $Z_V(f_1, \ldots, f_{i-1})$. By assumptions $(f_1, \ldots, f_{i-1}) \subset A(V)$ moreover, since f_i is not a zero-divisor modulo $(f_1, \ldots, f_{i-1})$, it follows that f_i is not a zero-divisor modulo $\sqrt{(f_1, \ldots, f_{i-1})}$. In particular, $f_i \notin \sqrt{(f_1, \ldots, f_{i-1})}$. If, by contradiction, $Z_V(f_i)$ contained an irreducible component of $Z_V(f_1, \ldots, f_{i-1})$, there would exist $g_i \in A(V) \setminus \sqrt{(f_1, \ldots, f_{i-1})}$ such that $Z_V(f_1, \ldots, f_{i-1}) \subset Z_V(f_i) \cup Z_V(g_i) = Z_V(f_i g_i)$, i.e. $f_i g_i \in \sqrt{(f_1, \ldots, f_{i-1})}$ which is a contradiction. □

Definition 10.4.3.

(i) Let $V \subset \mathbb{A}^n$ be an affine variety. Assume that $I_a(V) = (f_1, \ldots, f_s)$, where $f_1, \ldots, f_s \in A^{(n)}$ is a regular sequence. Then V is said to be (*affine*) *complete intersection* of the s hypersurfaces $Z_a(f_1), \ldots, Z_a(f_s)$ of $\mathbb{A}^n$. If $V = Z_a(f_1, \ldots, f_s)$, where $f_1, \ldots, f_s \in A^{(n)}$ is a regular sequence (namely $I_a(V) = \sqrt{(f_1, \ldots, f_s)}$) then V is said to be (*affine*) *set-theoretically complete intersection* of the s hypersurfaces $Z_a(f_1), \ldots, Z_a(f_s)$ of $\mathbb{A}^n$.

(ii) Let $V \subset \mathbb{P}^n$ be a projective variety such that $I_p(V) = (F_1, \ldots, F_s)$, where $F_1, \ldots, F_s \in S^{(n)}$ is a collection of homogeneous elements forming a regular sequence. Then V is said to be (*projective*) *complete intersection* of the s hypersurfaces $Z_p(F_1), \ldots, Z_p(F_s)$ of $\mathbb{P}^n$. If $V = Z_p(F_1, \ldots, F_s)$, where $F_1, \ldots, F_s \in S^{(n)}$ is a collection of homogeneous elements forming a regular sequence (namely $I_p(V) = \sqrt{(F_1, \ldots, F_s)}$) then V is said to be (*projective*) *set-theoretically complete intersection* of the s hypersurfaces $Z_p(F_1), \ldots, Z_a(F_s)$ of $\mathbb{P}^n$.

Note that, if V is either set-theoretically complete intersection or complete intersection as above, in both cases $\dim(V) = n - s$, i.e. V has codimension s in the ambient space. Moreover, it is clear from the definition that if V is a complete intersection then it is also set-theoretically complete intersection. The converse does not hold in general, as we shall discuss in the following clarifying examples.

Example 10.4.4.

(i) Consider $C \subset \mathbb{A}^3$ the affine twisted cubic as in Section 3.3.12. From (3.33) we know that $I_a(C) = (x_2 - x_1^2, \ x_3 - x_1 x_2)$. Set $f_1 := x_2 - x_1^2$, $f_2 := x_3 - x_1 x_2 \in A^{(3)}$ and note that f_1, f_2 is a regular sequence: indeed, $R := \frac{A^{(3)}}{(f_1)} \cong \mathbb{K}[x_1, x_3]$ and the image $\pi_1(f_2)$ under the quotient epimorphism $\pi_1 : A^{(3)} \twoheadrightarrow \frac{A^{(3)}}{(f_1)} = R$ is $\pi_1(f_2) = x_3 - x_1^3$ which is a non-zero divisor in R since $\frac{R}{(\pi_1(f_2))} \cong \mathbb{K}[x_1]$ is an integral domain. Therefore, the affine twisted cubic is (affine) complete intersection of the two quadrics $Z_a(f_1)$ and $Z_a(f_2)$ in $\mathbb{A}^3$; it is therefore also set-theoretically complete intersection of the given two quadrics and $\operatorname{codim}_{\mathbb{A}^3}(C) = 2$ which equals the number of generators of its radical ideal and the number of hypersurfaces which cut-out C in $\mathbb{A}^3$.

(ii) Let us consider $\Gamma := \overline{C}^p \subset \mathbb{P}^3$ the projective twisted cubic, defined as the projective closure in $\mathbb{P}^3$ of the affine twisted cubic C as in (i), when $\mathbb{A}^3$ is identified with the affine chart $U_0 \subset \mathbb{P}^3$ (cf. Claim 3.3.18). We proved therein that $I_p(\Gamma) = (F_1, F_2, F_3)$, where

$$F_1 := X_0 X_2 - X_1^2, \quad F_2 := X_0 X_3 - X_1 X_2, \quad F_3 := X_1 X_3 - X_2^2,$$

are three irreducible, homogeneous polynomials of degree 2 which are linearly independent in $\mathcal{S}_2^{(3)}$ and which arise as the maximal minors of the matrix of linear forms $A := \left(\begin{smallmatrix} X_0 & X_1 & X_2 \\ X_1 & X_2 & X_3 \end{smallmatrix} \right)$ (cf. (3.35), (3.36) and Claim 3.3.19). Note that, dehomogenizing w.r.t. the indeterminates X_0, one has

$$\delta_0(F_1) = f_1, \ \delta_0(F_2) = f_2, \ \delta_0(F_3) = f_3 := x_1 f_2 - x_2 f_1 \in I_a(C).$$

From Proposition 10.4.2-(ii), $F_1, F_2, F_3 \in \mathcal{S}^{(3)}$ cannot be a regular sequence since $\dim(\Gamma) = 1 > \dim(\mathbb{P}^3) - 3$, so Γ is not complete intersection in $\mathbb{P}^3$. On the other hand, consider the homogeneous cubic polynomial $G := X_2 F_3 - X_3 F_2 = 2 X_1 X_2 X_3 - X_2^3 - X_0 X_3^2 \in \mathcal{S}_3^{(3)}$. Note that, $F_1, G \in \mathcal{S}^{(3)}$ is a regular sequence: indeed $Z_p(F_1) \subset \mathbb{P}^3$ is a quadric cone which is irreducible (cf. Section 8.2.2) thus $\frac{\mathcal{S}^{(3)}}{(F_1)}$ is an integral domain and therefore G is not a zero-divisor modulo (F_1). From the definition of G, it is clear that $\Gamma \subseteq Z_p(F_1, G)$; we claim that one actually has $\Gamma = Z_p(F_1, G)$. This is a consequence of the following Linear Algebra result.

Lemma 10.4.5. *Let* $\mathbb{K}$ *be a field and let* $A := \left(\begin{smallmatrix} a_{11} & a_{12} & a_{13} \\ a_{21} & a_{22} & a_{23} \\ a_{31} & a_{32} & a_{33} \end{smallmatrix} \right)$ *be a square matrix of order* 3, *where* $a_{ij} \in \mathbb{K}$. *Set*

$$A_1 := \begin{pmatrix} a_{11} & a_{12} \\ a_{21} & a_{22} \end{pmatrix} \quad A_2 := \begin{pmatrix} a_{11} & a_{12} & a_{13} \\ a_{21} & a_{22} & a_{23} \end{pmatrix} \quad A_3 := \begin{pmatrix} a_{11} & a_{12} \\ a_{21} & a_{22} \\ a_{31} & a_{32} \end{pmatrix}$$

as minors of A. *If* $\det(A) = \det(A_1) = 0$, *then either* $\operatorname{rank}(A_2) < 2$ *or* $\operatorname{rank}(A_3) < 2$.

Proof. From the assumptions $\operatorname{rank}(A) \leqslant 2$. If $\operatorname{rank}(A_2) < 2$, there is nothing else to prove. Assume therefore $\operatorname{rank}(A_2) = 2$; thus, since $\det(A) = 0$, the homogeneous linear system having A as coefficient matrix is equivalent to that having A_2 as coefficient matrix. Solutions of $A_2 \, \underline{x} = \underline{0}$ depend on one parameter in $\mathbb{K}$ and are proportional to the determinants, up to suitable signs $\pm$, of the 2×2 minors of A_2. On the other hand, one of this minors is A_1 whose determinant is by assumption equal to 0. Therefore, one has $\underline{x} = t \begin{pmatrix} b_1 \\ b_2 \\ 0 \end{pmatrix}$ where $t \in \mathbb{K}$ a parameter and where b_1 and b_2 the determinants of the other two minors (with sign) of A_2 and at least one of the b_i's is non-zero. One has therefore that

$$\begin{cases} a_{11}b_1 + a_{12}b_2 = 0 \\ a_{21}b_1 + a_{22}b_2 = 0 \\ a_{31}b_1 + a_{32}b_2 = 0 \end{cases}$$

holds, which implies that the homogeneous linear system of three equations in two indeterminates having A_3 as coefficient matrix has non-zero solutions, therefore $\mathrm{rank}(A_3) < 2$. $\square$

With above set-up, consider the following matrices:

$$A := \begin{pmatrix} X_0 & X_1 & X_2 \\ X_1 & X_2 & X_3 \\ X_2 & X_3 & 0 \end{pmatrix}, \quad A_1 := \begin{pmatrix} X_0 & X_1 \\ X_1 & X_2 \end{pmatrix},$$

$$A_2 := \begin{pmatrix} X_0 & X_1 & X_2 \\ X_1 & X_2 & X_3 \end{pmatrix}, \quad A_3 := \begin{pmatrix} X_0 & X_1 \\ X_1 & X_2 \\ X_2 & X_3 \end{pmatrix}$$

containing linear forms in $\mathcal{S}_1^{(3)}$. Note that $\det(A) = G$, $\det(A_1) = F_1$ and that $A_2 = A_3^t$. For any point $P \in \mathbb{P}^3$, denote by $A(P)$ (equivalently, $A_i(P)$) the matrix obtained by evaluating the linear forms at the homogeneous coordinates of P. Therefore, $P \in Z_p(F_1, G)$ if and only if $\det(A(P)) = 0 = \det(A_1(P))$ which, from Lemma 10.4.5, implies $\mathrm{rank}(A_2(P)) = \mathrm{rank}(A_3(P)) < 2$; since the vanishing of the maximal minors of A_2 (equivalently of A_3) are exactly the generators of $I_p(\Gamma)$ (cf. (3.36)), this means that $P \in \Gamma$, i.e. $\Gamma = Z_p(F_1, G)$ which shows that Γ is set-theoretically complete intersection of a quadric and a cubic in $\mathbb{P}^3$.

(iii) Any irreducible hypersurface $Z_p(f) \subset \mathbb{P}^n$ obviously is a complete intersection. In particular any line in $\mathbb{P}^2$ is. The same holds for any line $\ell \subset \mathbb{P}^n$, for any $n \geqslant 2$, since its homogenous ideal is generated by $n - 1$ linearly independent linear forms. Thus, take the line ℓ in $\mathbb{P}^3$ such that $I_p(\ell) = (X_0, X_1)$. Let $\mathfrak{Q} = Z_p(X_0 X_3 - X_1 X_2)$ the rank-four quadric in $\mathbb{P}^3$ which is the isomorphic image under the Segre morphism

$\sigma_{2,2}$ of $\mathbb{P}^1 \times \mathbb{P}^1$ (cf. Section 8.2.1). From the equation of $\mathfrak{Q}$ one note that $\ell \subset \mathfrak{Q}$ and ℓ is a hypersurface in $\mathfrak{Q}$. Then ℓ is neither complete intersection nor set-theoretically complete intersection in $\mathfrak{Q}$ even if ℓ is complete intersection in $\mathbb{P}^3$ (cf. Exercise 10.4).

Exercises

Exercise 10.1. Let V and W be algebraic varieties and let $\varphi \in \text{Morph}(V, W)$ be a finite morphism (cf. Exercise 9.4). Show that $\dim(V) = \dim(W)$.

Exercise 10.2. Let $V \subset \mathbb{P}^n$ be a projective variety. Let s be the maximum among the dimension of linear subspaces $\Lambda \subset \mathbb{P}^n$ such that $\dim(\Lambda) = s$ and $\Lambda \cap V = \emptyset$. Deduce that $\dim(V) \leqslant n - s - 1$.

Exercise 10.3. Let $V := Z_a(x_1 x_3 - x_2) \subset \mathbb{A}^3$. Show that V is an irreducible affine hypersurface. Then, let $\varphi \in \text{Morph}(V, \mathbb{A}^2)$ be the restriction of the projection $\pi_I : \mathbb{A}^3 \to \mathbb{A}^2$, with $I = (1, 2)$. Prove that the dimension of the fibers of φ is not constant. Find the maximal open set where the fiber dimension is constant. Deduce that φ cannot be a finite morphism.

Exercise 10.4. Show that the line $\ell = Z_p(X_0, X_1) \subset \mathbb{P}^3$ is contained in the rank-four quadric surface

$$\mathfrak{Q} = Z_p(X_0 X_3 - X_1 X_2) \subset \mathbb{P}^3$$

and it is neither complete intersection nor set-theoretically complete intersection in $\mathfrak{Q}$.

Exercise 10.5. Show that the line $\ell = Z_p(X_0, X_1) \subset \mathbb{P}^3$ is contained in the rank-three quadric surface (i.e. *quadric cone*)

$$\mathfrak{Q}' = Z_p(X_0 X_2 - X_1^2) \subset \mathbb{P}^3$$

and it is not complete intersection but set-theoretically complete intersection in $\mathfrak{Q}'$.

Chapter 11

Fiber-Dimension: Semicontinuity

In this chapter, we study dimensional behavior of the *fibers* of a given dominant morphism $\varphi \in \mathrm{Morph}(V, W)$ of algebraic varieties, namely the dimension of the subsets $\varphi^{-1}(P) \subseteq V$, for $P \in \mathrm{Im}(\varphi)$ varying. For further reading, we refer to, e.g. Shafarevich (1994, § I.6.3, pp. 76–78).

11.1 Fibers of a Dominant Morphism

If V and W are algebraic varieties and if $\varphi \in \mathrm{Morph}(V, W)$ is a dominant morphism, recall that from Proposition 10.1.2-(v) the integer $r := \dim(V) - \dim(W)$ is non-negative.

Theorem 11.1.1. *Let V and W be algebraic varieties and $\varphi \in \mathrm{Morph}(V, W)$ be a dominant morphism. Set $r := \dim(V) - \dim(W)$. For any point $P \in \mathrm{Im}(\varphi) \subseteq W$ consider the set*

$$\varphi^{-1}(P) := \{Q \in V \mid \varphi(Q) = P\} \subseteq V,$$

which is called the fiber *of the morphism φ over the point P. Then $\varphi^{-1}(P)$ is a closed algebraic subset of V. Moreover:*

(i) *for any irreducible component F_P of $\varphi^{-1}(P)$, one has $\dim(F_P) \geqslant r$;*

(ii) *there exists a non-empty, open subset U of W such that $U \subseteq \mathrm{Im}(\varphi)$ and, for any point $P \in U$ and for any irreducible component $F_P \subseteq \varphi^{-1}(P)$, one more precisely has $\dim(F_P) = r$.*

Proof. $\varphi^{-1}(P)$ is closed in V since $P \in \mathrm{Im}(\varphi) \subseteq W$ is closed and φ is a continuous map.

To prove (i), since V is irreducible then $\operatorname{Im}(\varphi)$ is an algebraic subvariety of W. From Proposition 6.4.2, let $U \subseteq \operatorname{Im}(\varphi)$ be any affine open set of $\operatorname{Im}(\varphi)$. Similarly, let $U' \subseteq \varphi^{-1}(U)$ be any affine open set of $\varphi^{-1}(U)$. Note that, by the dominance of φ and the fact that U is an open set of $\operatorname{Im}(\varphi)$, one has $\overline{U} = \overline{\operatorname{Im}(\varphi)} = W$ so the composition $\varphi' := \varphi \circ \iota_{U'} \in \operatorname{Morph}(U', U)$ is dominant too. Therefore, with no loss of generality, we may assume V and W to be affine varieties. In this set-up, since $\varphi \in \operatorname{Morh}(V, W)$ is dominant and V and W are affine varieties, by Corollary 6.2.8 the map $\varphi^{\#} \in \operatorname{Hom}_{\mathbb{K}}(A(W), A(V))$ is an injective $\mathbb{K}$-algebra homomorphism. Set $d := \dim(W)$; from Proposition 10.3.3 it follows that there exist $f_1, \ldots, f_d \in A(W)$ such that $\{P\}$ is an irreducible component of $Z_W(f_1, \ldots, f_d)$. Then

$$\varphi^{-1}(P) \subseteq Z_V(\varphi^{\#}(f_1), \ldots, \varphi^{\#}(f_d)),$$

where $0 \neq \varphi^{\#}(f_i) \in A(V)$ by injectivity of $\varphi^{\#}$ and moreover, from the fact that $\{P\}$ is an irreducible component of $Z_W(f_1, \ldots, f_d)$, it follows that $\varphi^{-1}(P)$ is a connected component of $Z_V(\varphi^{\#}(f_1), \ldots, \varphi^{\#}(f_d))$. Thus any irreducible component F_p of $\varphi^{-1}(P)$ is also an irreducible component of $Z_V(\varphi^{\#}(f_1), \ldots, \varphi^{\#}(f_d))$. From Theorem 10.3.1-(i) and from the injectivity of $\varphi^{\#}$ we have that $\dim(F_p) \geqslant \dim(V) - d = \dim(V) - \dim(W) = r$.

To prove (ii), as above we may assume that V and W are affine varieties. By the dominance of φ, $A(W)$ is a subalgebra of $A(V)$ via the monomorphism $\varphi^{\#}$. We may assume $\dim(V) > 0$ otherwise $\dim(W) = \dim(V) = 0$ and φ constant and there is nothing else to prove. Let $v_1, \ldots, v_n \in A(V)$ such that $A(V) = A(W)[v_1, \ldots, v_n]$ namely, taking $x_1, \ldots, x_n$ indeterminates over $A(W)$ and setting $\underline{x} := (x_1, \ldots, x_n)$, there exist polynomials $f_1(\underline{x}), \ldots, f_k(\underline{x}) \in A(W)[\underline{x}]$ such that $A(V) \cong \frac{A(W)[\underline{x}]}{(f_1(\underline{x}), \ldots, f_k(\underline{x}))}$. For any point $P \in \operatorname{Im}(\varphi) \subseteq W$ and for any irreducible component $F_P \subseteq \varphi^{-1}(P)$, the coordinate ring $A(F_P)$ is a quotient ring of the $\mathbb{K}$-algebra

$$\frac{\mathbb{K}[\underline{x}]}{(f_1(P)(\underline{x}), \ldots, f_k(P)(\underline{x}))} := \mathbb{K}[\widetilde{v}_1, \ldots, \widetilde{v}_n].$$

Since $\varphi^{\#}$ extends to a field homomorphism $K(W) \overset{\varphi^{*}}{\hookrightarrow} K(V)$, by the transitivity of transcendence degree (cf. Proposition 1.8.5), we have

$$\operatorname{trdeg}_{K(W)}(K(V)) = \operatorname{trdeg}_{\mathbb{K}}(K(V)) - \operatorname{trdeg}_{\mathbb{K}}(K(W))$$
$$= \dim(V) - \dim(W) = r \geqslant 0.$$

Thus, from the fact that $\mathbb{K}(V) \cong \mathbb{K}(W)(v_1, \ldots, v_n)$, up to reordering generators, we may assume that $v_1, \ldots, v_r \in A(V)$ are a transcendence basis of $\mathbb{K}(V)$ over $\mathbb{K}(W)$ so they are algebraically independent over $A(W)$. This implies that any element of $A(V)$ is algebraic over $A(W)[v_1, \ldots, v_r]$. Therefore, for any $r + 1 \leqslant j \leqslant n$, there exist polynomials $g_j(x_1, \ldots, x_r, x_j) \in A(W)[x_1, \ldots, x_r, x_j]$ which are non-constant w.r.t. the indeterminate x_j and such that $g_j(v_1, \ldots, v_r, v_j) = 0$, i.e. $g_j(x_1, \ldots, x_r, x_j) \in (f_1(\underline{x}), \ldots, f_k(\underline{x}))$ for any $r + 1 \leqslant j \leqslant n$.

Let $l.c._{x_j}(g_j) \in A(W)[x_1, \ldots, x_r]$ be the leading-coefficient polynomial of $g_j(x_1, \ldots, x_r, x_j)$ w.r.t. the indeterminate x_j, $r + 1 \leqslant j \leqslant n$. By the assumptions $U_j := Z_W(l.c._{x_j}(g_j))^c$ is a non-empty, principal open subset of W, for any $r + 1 \leqslant j \leqslant n$. Set

$$U := \bigcap_{j=r+1}^{n} U_j, \tag{11.1}$$

which is a non-empty open subset of W. For any point $P \in U$, the polynomials $g_j(P)(x_1, \ldots, x_r, x_j) \in \mathbb{K}[x_1, \ldots, x_r, x_j]$ are not identically zero (because of the leading coefficients w.r.t. the indeterminate x_j and the definition of U_j, for any $r + 1 \leqslant j \leqslant n$) and are such that $g_j(P)(\tilde{v}_1, \ldots, \tilde{v}_r, \tilde{v}_j) = 0$, i.e. $g_j(P)(x_1, \ldots, x_r, x_j) \in (f_1(P)(\underline{x}), \ldots, f_k(P)(\underline{x})) \subset \mathbb{K}[\underline{x}]$, for any $r + 1 \leqslant j \leqslant n$. This implies that, for any irreducible component $F_P \subseteq \varphi^{-1}(P)$ the elements $\tilde{v}_{r+1}, \ldots, \tilde{v}_n \in A(F_P)$ are algebraically dependent over $\mathbb{K}[\tilde{v}_1, \ldots, \tilde{v}_r]$. Thus $\dim(F_P) \leqslant r$. From Theorem 11.1.1-(i) one concludes that $\dim(F_P) = r$. □

Another fundamental property, which directly follows from the previous results, deals with what mentioned in Remark 6.1.6. Indeed, we are now in position to deduce the following.

Corollary 11.1.2. *Let V and W be algebraic varieties and let $\varphi \in \mathrm{Morph}(V, W)$ be a dominant morphism. Then $\mathrm{Im}(\varphi)$ always contains a non-empty open subset U of W. Moreover for all the points P in such an open set U, setting $r = \dim(V) - \dim(W)$, the fiber $\varphi^{-1}(P)$ is a non-empty, pure, r-dimensional algebraic subset of V.*

Proof. The open set U is that in (11.1). □

Similarly, recalling Remark 6.1.10, we have the following.

Corollary 11.1.3 (Chevalley's theorem). *Let V and W be algebraic varieties and let $\varphi \in \mathrm{Morph}(V, W)$ be any morphism. The image of φ is a* constructible set, *namely a union of finitely many locally-closed subsets of W. More generally, any $\varphi \in \mathrm{Morph}(V, W)$ maps constructible sets in V to constructible sets in W.*

Proof. The second part of the statement follows directly from the first. To prove the first part of the statement, we use induction on $\dim(W)$. If $\dim(W) = 0$, the morphism φ is constant and surjective. Therefore, we may assume $\dim(W) := m > 0$ and that the statement holds true in dimension less than m. To proceed with the proof, there are two cases to be considered.

If φ is not dominant, since $\mathrm{Im}(\varphi)$ is irreducible in W then $Z := \overline{\mathrm{Im}(\varphi)}$ is a closed subvariety properly contained in W, $\varphi \in \mathrm{Morph}(V, Z)$ is dominant and $\dim(Z) < \dim(W)$ from Proposition 10.1.2-(vi). By induction, we are done in this case.

Assume therefore $\varphi \in \mathrm{Morph}(V, W)$ to be dominant. Let U be a non-empty subset of W such that $U \subseteq \mathrm{Im}(\varphi)$, as in (11.1). If $W = U$, i.e. φ is surjective, there is nothing else to prove. If otherwise $U \subset W$, then U^c is a non-empty, proper closed subset of W. Denote by $Z_1, \ldots, Z_k$ the irreducible components of U^c and let $Y_{i,1}, \ldots, Y_{i,t_i}$ be irreducible components of $\varphi^{-1}(Z_i)$, $1 \leqslant i \leqslant k$. For any (i, j_i), $1 \leqslant i \leqslant k$, $1 \leqslant j_i \leqslant t_i$, let $\iota_{Y_{i,j_i}}$ be the closed immersion of Y_{i,j_i} into V and let $\varphi_{i,j_i} := \varphi \circ \iota_{Y_{i,j_i}} \in \mathrm{Morph}(Y_{i,j_i}, Z_i)$. Since $\dim(Z_i) < \dim(W)$ by Proposition 10.1.2-(vi), for any $1 \leqslant i \leqslant k$, by inductive assumptions $\varphi_{i,j_i}(Y_{i,j_i})$ is constructible in Z_i, which is closed in W. Thus, $\mathrm{Im}(\varphi) = U \cup (\cup_{i,j_i} \varphi_{i,j_i}(Y_{i,j_i}))$ is constructible in W. $\square$

11.2   Semicontinuity

Focusing on projective varieties, another interesting and very useful consequence of Theorem 11.1.1 is given by the following result.

Proposition 11.2.1. *Let V, W be projective varieties, $\varphi \in \mathrm{Morph}(V, W)$ be a surjective morphism and $r = \dim(V) - \dim(W)$. For any integer j such that $r \leqslant j \leqslant \dim(V)$, the sets*

$$W_j := \{P \in W \mid \varphi^{-1}(P) \text{ has an irreducible component } F_P \text{ s.t. } \dim(F_P) \geqslant j\}$$

are closed subsets of W.

Proof. From Theorem 11.1.1 and from the surjectivity of φ, it follows that $W_r = W$ and moreover there exists a closed subset $X \subset W$ for which $W_j \subseteq X$, for any $j > r$. Assume therefore, $j > r$ and $W_j \neq \emptyset$. To prove that W_j is closed in W it suffices to proving that the intersection of W_j

with any of the irreducible component of X, which is closed in W, is closed in X.

We therefore may assume X to be irreducible and let $Y_1, \ldots, Y_h$ be the irreducible components of $\varphi^{-1}(X)$ for which $\varphi_i := \varphi \circ \iota_{Y_i} \in \mathrm{Morph}(Y_i, X)$ is dominant, for any $1 \leqslant i \leqslant h$. Since V is projective and $Y_i \subset V$ is closed in V, then Y_i is a projective variety, for any $1 \leqslant i \leqslant h$. From Theorem 9.2.1 and Corollary 9.1.2, it follows that any φ_i is a surjective morphism. If $j \leqslant \dim(Y_i) - \dim(X)$, for some i, then from Theorem 11.1.1-(i) we get that $W_j = X$ which is closed in W. If otherwise $j > \dim(Y_i) - \dim(X)$, for any $1 \leqslant i \leqslant h$, then W_j is contained in a proper, closed subset of X. Applying once again the same arguments, one concludes. $\qquad\square$

To sum-up, for projective varieties the content of the previous results is the following: for any surjective morphism between projective varieties V and W, there exists a non-empty open set U of W over which all the fibers of the morphism are pure algebraic subsets of V of *minimal possible dimension*, which is $\dim(V) - \dim(W)$ (cf. Theorem 11.1.1-(ii)). There could also exist possible subsets of W, which are in the complementary of the open set U, where the dimension of the fibers may jump and these subsets are closed in W (cf. Proposition 11.2.1). This behavior is known as *(upper)-semicontinuity of fiber dimension* according to the following.

Theorem 11.2.2 (Upper-semicontinuity of the fiber dimension). *Let V and W be projective varieties, $\varphi \in \mathrm{Morph}(V, W)$ be a surjective morphism. For all $Q \in V$ define*

$$e(Q) := \big\{ \max(\dim(Z)) \mid Z \text{ is an irreducible component of } \varphi^{-1}(\varphi(Q)) \text{ containing } Q \big\}.$$

The function $e : V \to \mathbb{Z}_{\geqslant 0}$ is upper-semicontinuous, i.e. for all $j \in \mathbb{Z}_{\geqslant 0}$, the set $\Sigma_j(\varphi) := \{ Q \in V \mid e(Q) \geqslant j \}$ is closed in W.

Proof. The proof is based on Proposition 11.2.1. Indeed, set $r := \dim(V) - \dim(W)$ which is non-negative. If $j \in \mathbb{Z}_{\geqslant 0}$ is such that $j \leqslant r$ then, from Theorem 11.1.1-(i), one has that $\Sigma_j(\varphi) = \varphi^{-1}(W) = V$ and we are done. For those $j \in \mathbb{Z}_{\geqslant 0}$ for which $j > r$ one has either $\Sigma_j(\varphi) = \emptyset$ (e.g. for those $j > \dim(V)$) or $\Sigma_j(\varphi) = \varphi^{-1}(W_j)$, for those $\emptyset \neq W_j \subset W$ defined as in Proposition 11.2.1. Since W_j is closed in W and φ is continuous, we deduce that $\Sigma_j(\varphi)$ is a proper, closed subset of V. $\qquad\square$

A nice consequence of the previous result is the following.

Corollary 11.2.3 (Irreducibility of a fibration with irreducible fibers of constant dimension and with irreducible base variety). *Let*

V be a projective algebraic set, W be a projective variety with $\dim(W) = m$ and let $\varphi \in \mathrm{Morph}(V, W)$ be a surjective morphism. Assume that, for any $P \in W$, the fiber $\varphi^{-1}(P)$ is irreducible and $\dim(\varphi^{-1}(P)) = d$. Then V is irreducible, i.e. V is a projective variety, of dimension $\dim(V) = m + d$.

Proof. Write $V = \cup_{i=1}^n V_i$ as an irredundant union of irreducible components of V. To prove irreducibility of V, we want to show that $i = 1$ and that $V = V_1$. Since V is projective, so any V_i is, and φ is a surjective morphism there is at least one $i \in \{1, \ldots, n\}$ for which $\varphi(V_i) = W$ otherwise, from Corollary 9.1.2, W would be reducible against assumption. Up to reordering the irreducible components of V, assume that this occurs for, e.g. $i = 1$. Let ι_{V_i} be the closed immersion of V_i in V and let $\varphi_i := \varphi \circ \iota_{V_i} \in \mathrm{Morph}(V_i, W)$, $1 \leqslant i \leqslant n$.

For any $P \in W$ set $\lambda_i(P) := \dim(\varphi_i^{-1}(P))$. Thus, for any $P \in W$, one has $d = \dim(\varphi^{-1}(P)) = \max_{1 \leqslant i \leqslant n}\{\lambda_i(P)\}$. Since any λ_i is upper-semicontinuous, from Theorem 11.2.2 there exists an $i_0 \in \{1, \ldots, n\}$ such that $\lambda_{i_0}(P) = d$, for any $P \in W$; in particular $\varphi_{i_0} : V_{i_0} \to W$ is surjective. For such an index i_0 one has that $\varphi_{i_0}^{-1}(P) = \varphi^{-1}(P) \cap V_i$ is a closed subset of $\varphi^{-1}(P)$ of the same dimension. Therefore, one has $\varphi_{i_0}^{-1}(P) = \varphi^{-1}(P)$, for any $P \in W$, as it follows from Proposition 10.1.2-(vi) and from the irreducibility of the fiber $\varphi^{-1}(P)$ for any $P \in W$. Therefore, $V_{i_0} = V_1 = V$ and the irredundant decomposition consists of only V, namely V is irreducible.

Finally since the open set U, as in Theorem 11.1.1-(ii), where the upper-semicontinuous function $\lambda := \lambda_1$ reaches its minimum d coincides with the whole variety W, by Theorem 11.1.1-(ii) one concludes that $\dim(V) = m + d = \dim(W) + \dim(\varphi^{-1}(P))$, for any $P \in W$. $\square$

Exercises

Exercise 11.1. Let V and W be algebraic varieties. Give an example of a morphism $\varphi \in \mathrm{Morph}(V, W)$ which is dominant, where all the φ-fibers have the same dimension and moreover there exists a suitable non-empty open set $U \subseteq W$ such that, for each point $P \in U$ the fiber $\varphi^{-1}(P)$ is reducible nonetheless with V irreducible.

Exercise 11.2. Give an example of an algebraic variety $V \subset \mathbb{P}^2 \times \mathbb{A}^1$ such that, if φ denotes the restriction to V of the second projection onto $\mathbb{A}^1$, then φ is surjective, not all the φ-fibers are irreducible nonetheless V is irreducible and all the φ-fibers have the same dimension.

Exercise 11.3. Let $\varphi : V \to W$ be a morphism between projective varieties of positive dimensions for which there exists a point $P \in \mathrm{Im}(\varphi)$ such that $\varphi^{-1}(P)$ consists of finitely many points. Show that $\dim(\overline{\mathrm{Im}(\varphi)}) = \dim(V)$.

Exercise 11.4. Let V be an algebraic variety and let $W \subset V$ be a proper subvariety. Prove that $K \dim(\mathcal{O}_{V,W}) = \mathrm{codim}_V(W)$.

Exercise 11.5. Let V be an algebraic variety and let $P \neq Q \in V$ be any pair of distinct points. Show that there always exists an irreducible hypersurface $W \subset V$ such that $P \in W$ but $Q \notin W$.

Chapter 12

Tangent Spaces: Smoothness of Algebraic Varieties

In this chapter, we consider the *local study* of an algebraic variety V, namely we examine the structure of V locally at a given point $P \in V$. Throughout this chapter the field $\mathbb{K}$ is taken to be not only algebraically closed but also of characteristic 0.

12.1 Tangent Space at a Point of an Affine Variety: Smoothness

We start as usual with the affine case. Up to isomorphism, we may assume $V \subseteq \mathbb{A}^r$ is an affine variety, for some non-negative integer r. Let $I_a(V) := (f_1, \ldots, f_s) \subseteq A^{(r)}$ be its radical ideal and let $P = (p_1, \ldots, p_r) \in V$ be a point.

Definition 12.1.1. The *affine tangent space at the point P of V* is the affine subspace of $\mathbb{A}^r$ defined by the system of linear equations $\Sigma_{i=1}^r \frac{\partial f_j}{\partial x_i}(P)(x_i - p_i) = 0$, $1 \leqslant j \leqslant s$. It will be denoted by $T_{V/\mathbb{A}^r, P}$.

By Leibniz's rule, it is easy to observe that the previous definition depends only on the point $P \in V$ and on the ideal $I_a(V)$, i.e. it does not depend on the choice of generators of $I_a(V)$. When in particular V is an affine subspace of $\mathbb{A}^r$, then at any $P \in V$, we have $T_{V/\mathbb{A}^r, P} = V$.

For $V \subset \mathbb{A}^r$ an affine variety with $I_a(V) := (f_1, \ldots, f_s)$ one considers

$$J := J(f_1, \ldots, f_s) = \left(\frac{\partial f_j}{\partial x_i} \right)_{1 \leqslant i \leqslant r, \, 1 \leqslant j \leqslant s} \tag{12.1}$$

the *Jacobian matrix* of the equations defining V, which is a $s \times r$ matrix with polynomial entries. For any integer $h \in \{0, \ldots, r\}$, let $J_h \subset A^{(r)}$ be the ideal generated by the minors of order $r - h + 1$ of J. One can consider the set $V \cap Z_a(J_h) := \{P \in V \mid \mathrm{rk}(J(P)) \leqslant r - h\}$, i.e.

$$V \cap Z_a(J_h) = \{ P \in V \mid \dim(\mathrm{T}_{V/\mathbb{A}^r, P}) \geqslant h \}, \tag{12.2}$$

where $\dim(\mathrm{T}_{V/\mathbb{A}^r, P})$ denotes the dimension as a (classical) affine subspace, i.e. the dimension of the $\mathbb{K}$-vector space determined by the direction of $\mathrm{T}_{V/\mathbb{A}^r, P}$.

From above, the right-hand side member of (12.2) is in particular a closed subset of V. Thus, $\dim(\mathrm{T}_{V/\mathbb{A}^r, P})$ reaches its minimum over a non-empty, open subset U of V (possibly $U = V$, as we will see below). The open set U consists of points $P \in V$ where the Jacobian matrix evaluated at P, namely $J(P)$, has maximal rank. From now on we will denote such an open set U as $\mathrm{Sm}(V) := U$ according to the following.

Definition 12.1.2. A point $P \in \mathrm{Sm}(V)$ is said to be a *smooth point* (also *simple* or *non-singular point*) for V. Otherwise, $P \in V \setminus \mathrm{Sm}(V)$ is said to be a *singular point* for V.

The non-empty open set $\mathrm{Sm}(V)$ is called the *smooth locus* of V whereas the (possibly empty) proper, closed subset $V \setminus \mathrm{Sm}(V)$ will be denoted by $\mathrm{Sing}(V)$ and called the *singular locus* of V. If $\mathrm{Sing}(V) = \emptyset$, then V is said to be a *smooth affine variety* (or also a *non-singular affine variety*), otherwise V is said to be *singular*.

Example 12.1.3.

(i) Let $V = Z_a(x_2 - x_1^2) \subset \mathbb{A}^2$ be the parabola. Setting $f := x_2 - x_1^2$, its Jacobian matrix is $J = (-2x_1 \ \ 1)$. Note that $Z_a(J_0) = Z_a(J_1) = \mathbb{A}^2$. This implies that $\dim(\mathrm{T}_{V/\mathbb{A}^2, P}) \geqslant 1$ at any point $P \in V$. Since one has $J_2 = (2x_1, 1)$, then $Z_a(J_2) = \emptyset$ so the previous inequality is an equality for any $P \in V$. It follows that V is a smooth irreducible curve and that at any point $P \in V$ one has $\dim(\mathrm{T}_{V/\mathbb{A}^2, P}) = 1 = \dim(V)$. Similarly, one can easily check that the "standard" affine twisted cubic $V = \{(t, t^2, t^3) \mid t \in \mathbb{A}^1\} \subset \mathbb{A}^3$ (as well as more generally the "standard" affine rational normal curve of degree d in

$\mathbb{A}^d$, $V = \{(t, t^2, t^3, \ldots, t^d) \mid t \in \mathbb{A}^1\} \subset \mathbb{A}^d)$ is smooth at any of its points.

(ii) Consider the semi-cubic parabola, i.e. the cuspidal plane cubic $V = Z_a(x_2^2 - x_1^3) \subset \mathbb{A}^2$ (cf., Example 5.3.24). We have $J = (-3x_1^2 \quad 2x_2)$ and, as above, $Z_a(J_0) = Z_a(J_1) = \mathbb{A}^2$ so any $P \in V$ is s.t. $\dim(T_{V/\mathbb{A}^2, P}) \geqslant 1$. For $h = 2$, $J_2 = (-3x_1^2, 2x_2) = (x_1^2, x_2)$ so $V \cap Z_a(J_2) = \{O = (0, 0)\}$. Thus, for any $P \in V \setminus \{O\}$ one has $\dim(T_{V/\mathbb{A}^2, P}) = 1 = \dim(V)$. On the contrary, at the origin one has $\dim(T_{V/\mathbb{A}^2, O}) \geqslant 2$. Since $T_{V/\mathbb{A}^2, O}$ has to be an affine subspace of $\mathbb{A}^2$, we must have $T_{V/\mathbb{A}^2, O} = \mathbb{A}^2$ and the previous inequality is an equality. Thus, V is singular; more precisely $\mathrm{Sing}(V) = \{O = (0, 0)\}$ whereas its smooth locus is $\mathrm{Sm}(V) = V \setminus \{O\}$.

(iii) Identical conclusions as in (ii) hold for the plane nodal cubic $V = Z_a(x_1^3 + x_1^2 - x_2^2) \subset \mathbb{A}^2$ which is monoid of degree 3 in $\mathbb{A}^2$ (cf. Example 8.2.12-(ii)). In this case $J = (3x_1^2 + 2x_1 - 2x_2)$ and once again $V \cap Z_a(J_2) = \{O = (0, 0)\}$ which is the singular locus of V.

(iv) More generally, if V is an (irreducible) hypersurface $V = Z_a(f) \subset \mathbb{A}^r$, the Jacobian matrix is simply $J = (\frac{\partial f}{\partial x_1} \quad \frac{\partial f}{\partial x_2} \quad \cdots \quad \frac{\partial f}{\partial x_r})$. For any $0 \leqslant h \leqslant r - 1$, one has $Z_a(J_h) = \mathbb{A}^r$ therefore any $P \in V$ is such that $\dim(T_{V/\mathbb{A}^r, P}) \geqslant r - 1 = \dim(V)$. The only non-trivial information is contained in $V \cap Z_a(J_r) = \mathrm{Sing}(V)$. In particular, V is smooth if and only if $J(P) \neq (0 \ 0 \ldots \ 0)$ for any $P \in V$. Otherwise any $P \in V \cap Z_a(J_r)$ is a singular point. Note that $\mathrm{Sing}(V) = V \cap Z_a(J_r)$ is defined by the ideal $(f, \frac{\partial f}{\partial x_1}, \frac{\partial f}{\partial x_2}, \ldots, \frac{\partial f}{\partial x_r})$; this confirms that $\mathrm{Sing}(V)$ is a proper, closed subset in V, since $I_a(V) = (f)$ is a principal ideal.

To sum-up, at any smooth point $Q \in \mathrm{Sm}(V)$ one has

$$\dim(T_{V/\mathbb{A}^r, Q}) = r - 1 = \dim(V). \tag{12.3}$$

If otherwise $P \in \mathrm{Sing}(V)$, then $\dim(T_{V/\mathbb{A}^r, P}) = r = \dim(\mathbb{A}^r)$.

(v) One word of warning concerning the singular locus of an affine hypersurface. Recall that $\mathrm{Sing}(V) = V \cap Z_a(J_r)$. Indeed, it can happen that $Z_a(J_r) \neq \emptyset$ even if $V \cap Z_a(J_r) = \emptyset$ and in such a case V is a smooth hypersurface. Consider, e.g. the quadric hypersurface $V = Z_a(x_1^2 + x_2^2 + \cdots + x_r^2 - 1) \subset \mathbb{A}^r$. In this case, $Z_a(J_r) = \{O = (0, 0, \ldots, 0)\}$ but $\mathrm{Sing}(V) = \emptyset$ as $O \notin V$.

Recalling Example 6.5.13, we can give a geometric interpretation of the affine tangent space at a point of an affine variety. Let $P \in V$ and let $\underline{b} := (b_1, \ldots, b_r) \in \mathbb{K}^r$ be a non-zero vector. Consider parametric equations of the line $\ell_{P,\underline{b}}$ passing through P and with direction $\underline{b}$, i.e.

$$\ell_{P,\underline{b}} : \quad x_i = b_i \, t + p_i, \quad 1 \leqslant i \leqslant r, \ t \in \mathbb{K}.$$

Points of $V \cap \ell_{P,\underline{b}}$ corresponds to the solutions of the system of equations $\phi_1(t) = \cdots = \phi_s(t) = 0$, where $\phi_i(t) := f_i(b_1 \, t + p_1, \ldots, b_r \, t + p_r)$, $1 \leqslant i \leqslant s$, and where $I_a(V) = (f_1, \ldots f_s)$ as above. If one takes

$$\phi(t) = \text{g.c.d.}(\phi_1(t), \ldots, \phi_s(t)), \tag{12.4}$$

the points of $\ell_{P,\underline{b}} \cap V$ bijectively correspond to the roots of the polynomial $\phi(t) \in \mathbb{K}[t]$. The root $t = 0$ of $\phi(t)$, corresponding to $P \in \ell_{P,\underline{b}} \cap V$, will occur with a certain multiplicity $\mu \geqslant 1$.

Definition 12.1.4. The *intersection multiplicity* between $\ell := \ell_{P,\underline{b}}$ and V at the point P is the multiplicity μ of the root $t = 0$ of the polynomial $\phi(t)$. It is denoted by $\mu(V, \ell; P)$.

It is easy to check that the previous definition is independent from the chosen parametric equations of $\ell_{P,\underline{b}}$ and from the chosen generators of $I_a(V)$.

Remark 12.1.5.

(i) $\phi(t)$ is identically zero if and only if $\ell_{P,\underline{b}}$ is entirely contained in V; in such a case we will set $\mu(V, \ell; P) = +\infty$.
(ii) If a line ℓ does not pass through P, we will put $\mu(V, \ell; P) = 0$.
(iii) It $t = 0$ is a *simple root* of $\phi(t)$, i.e. it has algebraic multiplicity 1, then $\mu(V, \ell; P) = 1$ and we will say that the intersection at P is *simple* or even *transverse*. When otherwise $\mu(V, \ell; P) \geqslant 2$, we will say that the line has *contact of order greater than* 1 at P with V.

Proposition 12.1.6. *Let $V \subset \mathbb{A}^r$ be an affine variety and let $P \in \text{Sm}(V)$ be a smooth point. The affine tangent space $T_{V/\mathbb{A}^r, P}$ is the union of all the lines in $\mathbb{A}^r$ which have contact of order greater that 1 at P with V.*

Proof. As above, let $I_a(V) = (f_1, \ldots, f_s)$. To simplify notation, let $\underline{x} := (x_1, \ldots x_r)$ be the vector of indeterminates and let $\underline{p} = (p_1, \ldots, p_r)$ be the coordinate vector of the point $P \in V$. For any $j \in \{1, \ldots, s\}$, by Taylor expansion, we set

$$f_j(\underline{x}) = f_j((\underline{x} - \underline{p}) + \underline{p}) = h_j(\underline{x} - \underline{p}) + g_j(\underline{x} - \underline{p}), \qquad (12.5)$$

where h_j is a linear form and where g_j is a polynomial whose homogeneous factors have at least degree two. For any $1 \leqslant j \leqslant s$, one has

$$f_j(\underline{b}\, t + \underline{p}) = h_j(\underline{b}\, t) + g_j(\underline{b}\, t) = t\, h_j(\underline{b}) + t^2\, q_j(t),$$

for suitable $q_j \in \mathbb{K}[t]$. The line $\ell_{P,\underline{b}}$ has contact of order greater than 1 at the point P with V if and only if

$$h_1(\underline{b}) = h_2(\underline{b}) = \cdots = h_s(\underline{b}) = 0. \qquad (12.6)$$

On the other hand, from Taylor's formula applied to (12.5), one has

$$h_j(\underline{x} - \underline{p}) = \sum_{i=1}^{r} \frac{\partial f_j}{\partial x_i}(\underline{p})(x_i - p_i),$$

so (12.6) establishes that $\underline{b}$ belongs to the direction of $T_{V/\mathbb{A}^r,P}$, which is the sub-vector space of $\mathbb{K}^r$ defined by a homogeneous linear system whose coefficient matrix is precisely $J(P)$.   □

12.2   Tangent Space at a Point of a Projective Variety: Smoothness

The case of a projective variety is similar to the affine case; we shall briefly mention to it. Up to isomorphism, let $V \subset \mathbb{P}^r$ be a projective variety and let $P = [p_0, p_1, \ldots, p_r] \in V$ be a point. Let $I_p(V) = (F_1, \ldots, F_s)$ be the homogeneous radical ideal of V.

Definition 12.2.1. The *projective tangent space to V at the point P* is the linear subspace of $\mathbb{P}^r$ defined by the homogeneous linear system $\sum_{i=0}^{r} \frac{\partial F_j}{\partial X_i}(P)X_i = 0$, $1 \leqslant j \leqslant s$. It will be denoted by $T_{V/\mathbb{P}^r,P}$.

As in the affine case, the definition is independent from the choice of the generators of $I_p(V)$ and one can similarly define the set $\mathrm{Sm}(V)$, $\mathrm{Sing}(V)$ as well as the concepts of *smooth projective variety* (or also *non-singular projective variety*) and of *singular projective variety*.

If $p_0 \neq 0$, i.e. if $P \in V \cap U_0 := V_0$, the projective hyperplanes defining $T_{V/\mathbb{P}^r,P}$ are the projective closures of the affine hyperplanes in Definition 12.1.1 defining $T_{V_0/\mathbb{A}^r,P}$. Thus, from Section 3.3.5, the linear projective subspace $T_{V/\mathbb{P}^r,P}$ turns out to be the projective closure of the affine space $T_{V_0/\mathbb{A}^r,P}$. Same conclusion holds for any other non-empty affine chart $V_i = V \cap U_i \neq \emptyset$, $1 \leqslant i \leqslant r$.

Let $L_P \subset \mathbb{P}^r$ be any line passing through the point $P \in V$; let moreover $Q = [q_0, q_1, \ldots, q_r] \neq P$ be any other point of L_P. To simplify notation, we pose $\underline{p} := (p_0, p_1, \ldots, p_r)$, $\underline{q} = (q_0, q_1, \ldots, q_r)$ and $\underline{X} = (X_0, \ldots, X_r)$ the row-vector of homogeneous coordinates in $\mathbb{P}^r$. Thus, parametric equations of L_P are given by $\underline{X} = \lambda \underline{p} + \mu \underline{q}$, $[\lambda, \mu] \in \mathbb{P}^1$. Points of $V \cap L_P$ corresponds to the solutions $[\lambda, \mu] \in \mathbb{P}^1$ of the system of equations $F_1(\lambda \, \underline{p} + \mu \, \underline{q}) = F_2(\lambda \, \underline{p} + \mu \, \underline{q}) = \cdots = F_s(\lambda \, \underline{p} + \mu \, \underline{q}) = 0$. The previous system of equations is equivalent to a unique homogeneous equation $F(\lambda, \mu) = 0$, where

$$F(\lambda, \mu) := \text{g.c.d.}(F_1(\lambda \, \underline{p} + \mu \, \underline{q}), \ldots, F_s(\lambda \, \underline{p} + \mu \, \underline{q})).$$

The multiplicity of the root $[1, 0]$ (corresponding to the point P) will be called *intersection multiplicity* between L_P and V at the point P and will be denoted by $\mu(V, L_P; P)$.

Similarly to the affine case, $\mu(V, L; P) = 0$ if the line L does not pass through P, whereas $\mu(V, L; P) = +\infty$ when the line L passes through P and it is entirely contained in V. Moreover, the definition of $\mu(V, L; P)$ is compatible with that given in Definition 12.1.4, i.e. if $P \in V_0 = V \cap U_0 \neq \emptyset$ and if $\ell := L \cap U_0$, then $\mu(V, L; P) = \mu(V_0, \ell; P)$ and $T_{V/\mathbb{P}^r,P}$ is the union of all projective lines $L \subset \mathbb{P}^r$ having contact of order greater than 1 at the point P with V, i.e. such that $\mu(V, L; P) \geqslant 2$.

Example 12.2.2.

(i) When $V = Z_p(F)$ is an (irreducible) hypersurface in $\mathbb{P}^r$, with $F \in \mathcal{S}_d^{(r)}$ for some positive integer d, its Jacobian matrix is $J = (\frac{\partial F}{\partial X_0} \, \frac{\partial F}{\partial X_1} \, \cdots \, \frac{\partial F}{\partial X_r})$. As in the affine case, for any $0 \leqslant h \leqslant r - 1$, one has $Z_p(J_h) = \mathbb{P}^r$ therefore any $P \in V$ is such that $\dim(T_{V/\mathbb{P}^r,P}) \geqslant r - 1 = \dim(V)$. When $h = r$, J_r is a homogeneous ideal generated by homogeneous polynomials of degree $(d - 1)$ and $\text{Sing}(V)$ is simply given by $Z_p(J_h)$, as it follows from Euler's identity (cf. Proposition 1.10.15-(i)). For any smooth point $P \in \text{Sm}(V)$ one has therefore $\dim(T_{V/\mathbb{P}^r,P}) = r - 1 = \dim(V)$. On the contrary, any $P \in Z_a(J_r)$ belongs to $\text{Sing}(V)$ and $\dim(T_{V/\mathbb{P}^r,P}) = r = \dim(\mathbb{P}^r)$. In particular, V is smooth if and only if $Z_p(J_h) = \emptyset$, i.e. $\sqrt{J_h} = \mathcal{S}_+^{(r)}$.

(ii) For $V = Z_p(F)$ an irreducible hypersurface in $\mathbb{P}^r$ as above, if $L \subset \mathbb{P}^r$ is a line not contained in V one has

$$\sum_{P \in L} \mu(V, L; P) = d, \tag{12.7}$$

since the left-hand side member above equals the sum of the (algebraic) multiplicities of the roots of the homogeneous polynomial $F(\lambda \underline{p} + \mu \underline{q}) \in \mathbb{K}[\lambda, \mu]_d$, for any pair of points $P, Q \in L$.

12.3 Zariski Tangent Space of an Algebraic Variety: Intrinsic Definition of Smoothness

The previous definitions of tangent spaces take into account the embedding of the variety V either in an affine or in a projective space; therefore these definitions are not intrinsic. It is possible to intrinsically associate a $\mathbb{K}$-vector space to any point P of any algebraic variety V in such a way that, under this association, one recovers the affine tangent space in Definition 12.1.1 (viewed as a vector space whose origin is P) when V is affine (cf. also Mumford, 1995, Chapter 1, pp. 3–5).

Take V any algebraic variety and let $P \in V$ be any point. From Theorem 5.3.10 and Example 5.3.22, we know that $(\mathcal{O}_{V,P}, \mathfrak{m}_{V,P})$ is a local ring, with residue field $\mathbb{K}$. To simplify notation, in what follows we will simply denote by $\mathfrak{m}$ the maximal ideal $\mathfrak{m}_{V,P}$. The $\mathcal{O}_{V,P}$-module $\frac{\mathfrak{m}}{\mathfrak{m}^2}$ is annihilated by $\mathfrak{m}$, therefore it is a $\frac{\mathcal{O}_{V,P}}{\mathfrak{m}}$-module, i.e. it is a $\mathbb{K}$-vector space.

Lemma 12.3.1. *For any* $P \in V$, *the* $\mathbb{K}$-*vector space* $\frac{\mathfrak{m}}{\mathfrak{m}^2}$ *is finite dimensional.*

Proof. From Corollary 5.3.20, for any open set $U \subseteq V$ containing P one has an isomorphism of local rings $\mathcal{O}_{U,P} \cong \mathcal{O}_{V,P}$. Thus, by Proposition 6.4.2, with no loss of generality we may assume that V is an affine variety. In particular we can consider $V \subseteq \mathbb{A}^r$, for some integer r, as an irreducible, closed subset. In such a case, from Claim 5.3.15, we have $\mathcal{O}_{V,P} \cong A(V)_{\mathfrak{m}_V(P)}$ and $\mathfrak{m} \cong \mathfrak{m}_V(P)A(V)_{\mathfrak{m}_V(P)}$. Thus, $\frac{\mathfrak{m}}{\mathfrak{m}^2} \cong \frac{\mathfrak{m}_V(P)A(V)_{\mathfrak{m}_V(P)}}{\mathfrak{m}_V(P)^2 A(V)_{\mathfrak{m}_V(P)}}$. On the other hand, from Theorem 5.3.14-(b), $\mathfrak{m}_V(P) \cong I_{a,V}(P) = \frac{\mathfrak{m}_P}{I_a(V)}$, where $\mathfrak{m}_P = (x_1 - p_1, \ldots, x_r - p_r)$ is the maximal ideal in $A^{(r)}$ corresponding to $P \in \mathbb{A}^r$ and $I_a(V) \subseteq A^{(r)}$ the radical ideal of $V \subseteq \mathbb{A}^r$. Therefore, $\dim\left(\frac{\mathfrak{m}}{\mathfrak{m}^2}\right) \leqslant \dim\left(\frac{\mathfrak{m}_P}{\mathfrak{m}_P^2}\right) = r$, as desired. $\square$

Definition 12.3.2. For any point $P \in V$, the $\mathbb{K}$-vector space $\left(\frac{\mathfrak{m}}{\mathfrak{m}^2}\right)^* := \mathrm{Hom}_{\mathbb{K}}\left(\frac{\mathfrak{m}}{\mathfrak{m}^2}, \mathbb{K}\right)$ is called the *Zariski tangent space of V at P* and it is denoted by $T_{V,P}$.

Note that the definition of $T_{V,P}$ is of *local nature*, i.e. it depends only on the local ring $\mathcal{O}_{V,P}$. In particular, for any open neighborhood $U \subseteq V$ containing P one has $T_{V,P} \cong T_{U,P}$ (cf. Shafarevich, 1994, § II.1.3, pp. 86–92).

Definition 12.3.3. A $\mathbb{K}$-linear map $D : \mathcal{O}_{V,P} \to \mathbb{K}$ is called a $\mathbb{K}$-*derivation from $\mathcal{O}_{V,P}$ to $\mathbb{K}$* if, for any $\Phi, \Psi \in \mathcal{O}_{V,P}$ and for any $\lambda \in \mathbb{K}$, one has

$$D(\lambda) = 0 \quad \text{and} \quad D(\Phi\Psi) = \Phi(P)\, D(\Psi) + \Psi(P)\, D(\Phi).$$

The set of all $\mathbb{K}$-derivations from $\mathcal{O}_{V,P}$ to $\mathbb{K}$ determines a sub-vector space of $\mathrm{Hom}_{\mathbb{K}}(\mathcal{O}_{V,P}, \mathbb{K})$, which will be denoted by $\mathrm{Der}_{\mathbb{K}}(\mathcal{O}_{V,P}, \mathbb{K})$.

Proposition 12.3.4.

(i) *For any algebraic variety V and any point $P \in V$, there exists a canonical isomorphism $T_{V,P} \cong \mathrm{Der}_{\mathbb{K}}(\mathcal{O}_{V,P}, \mathbb{K})$ of $\mathbb{K}$-vector spaces.*

(ii) *If $V \subseteq \mathbb{A}^r$ is an affine variety and $P \in V$, there exists a canonical isomorphism of $\mathbb{K}$-vector spaces $T_{V,P} \cong T_{V/\mathbb{A}^r,P}$, where $T_{V/\mathbb{A}^r,P}$ is considered as a $\mathbb{K}$-vector space with origin at the point P.*

Proof.

(i) If $D \in \mathrm{Der}_{\mathbb{K}}(\mathcal{O}_{V,P}, \mathbb{K})$, for any $\Phi, \Psi \in \mathfrak{m}$ we have $D(\Phi\,\Psi) = 0$, i.e. $D|_{\mathfrak{m}^2} = 0$. This implies that D induces a $\mathbb{K}$-linear homomorphism $L_D : \left(\frac{\mathfrak{m}}{\mathfrak{m}^2}\right) \to \mathbb{K}$, i.e. $L_D \in \left(\frac{\mathfrak{m}}{\mathfrak{m}^2}\right)^* = T_{V,P}$.

Conversely, given $L \in T_{V,P}$, define $D_L : \mathcal{O}_{V,P} \to \mathbb{K}$, $\Phi \to L(\overline{\Phi - \Phi(P)})$, where $\overline{\Phi - \Phi(P)} \in \frac{\mathfrak{m}}{\mathfrak{m}^2}$. For any $\lambda \in \mathbb{K}$, one has $D_L(\lambda) = L(\overline{\lambda - \lambda}) = L(0) = 0$; for any $\Phi, \Psi \in \mathcal{O}_{V,P}$, one has

$$D_L(\Phi\,\Psi) = D_L\left((\Phi - \Phi(P) + \Phi(P))\,((\Psi - \Psi(P) + \Psi(P))\right)$$
$$= D_L\left((\Phi - \Phi(P))\,(\Psi - \Psi(P)) + (\Phi - \Phi(P))\,\Psi(P)\right.$$
$$\left. + (\Psi - \Psi(P))\,\Phi(P) + \Phi(P)\,\Psi(P)\right).$$

Since L is $\mathbb{K}$-linear, by its definition D_L is also $\mathbb{K}$-linear. Using that $\Phi(P)\,\Psi(P) \in \mathbb{K}$ and $(\Phi - \Phi(P))\,(\Psi - \Psi(P)) \in \mathfrak{m}^2$, the latter expression equals $D_L \left((\Phi - \Phi(P)) \, \Psi(P) + (\Psi - \Psi(P)) \, \Phi(P) \right)$ which, by definition of D_L, is

$$
L \left(\overline{(\Phi - \Phi(P)) \, \Psi(P) + (\Psi - \Psi(P)) \, \Phi(P)} \right)
$$
$$
= \Psi(P) \, L \left(\overline{\Phi - \Phi(P)} \right) + \Phi(P) \, L \left(\overline{\Psi - \Psi(P)} \right);
$$

this reads $\Psi(P) \quad D_L(\Phi) + \Phi(P) \quad D_L(\Psi)$. In other words, $D_L \in \mathrm{Der}_{\mathbb{K}}(\mathcal{O}_{V,P}, \mathbb{K})$. Since the maps $D \to L_D$ and $L \to D_L$ are one the inverse of the other, we conclude.

(ii) Let $V \subset \mathbb{A}^r$ be an affine variety; from Theorem 5.3.14-(c), for any $P \in V$ one has $\mathcal{O}_{V,P} \cong A(V)_{\mathfrak{m}_V(P)}$. With small abuse of notation, set $x_1, \ldots x_r \in \mathcal{O}_{V,P}$ to be the images of the indeterminates of $A^{(r)}$. For any $D \in \mathrm{Der}_{\mathbb{K}}(\mathcal{O}_{V,P}, \mathbb{K})$, let

$$
\lambda_i = D(x_i), \quad 1 \leqslant i \leqslant r. \tag{12.8}
$$

For any $f \in I_a(V)$, by the $\mathbb{K}$-linearity of D one has $0 = D(f(x_1, \ldots, x_r)) = \Sigma_{i=1}^r \frac{\partial f_j}{\partial x_i}(P)\lambda_i$, i.e. $(\lambda_1, \ldots, \lambda_r) \in \mathbb{K}^r$ lies in the direction of $\mathrm{T}_{V/\mathbb{A}^r, P}$, i.e. $(p_1, \ldots, p_r) + (\lambda_1, \ldots, \lambda_r) \in \mathrm{T}_{V/\mathbb{A}^r, P}$.

Conversely, for $(p_1, \ldots, p_r) + (\lambda_1, \ldots, \lambda_r) \in \mathrm{T}_{V/\mathbb{A}^r, P}$, posing as in (12.8) $D(x_i) := \lambda_i$, we can define a derivation $D \in \mathrm{Der}_{\mathbb{K}}(\mathcal{O}_{V,P}, \mathbb{K})$ as follows:

$$
D(g) := \sum_{i=1}^r \frac{\partial f_j}{\partial x_i}(P)\lambda_i, \quad \forall \, g \in A(V)
$$

and

$$
D \left(\frac{g}{h} \right) = \frac{D(g) \, h(P) - D(h) \, g(P)}{h(P)^2}, \quad \forall \, \frac{g}{h} \in \mathcal{O}_{V,P}.
$$

The map $\mathrm{T}_{V/\mathbb{A}^r, P} \to \mathrm{Der}_{\mathbb{K}}(\mathcal{O}_{V,P}, \mathbb{K})$ just constructed is well-defined, $\mathbb{K}$-linear and it is the inverse of the previous one; so $\mathrm{T}_{V/\mathbb{A}^r, P} \cong \mathrm{Der}_{\mathbb{K}}(\mathcal{O}_{V,P}, \mathbb{K})$. We then conclude by part (i). $\qquad \square$

As in the affine case, a point $P \in V$ is said to be *smooth* (*or simple, or non-singular*) if $\dim(\mathrm{T}_{V,P}) = \min_{Q \in V} \dim(\mathrm{T}_{V,Q})$. Otherwise P is said to be a *singular point*. Applying Proposition 12.3.4-(ii) to any affine open set of an affine open covering of V and using (12.2) for any such affine open set, one deduces that the set of non-singular points in V is a non-empty, dense open set in V which is denoted by $\mathrm{Sm}(V)$. Thus, $\mathrm{Sing}(V) :=$

$V \setminus \mathrm{Sm}(V)$, which is the set of *singular points* of V, is a proper closed subset of V. The algebraic variety V is said to be *singular* (respectively, *smooth*) if $\mathrm{Sing}(V) \neq \emptyset$ (respectively, $\mathrm{Sing}(V) = \emptyset$).

There is another interpretation of the Zariski tangent space at a point of an algebraic variety, which is extremely useful when, e.g. one is concerned with the study of local properties of algebraic varieties, like *Hilbert schemes*, which are also "parameter spaces" for families of algebraic varieties (cf., e.g. Sernesi, 2006), or like *moduli spaces*, parametrizing isomorphism classes of projective varieties (cf., e.g. Caporaso, 2004, for moduli spaces of smooth projective curves, whereas Catanese, 1986, § 19, for moduli spaces of some families of surfaces with given invariants). To discuss this further definition, we first need some algebraic preliminaries.

Let $(A, \mathfrak{m})$ be a local $\mathbb{K}$-algebra with residue field $\mathbb{K}$ (cf. Section 1.11.1). Similarly as above, one observes that $\frac{\mathfrak{m}}{\mathfrak{m}^2}$ is a $\mathbb{K}$-vector space. If furthermore A is Noetherian, then $\frac{\mathfrak{m}}{\mathfrak{m}^2}$ obviously has finite dimension over $\mathbb{K}$. Thus, one can define $\left(\frac{\mathfrak{m}}{\mathfrak{m}^2}\right)^*$ to be the *Zariski tangent space of the local Noetherian* $\mathbb{K}$-*algebra* $(A, \mathfrak{m})$. Recall that, for any algebraic variety V and any $P \in V$ the local $\mathbb{K}$-algebra $\mathcal{O}_{V,P}$ is Noetherian (cf. Proposition 10.2.3); therefore, we can think about $(A, \mathfrak{m})$ in the above definitions as it were the local ring at a point of an algebraic variety. Applying verbatim the same strategy as in the proof of Proposition 12.3.4-(i), one shows that

$$\left(\frac{\mathfrak{m}}{\mathfrak{m}^2}\right)^* \cong \mathrm{Der}_{\mathbb{K}}(A, \mathbb{K}). \tag{12.9}$$

Consider now the polynomial $\mathbb{K}$-algebra $\mathbb{K}[t]$, where t an indeterminate over $\mathbb{K}$, and the quotient ring $\frac{\mathbb{K}[t]}{(t^2)}$ which is called the *ring of dual numbers over* $\mathbb{K}$ and denoted by $\mathbb{K}[\epsilon]$, where ϵ denotes the class of t modulo the ideal (t^2); in particular $\epsilon^2 = 0$ in $\mathbb{K}[\epsilon]$. By its definition, $\mathbb{K}[\epsilon]$ is a finite $\mathbb{K}$-algebra of dimension 2, whose elements are formal expressions $a + b\epsilon$, $a, b \in \mathbb{K}$, with the following operations:

$$(a + b\epsilon) + (c + d\epsilon) := (a + c) + (b + d)\epsilon, \ \lambda(a + b\epsilon) := \lambda a + \lambda b\epsilon$$

$$(a + b\epsilon) \cdot (c + d\epsilon) := ac + (ad + bc)\epsilon, \ \forall \, a, b, c, d, \lambda \in \mathbb{K},$$

where the last equality follows from the fact that $\epsilon^2 = 0$. Furthermore, $\mathbb{K}[\epsilon]$ is a Noetherian local ring, of maximal ideal (ϵ) and residue field $\mathbb{K}$. Let $(A, \mathfrak{m})$ be any Noetherian local $\mathbb{K}$-algebra with residue field $\mathbb{K}$ and let $\phi : A \to \mathbb{K}[\epsilon]$ be any local homomorphism of local rings. Then, for any

$a \in A$, one has

$$\phi(a) := a_0 + d_\phi(a)\epsilon, \qquad (12.10)$$

where $a_0 \in \frac{A}{\mathfrak{m}}$ is the class of a modulo $\mathfrak{m}$ and where $d_\phi(a) \in \mathbb{K}$ is determined by ϕ. Conversely given a $d \in \mathrm{Hom}_{\mathbb{K}}(A, \mathbb{K})$ then it recovers a local homomorphism $\phi = \phi_d$ as in (12.10).

Given $\phi : A \to \mathbb{K}[\epsilon]$ a local homomorphism as above, we want to show that $d := d_\phi$ is a $\mathbb{K}$-derivation from A to $\mathbb{K}$, i.e. an element in $\mathrm{Der}_{\mathbb{K}}(A, \mathbb{K})$. Since ϕ is a local homomorphism of $\mathbb{K}$-algebras, for any $a, b \in A$ and any $\lambda \in \mathbb{K}$, one has

$$a_0 + b_0 + d(a + b)\,\epsilon = \phi(a + b) = \phi(a) + \phi(b) = a_0 + d(a)\,\epsilon + b_0 + d(b)\,\epsilon,$$

$$\lambda\,a_0 + d(\lambda\,a)\epsilon = \phi(\lambda\,a) = \lambda\,\phi(a) = \lambda\,a_0 + \lambda\,d(a)\,\epsilon,$$

$$a_0\,b_0 + d(ab)\epsilon = \phi(a\,b) = \phi(a)\,\phi(b) = (a_0\,b_0) + (a_0\,d(b) + b_0\,d(a))\,\epsilon.$$

Thus, for any $a, b \in A$ and any $\lambda \in \mathbb{K}$, one has

$$d(a + b) = d(a) + d(b), \; d(\lambda\,a) = \lambda\,d(a), \; d(ab) = a_0\,d(b) + b_0\,d(a),$$

which proves that $d = d_\phi$ is a $\mathbb{K}$-derivation. With this set-up and recalling the definition of the vector space of $\mathbb{K}$-algebra homomorphisms as in (6.8), one has the following

Proposition 12.3.5. *Let $(A, \mathfrak{m})$ be a Noetherian local $\mathbb{K}$-algebra with residue field $\mathbb{K}$. Then the Zariski tangent space $\left(\frac{\mathfrak{m}}{\mathfrak{m}^2}\right)^*$ of $(A, \mathfrak{m})$ is isomorphic to the vector space of $\mathbb{K}$-algebra homomorphisms $\mathrm{Hom}_{\mathbb{K}}(A,\ \mathbb{K}[\epsilon])$. In particular, if V is an algebraic variety and if $P \in V$ is any point, then the Zariski tangent space $\mathrm{T}_{V,P}$ of V at P is isomorphic to $\mathrm{Hom}_{\mathbb{K}}(\mathcal{O}_{V,P}, \mathbb{K}[\epsilon])$.*

Proof. From (12.9), we know that $\left(\frac{\mathfrak{m}}{\mathfrak{m}^2}\right)^* \cong \mathrm{Der}_{\mathbb{K}}(A, \mathbb{K})$. Previous computation shows that $\mathrm{Der}_{\mathbb{K}}(A, \mathbb{K}) \cong \mathrm{Hom}_{\mathbb{K}}(A,\ \mathbb{K}[\epsilon])$. The last part of the statement follows by replacing A with $\mathcal{O}_{V,P}$. $\qquad \square$

Due to the local nature of the Zariski tangent space at a point, we describe here how this local definition reflects on morphisms among algebraic varieties.

Definition 12.3.6. Let V and W be algebraic varieties, $P \in V$ be a point and $\varphi : V \to W$ be a morphism. The local $\mathbb{K}$-algebra homomorphism $\varphi_P : \mathcal{O}_{W,\varphi(P)} \to \mathcal{O}_{V,P}$ induces a homomorphism of $\mathbb{K}$-vector spaces, denoted by $d\varphi_P : \mathrm{T}_{V,P} \longrightarrow \mathrm{T}_{W,\varphi(P)}$, which is called the *differential of φ at the point P.*

If $\varphi : V \to W$ and $\psi : W \to Z$ are morphisms, for any point $P \in V$ one has

$$d(\psi \circ \varphi)_P = (d\psi_{\varphi(P)}) \circ (d\varphi_P) \quad \text{and} \quad (d\,\mathrm{Id}_V)_P = \mathrm{Id}_{T_{V,P}}. \tag{12.11}$$

Corollary 12.3.7. *If $\varphi : V \to W$ is an isomorphism of algebraic varieties then, for any point $P \in V$, $d\varphi_P$ is an isomorphism of $\mathbb{K}$-vector spaces.*

Proof. It immediately follows from (12.11). □

Using differentials, Theorem 12.3.9 gives an extra *alternative definition* for the "dimension" of an algebraic variety, with no use of either function or Krull-dimension or topological dimension. We first need an auxiliary result.

Proposition 12.3.8. *Any algebraic variety V is birationally equivalent to a hypersurface in some affine or projective space.*

Proof. It suffices to prove, e.g. the second part of statement. Let $\dim(V) = v$ and let $\{s_1, \ldots, s_v\}$ be a transcendence basis of $K(V)$ over $\mathbb{K}$. Since $\mathrm{char}(\mathbb{K}) = 0$, by the *Primitive Element Theorem* (cf., e.g. Lang, 2002, Theorem 4.6, p. 243), there exists $\theta \in \mathbb{K}(V)$ such that $K(V) \cong \mathbb{K}(s_1, \ldots, s_v, \theta)$. By the assumption on $\{s_1, \ldots, s_v\}$, the elements $s_1, \ldots, s_v, \theta \in K(V)$ are algebraically dependent over $\mathbb{K}$. There exists therefore a non-zero polynomial $f \in \mathbb{K}[x_1, \ldots, x_v, x_{v+1}]$ (we may assume to be irreducible) s.t. $f(s_1, \ldots, s_v, \theta) = 0$. Since $\{s_1, \ldots, s_v\}$ is a transcendence basis of $K(V)$ over $\mathbb{K}$, f is not constant with respect to the indeterminates x_{v+1}.

Let $W := Z_a(f) \subset \mathbb{A}^{v+1}$. Then the affine coordinate ring of W is an integral $\mathbb{K}$-algebra of the form $A(W) = \frac{A^{(v+1)}}{I_a(W)} = \frac{\mathbb{K}[x_1, x_2, \ldots, x_{v+1}]}{I_a(W)}$, where x_i's indeterminates. Since $I_a(W) = (f)$ is principal, as in the proof of Proposition 10.1.5, the elements $\overline{x}_1, \ldots, \overline{x}_v \in A(W) \subset Q(A(W)) \cong K(W)$ are algebraically independent, where $\overline{x}_i$ denotes the image of the indeterminate x_i, $1 \leqslant i \leqslant v + 1$. Therefore, $K(W) \cong \mathbb{K}(\overline{x}_1, \ldots, \overline{x}_v, \overline{x}_{v+1}) \cong \mathbb{K}(\overline{x}_1, \ldots, \overline{x}_v)[\overline{x}_{v+1}]$, where the second isomorphism follows from the fact that $\overline{x}_{v+1}$ is algebraic over $\mathbb{K}(\overline{x}_1, \ldots, \overline{x}_v)$. If we let y to be an indeterminate, then $\mathbb{K}(\overline{x}_1, \ldots, \overline{x}_v)[\overline{x}_{v+1}] \cong \frac{\mathbb{K}(\overline{x}_1, \ldots, \overline{x}_v)[y]}{(f(\overline{x}_1, \ldots, \overline{x}_v, y))}$. Since $\{s_1, \ldots, s_v\}$ is a transcendence basis over $\mathbb{K}$, then one has $\frac{\mathbb{K}(\overline{x}_1, \ldots, \overline{x}_v)[y]}{(f(\overline{x}_1, \ldots, \overline{x}_v, y))} \cong \frac{\mathbb{K}(s_1, \ldots, s_v)[y]}{(f(s_1, \ldots, s_v, y))}$.

By the definition of θ, $\frac{\mathbb{K}(s_1, \ldots, s_v)[y]}{(f(s_1, \ldots, s_v, y))} \cong \mathbb{K}(s_1, \ldots, s_v)[\theta] \cong \mathbb{K}(s_1, \ldots, s_v, \theta) \cong K(V)$. To sum up, $K(V) \cong K(W)$ and one concludes by using Corollary 8.1.7-(i). □

We remark that the assumption $char(\mathbb{K}) = 0$ can be replaced by the weaker assumption that $\mathbb{K}$ is a *perfect field* (see, e.g. Dolgachev, 2013, Remark 4.11, p. 27).

Theorem 12.3.9. *Let V be an algebraic variety and let $P \in \mathrm{Sm}(V)$. Then*

$$\dim(V) = \dim(T_{V,P}). \tag{12.12}$$

Proof. Note first that if V and W are birational algebraic varieties then equality (12.12) holds true for V if and only if it holds true for W: indeed there exist non-empty, open sets $U_V \subseteq V$ and $U_W \subseteq W$ which contain smooth points of V and W, respectively, and which are isomorphic so we can we apply Corollary 12.3.7. Moreover, from (12.3), equality (12.12) holds true for V a hypersurface in $\mathbb{A}^r$. Thus, from Proposition 12.3.8, any algebraic variety V is birationally equivalent to an affine hypersurface in some affine space for which the equality (12.12) holds true, therefore the proof is complete. $\square$

The above results show that the concept of Zariski tangent space at a point P of an algebraic variety V deals with the local ring $\mathcal{O}_{V,P}$. It is natural to expect that the same occurs for the concept of smoothness; this is what we actually show below. Let $(A, \mathfrak{m})$ be any Noetherian local $\mathbb{K}$-algebra with residue field $\mathbb{K}$. It is a standard result in Commutative Algebra (cf., e.g. Matsumura, 1980, p. 78) that

$$\dim\left(\frac{\mathfrak{m}}{\mathfrak{m}^2}\right) \geqslant K \dim(A), \tag{12.13}$$

where the dimension on the left is meant as $\mathbb{K}$-vector space whereas that on the right is the Krull-dimension of the ring. The local ring $(A, \mathfrak{m})$ is said to be *regular local ring* if and only if equality in (12.13) holds.

Theorem 12.3.10. *Let V be an algebraic variety and let $P \in V$ be a point. Then P is a smooth point of V if and only if $\mathcal{O}_{V,P}$ is a regular local $\mathbb{K}$-algebra.*

Proof. From Proposition 6.4.2, there always exists an affine open neighborhood $U \subset V$ of the point P; since $\mathcal{O}_{V,P} \cong \mathcal{O}_{U,P}$ then, with no loss of generality, we may assume V to be an affine variety. Moreover, up to an isomorphism, we may assume that $V \subset \mathbb{A}^r$, for some non-negative integer r. Let therefore $P = (p_1, p_2, \ldots, p_r) \in \mathbb{A}^r$ and let $\mathfrak{m}_P = (x_1 - p_1, x_2 - p_2, \ldots, x_r - p_r) \subset A^{(r)}$ be its corresponding maximal ideal.

Consider the $\mathbb{K}$-linear map $A^{(r)} \xrightarrow{\theta_P} \mathbb{K}^r$, defined as $f \xrightarrow{\theta_P} J(P)^t$, where $J = J(f)$ is the Jacobian row-matrix of f as in (12.1). Note that $\theta_P(x_i - p_i)$, $1 \leqslant i \leqslant r$, determine the canonical basis of $\mathbb{K}^r$ and that $\theta_P(\mathfrak{m}_P^2) = 0$. Therefore, θ_P induces a $\mathbb{K}$-linear map $\widehat{\theta}_P : \frac{\mathfrak{m}_P}{\mathfrak{m}_P^2} \to \mathbb{K}^r$.

Let now $I_a(V) = (f_1, \ldots, f_s) \subset A^{(r)}$ and let $J = J(f_1, \ldots, f_s)$ be the Jacobian matrix of the equations defining V as in (12.1). From the previous definition, one has $\dim(\theta_P(I_a(V))) = \dim(\theta_P(f_1, \ldots, f_s)) = \operatorname{rank}(J(P))$. Since $P \in V$, then $I_a(V) \subseteq \mathfrak{m}_P$ and moreover

$$\dim\left(\theta_P\left(I_a(V)\right)\right) = \dim\left(\widehat{\theta}_P\left(\frac{\mathfrak{m}_P^2 + I_a(V)}{\mathfrak{m}_P^2}\right)\right),$$

i.e. $\dim\left(\widehat{\theta}_P\left(\frac{\mathfrak{m}_P^2 + I_a(V)}{\mathfrak{m}_P^2}\right)\right) = \operatorname{rank}(J(P))$. Let now $\mathfrak{m}$ denote the maximal ideal of the local ring $\mathcal{O}_{V,P}$. We claim that there exists an isomorphism

$$\frac{\mathfrak{m}}{\mathfrak{m}^2} \cong \frac{\mathfrak{m}_P}{\mathfrak{m}_P^2 + I_a(V)}. \tag{12.14}$$

If (12.14) holds true then, from (12.13), one has

$$K\dim\left(\mathcal{O}_{V,P}\right) \leqslant \dim\left(\frac{\mathfrak{m}}{\mathfrak{m}^2}\right) = \dim\left(\frac{\mathfrak{m}_p}{\mathfrak{m}_P^2 + I_a(V)}\right)$$

$$= \dim\left(\frac{\mathfrak{m}_p}{\mathfrak{m}_P^2}\right) - \dim\left(\frac{\mathfrak{m}_P^2 + I_a(V)}{\mathfrak{m}_P^2}\right) = r - \operatorname{rank}(J(P)).$$

Therefore, we have $\operatorname{rank}(J(P)) \leqslant r - K\dim(\mathcal{O}_{V,P}) = r - \dim(V)$, where the second equality in the previous formula follows from Proposition 10.2.3. Now, from (12.12), the first inequality is an equality if and only if $P \in \operatorname{Sm}(V)$ equivalently if and only if $\mathcal{O}_{V,P}$ is a regular local ring.

We are therefore left to showing the existence of an isomorphism as in (12.14). Since V is affine, by Theorem 5.3.14-(c), we have $\mathcal{O}_{V,P} \cong A(V)_{\mathfrak{m}_V(P)}$, where $\mathfrak{m}_V(P) = \frac{\mathfrak{m}_P}{I_a(V)}$ the maximal ideal in $A(V) \cong \frac{A^{(r)}}{I_a(V)}$. Thus, $\mathfrak{m} = \mathfrak{m}_V(P) \cdot A(V)_{\mathfrak{m}_V(P)}$ and $\mathfrak{m}^2 = \mathfrak{m}_V^2(P) \cdot A(V)_{\mathfrak{m}_V(P)}$, so

$$\frac{\mathfrak{m}}{\mathfrak{m}^2} \cong \frac{\mathfrak{m}_V(P) \cdot A(V)_{\mathfrak{m}_V(P)}}{\mathfrak{m}_V^2(P) \cdot A(V)_{\mathfrak{m}_V(P)}} \cong \frac{\mathfrak{m}_V(P)}{\mathfrak{m}_V^2(P)} \cong \frac{\mathfrak{m}_P}{\mathfrak{m}_P^2 + I_a(V)},$$

which completes the proof. $\qquad\square$

Exercises

Exercise 12.1. Let $C \subset \mathbb{A}^3$ be the (reducible) affine curve given by $C = Z_a(x_1 x_2, x_1 x_3, x_2 x_3)$. Show that C cannot be isomorphic to any affine plane curve.

Exercise 12.2. Let $C := Z_p(X_0 X_1 - X_2^2) \subset \mathbb{P}^2$ which is a smooth projective conic. Set $F := X_0 X_1 - X_2^2 \in \mathcal{S}^{(2)}$ and consider the morphism

$$\varphi : C \to \mathbb{P}^2, \ P \xrightarrow{\ \varphi\ } \left[\frac{\partial F}{\partial X_0}(P), \ \frac{\partial F}{\partial X_1}(P), \ \frac{\partial F}{\partial X_2}(P) \right].$$

Then $\varphi(C) \subset \mathbb{P}^2$ is said to be the *dual curve* of C. Show that $\varphi(C)$ is a smooth curve. Compute its degree.

Exercise 12.3. Show that any projective monoid $Z \subset \mathbb{P}^n$ of degree d as in (8.5) has a point of multiplicity $d - 1$ at $P_0 = [1, 0, \ldots, 0]$, i.e. for any line $L \subset \mathbb{P}^n$ passing through P_0 one has $\mu(Z, \ell; P_0) \geqslant d - 1$.

Exercise 12.4. A hypersurface $Z \subset \mathbb{P}^n$ of degree d is said to be a *cone* if it has at least a point of multiplicity d. The set of points in Z of multiplicity d is called the *vertex* of the cone and denoted by $v(Z)$. Show that $v(Z)$ is a linear subspace of $\mathbb{P}^n$.

Exercise 12.5. If $Z \subset \mathbb{P}^n$ is a cone of degree d, show that Z is the union of lines passing through $P \in v(Z)$, for any point $P \in v(Z)$.

Solutions to Exercises

Here, we report solutions of all the exercises proposed at the end of each chapter.

Solution 1.1

(i) Assume that $I = I' \cap I''$. Note that $I \subseteq I'$ and $I \subseteq I''$. If by contradiction there existed $f' \in I' \setminus I$ and $f'' \in I'' \setminus I$, then $f' \cdot f'' \in I' \cdot I'' \subseteq I' \cap I'' = I$ which would contradict that I is a prime ideal. Therefore, either $I = I'$ or $I = I''$.

(ii) Any $a \in \sqrt{(0)}$ is such that $a^n = 0$, for some positive integer n. Thus, for any $\mathfrak{p} \in \operatorname{Spec}(R)$, $a \notin R \setminus \mathfrak{p}$ since $R \setminus \mathfrak{p}$ is a multiplicative system. This implies that $a \in \mathfrak{p}$, for any $\mathfrak{p} \in \operatorname{Spec}(R)$, namely $\sqrt{(0)} \subseteq \cap_{\mathfrak{p} \in \operatorname{Spec}(R)} \mathfrak{p}$. To prove the other inclusion, take any $f \in R \setminus \sqrt{(0)}$; we need to show that there exists a prime ideal $\mathfrak{p}$ such that $f \notin \mathfrak{p}$. To prove this note that, since f is not nilpotent, then $S := \{1, f, f^2, f^3, \ldots\}$ is a multiplicative system in R. From Corollary 1.11.8-(iii), any prime ideal $\mathfrak{q} \subset R_S$ is of the form $\mathfrak{q} = \mathfrak{p} \cdot R_S$, for some $\mathfrak{p} \in \operatorname{Spec}(R)$ such that $\mathfrak{p} \cap S = \emptyset$ and we are done. The previous arguments show that $\sqrt{(0)} = \cap_{\mathfrak{p} \in \operatorname{Spec}(R)} \mathfrak{p}$. Since the intersection of a family of ideals is an ideal, in particular one has also proved that $\sqrt{(0)}$ is an ideal in R.

(iii) For $a \in \sqrt{(0)}$ let $k := \operatorname{ord}(a)$, i.e. the smallest integer for which $a^k = 0$. Then, for any $u \in U(R)$ the element

$$b := u^{-1} - au^{-2} + a^2 u^{-3} - a^3 u^{-4} + \cdots + (-1)^{k-1} a^{k-1} u^{k-1} \in R$$

is such that $b = (u + a)^{-1}$ in R, namely $u + a \in U(R)$.

Solution 1.2

The fact that $(I : J)$ is an ideal is straightforward. Now, for any $i \in I$ and any $j \in J$, $i \cdot j \in I$ since I is an ideal; thus $i \in (I : J)$, i.e. $I \subseteq (I : J)$. Furthermore, $(I : J) \cdot J \ni r \cdot j \in I$, i.e. $(I : J) \cdot J \subseteq I$. Finally, for any $f \in \sqrt{I \cap J}$, there exists $m \in \mathbb{N}$ such that $f^m \in I \cap J \subseteq \sqrt{I} \cap \sqrt{J}$. Thus, there exist $k_1, k_2 \in \mathbb{N}$ such that $f^{mk_1} \in I$ and $f^{mk_2} \in J$. Therefore, $f \in \sqrt{I}$ and $f \in \sqrt{J}$, i.e. $f \in \sqrt{I} \cap \sqrt{J}$, so $\sqrt{I \cap J} \subseteq \sqrt{I} \cap \sqrt{J}$. To prove the other inclusion, note that for any $f \in \sqrt{I} \cap \sqrt{J}$ there exist $k_1, k_2 \in \mathbb{N}$ such that $f^{k_1} \in I$ and $f^{k_2} \in J$. Thus $f^{k_1+k_2} \in I \cap J$ and so $f \in \sqrt{I \cap J}$. This shows that $\sqrt{I} \cap \sqrt{J} \subseteq \sqrt{I \cap J}$.

Solution 1.3

(i) Note that $f(x) \in R[x]$ is invertible if and only if there exists $g(x) = b_0 + b_1 x + \cdots + b_m x^m \in R[x]$ such that

$$a_0 b_0 = 1, \ a_0 b_1 = -a_1 b_0, \ldots, a_0 b_k = -(b_{k-1} a_1 + \cdots + b_0 a_k), \ 2 \leqslant k \leqslant n.$$

Thus, $a_0 \in U(R)$ and $b_0 = a_0^{-1} \in U(R)$; moreover

$$(*) \quad b_1 = -a_1 a_0^{-2}, \ b_2 = a_1^2 a_0 - 2 - a_2 a_0^{-1}, \ldots.$$

Therefore, given $f(x) = a_0 + a_1 x + \cdots + a_n x^n \in R[X]$ such that $a_0 \in U(R)$ and $a_1, \ldots, a_n$ are nilpotent elements in R, then $g(x) \in R[x]$ such that $f(x) \cdot g(x) = 1$ can be explicitly and uniquely determined as a solution of the system of equations $(*)$ which, by the nilpotency of the element a_i, $1 \leqslant i \leqslant n$, has finitely many equations. Conversely, let $f(x) \in U(R[x])$. From above, one has $a_0 \in U(R)$. If $\deg(f(x)) = n$, then for $n = 0$ we are done. Assume therefore $n \geqslant 1$ and use induction on n. If $g(x) = b_0 + b_1 x + \cdots + b_m x^m \in R[x]$ such that $f(x) \cdot g(x) = 1$, in particular $a_n b_m = 0$. Moreover, multiplying $f(x) \cdot g(x) = 1$ by $f(x)^r$, one gets $f(x)^{r+1} \cdot g(x) = f(x)^r$ so $a_n^{r+1} b_{m-r} = 0$. As b_0 is a unit, then $a_n^{m+1} = 0$, i.e. a_n is nilpotent in R. This implies that $a_n x^n \in R[x]$ is nilpotent. Let

$$f(x) - a_n x^n = a_0 + a_1 x + \cdots + a_{n-1} x^{n-1}.$$

Since $f(x)$ is invertible and $a_n x^n$ is nilpotent, from Exercise 1.1-(iii), $f(x) - a_n x^n$ is invertible in $R[x]$ of degree less than n; by induction, $a_0 \in U(R)$ and $a_1, \ldots, a_{n-1}$ are nilpotent elements in R.

(ii) Since $f(x) \in R[x]$ is nilpotent and $1 \in U(R[x])$, then $1 + f(x)$ is invertible in $R[x]$ as it follows from Exercise 1.1-(iii). Therefore, by (i), $a_1, \ldots, a_n \in R$ are nilpotent elements whereas $1 + a_0 \in U(R)$. On the other hand, for t large enough, one has $f^t = 0$ thus in particular $a_0^t = 0$, i.e. a_0 is nilpotent in R. If conversely, $a_0, a_1, \ldots, a_n \in R$ are nilpotent elements, let $k_j := \mathrm{ord}(a_j)$ for $0 \leqslant j \leqslant n$. Set $\overline{k} := \Sigma_{j=0}^n k_j$, so that one gets $f(x)^{\overline{k}} = 0$. Indeed, $f(x)^{\overline{k}}$ is a linear combination with integral coefficients of products of the form $a_0^{r_0} a_1^{r_1} \cdots a_n^{r_n} x^{k_s}$ such that $\Sigma_{j=0}^n r_j = \overline{k}$, for each k_s. Since we cannot simultaneously have $r_j < k_j$, each of these products is zero.

(iii) $f(x) \in R[x]$ is a zero-divisor if there exists $g(x) \in R[x] \setminus \{0\}$ such that $f(x) \cdot g(x) = 0$. Take $g(x)$ of smallest degree w.r.t. this property, say $g(x) = b_0 + b_1 x + \cdots, b_m x^m$, where we may assume $b_0 \neq 0$. One has in particular $a_n \cdot b_m = 0$. Since $\deg(a_n g(x)) < \deg(g(x))$, by minimality of $g(x)$ it follows that $a_n \cdot g(x) = 0$ in $R[x]$. In particular $a_n \cdot b_0 = 0$, with $b_0 \neq 0$. Similarly, $a_{n-t} \cdot g(x) = 0$, for $1 \leqslant t \leqslant n$, so $a_{n-t} \cdot b_0 = 0$, for any $1 \leqslant t \leqslant n$, with $b_0 \neq 0$. Therefore, it suffices to considering $r := b_0 \in R \setminus \{0\}$.

Solution 1.4

For any $b \in R \setminus \{0\}$, there exists $b^{-1} \in \mathbb{K}$. We must show that b^{-1} is in R. Since $\mathbb{K}$ is a finite R-algebra and $R \subseteq R[b^{-1}] \subseteq \mathbb{K}$, then $R[b^{-1}]$ is a finite R-algebra. From Proposition 1.6.3, $b^{-1} \in \mathbb{K}$ is integral over R. Thus, there exist $r_0, \ldots, r_{n-1} \in R$ such that

$$b^{-n} + r_{n-1} b^{1-n} + \cdots + r_0 = 0.$$

Multiplying by b^{n-1} the previous equality gives

$$b^{-1} + r_{n-1} + \cdots + r_0 b^{n-1} = 0,$$

i.e. $b^{-1} = -\Sigma_{j=0}^{n-1} r_j b^{n-1-j} \in R$.

Solution 1.5

Note that for any a, $b \in \mathbb{Z}$, one has

$$\mathcal{S}_a \cdot \mathcal{M}(h)_b \cong \mathcal{S}_a \cdot \mathcal{M}_{b+h} \subseteq \mathcal{M}_{(a+b)+h} \cong \mathcal{M}(h)_{a+b}.$$

Therefore, $\mathcal{M}(h)$ is a graded $\mathcal{S}$-module. Moreover, $\mathcal{M}$ and $\mathcal{M}(h)$ are obviously isomorphic as $\mathcal{S}$-modules since they have the same elements but the graduation is simply shifted by h. On the other hand, $\mathcal{M}(h)_d \cong \mathcal{M}_{d+h}$ is not isomorphic to $\mathcal{M}_d$, if, e.g. $\mathcal{S} = \mathcal{M} = \mathbb{K}[X_0, \ldots, X_n] = \bigoplus_{d \geqslant 0} \mathbb{K}[X_0, \ldots, X_n]_d$.

Solution 2.1

First, consider the pencil of lines $\ell_t : x_2 - ix_1 - t = 0$ in $\mathbb{A}_{\mathbb{C}}^2$, where $t = t_1 + it_2 \in \mathbb{C}$. With the use of this pencil, one defines a bijective map $\psi : \mathbb{C} \setminus \{0\} \to Z_a(f) \subset \mathbb{A}_{\mathbb{C}}^2$, where

$$\psi(t) := \left(x_1(t) = -\frac{t^2 + 1}{2it}, \ x_2(t) = \frac{t^2 - 1}{2t} \right),$$

which is a (complex) *rational parametrization* of the conic $Z_a(f)$. Now, let us set-theoretically identify the affine plane $\mathbb{A}_{\mathbb{C}}^2$ with $\mathbb{R}^4$ via

$$x_1(t) := a(t_1, t_2) + ib(t_1, t_2) \ \text{ and } \ x_2(t) := c(t_1, t_2) + id(t_1, t_2), \quad i^2 = -1,$$

where $a(t_1, t_2), b(t_1, t_2), c(t_1, t_2), d(t_1, t_2) \in \mathbb{R}$. The equation defining $Z_a(f)$ reads in the given coordinates of $\mathbb{R}^4$ as

$$\begin{cases} a^2 + c^2 - b^2 - d^2 + 1 = 0 \\ ab + cd = 0. \end{cases}$$

The map ψ, given by the pencil of complex lines ℓ_t, can be considered as a map

$$\psi^* : \mathbb{R}^2 \setminus \{(0,0)\} \to Z_a(f) \subset \mathbb{R}^4.$$

Considering Euclidean topologies both in the domain and the target of the previous map, one observes that ψ^* is a homeomorphism.

Consider now the Euclidean topological space $\mathbb{R}^3$, with coordinates (t_1, t_2, t_3), in such a way that $\mathbb{R}^2$ above is identified with the coordinate plane $t_3 = 0$. In $\mathbb{R}^3$ consider the two-sphere Σ, centered at the point $C = (0, 0, \frac{1}{2})$ with radius $r = \frac{1}{2}$. Its North-pole is the point $N = (0, 0, 1)$ whereas its South-pole is $O = (0, 0, 0)$. Let $\pi : \Sigma \setminus \{N\} \to \mathbb{A}_{\mathbb{R}}^2$ be the *stereographic projection* of Σ from its North-pole N, which is well-known to be a homeomorphism (in the Euclidean topologies) between $\Sigma \setminus \{N\}$ and the coordinate plane $t_3 = 0$; its restriction $\pi_|$ to $\Sigma \setminus \{N, O\}$ is therefore a homeomorphism onto $\mathbb{R}^2 \setminus \{(0,0)\}$. Thus the map

$$\psi^* \circ \pi_| : \Sigma \setminus \{N, O\} \to Z_a(f) \subset \mathbb{R}^4$$

is the desired homeomorphism (in Euclidean topologies).

It is clear that Σ is homeomorphic, in the Euclidean topology of $\mathbb{R}^3$, to the unit 2-sphere $S^2 \subset \mathbb{R}^3$. This, endowed with a suitable two-chart atlas (cf., e.g. Miranda, 1995, Example 1.13, p. 3), has a natural structure of *complex* compact manifold of (complex) dimension one and it is called the *Riemann sphere* (cf. Miranda, 1995, Example 1.20, p. 4). The latter is homeomorphic to the projective complex line $\mathbb{P}^1_{\mathbb{C}}$ (see Miranda, 1995, Problem I.8.C, p. 12), thus $Z_a(f)$ is homeomorphic, in the Euclidean topology, to $\mathbb{P}^1_{\mathbb{C}}$ minus two points.

Solution 2.2

Note that $Y \subseteq Z_a(x_2 - x_1^2)$. On the other hand, any $P \in Z_a(x_2 - x_1^2)$ is of the form $P = (p_1, p_2)$ where $p_2 = p_1^2$, i.e. $P = (p_1, p_1^2)$, $p_1 \in \mathbb{K}$ varying. Thus, $Y = Z_a(x_2 - x_1^2)$ so it is an AAS. One has $I_a(Y) = (x_2 - x_1^2)$ is a principal ideal; moreover I is a prime ideal since $\frac{A^{(2)}}{I} \cong \mathbb{K}[x_1]$ is an integral domain.

Solution 2.3

(i) Consider the $\mathbb{R}$-algebra epimorphism

$$\mathbb{R}[x_1, x_2, x_3] \xrightarrow{\varphi} \mathbb{R}[t], \ \varphi(x_1) = t, \ \varphi(x_2) = t^2, \ \varphi(x_3) = t^3.$$

Since $\frac{\mathbb{R}[x_1,x_2,x_3]}{\mathrm{Ker}(\varphi)} \cong \mathbb{R}[t]$ is an integral domain, then $\mathrm{Ker}(\varphi)$ is a prime ideal; moreover $I \subseteq \mathrm{Ker}(\varphi)$. On the other hand, $g(x_1, x_2, x_3) \in \mathrm{Ker}(\varphi)$ if and only if $g(t, t^2, t^3) = 0$, for any $t \in \mathbb{R}$, i.e. $\mathrm{Ker}(\varphi) \subseteq I$. Therefore, $\frac{\mathbb{R}[x_1,x_2,x_3]}{I}$ is an integral domain.

(ii) Since $x_1(x_3 - x_1 x_2) - x_2(x_2 - x_1^2) = x_1 x_3 - x_2^2$ then $x_1 x_3 - x_2^2 \in I$.

(iii) Note that $Z_a(I) \subseteq Z_a(J)$. On the other hand, the line $Z_a(x_1, x_2)$ is contained in $Z_a(J)$ but not in $Z_a(I)$. Therefore, $Z_a(I) \subsetneq Z_a(J)$. For the last part, observe that in $\frac{\mathbb{R}[x_1,x_2,x_3]}{J}$ one has $x_2 = x_1^2$ and $x_1 x_3 = x_2^2$, in particular $x_1 x_3 = x_1^4$, namely $x_1(x_3 - x_1^3) = 0$ in $\frac{\mathbb{R}[x_1,x_2,x_3]}{J}$ even if $x_1 \neq 0$ and $x_3 - x_1^3 \neq 0$ in $\frac{\mathbb{R}[x_1,x_2,x_3]}{J}$. Therefore, the quotient ring cannot be integral.

Solution 2.4

(i) $I_a(Y_k) = I_a(Y') \cap I_a(Y_k'') = (x_1 x_2, \ x_1 x_3, \ x_2 x_3 - k x_2, \ x_3^2 - k x_3)$.

(ii) Note that, for $k \neq 0$, $Y' \cap Y_k'' = \emptyset$. On the other hand, $Y' \cap Y_0'' = \{O = (0, 0, 0)\}$. As for the ideals, I_k is radical by definition, for any

$k \neq 0$. On the other hand, I_k is not a prime ideal since $x_1, x_2 \notin I_k$ but $x_1 x_1 \in I_k$, for any $k \neq 0$.

(iii) By definition of I_0, one has $I_0 = (x_1 x_2, \ x_1 x_3, \ x_2 x_3, \ x_3^2)$. Note that $x_3 \in \sqrt{I_0}$, since $x_3^2 \in I_0$ but $x_3 \notin I_0$, since its generators are all quadratic polynomials. Therefore, I_0 is not radical, in particular it cannot be a prime ideal, and moreover $x_3 \in \sqrt{I_0} \setminus I_0$.

Solution 2.5

(i) Note that $f = x_1^2 + x_2^2 + 2x_1$ and $g := x_1^2 + x_2^2 - 2x_1$. One has $Z_a(J) = Z_a(\mathfrak{m}_O) = \{O = (0,0)\}$, where $\mathfrak{m}_O$ is a maximal ideal (cf. Proposition 2.1.11-(iv)).

(ii) and (iii) Note that $J \subset \mathfrak{m}_O$, where the inclusion is proper, since $x_2 \notin J$. On the other hand, $x_1 = \frac{1}{4}(f - g) \in J$ and $x_2^2 = f - x_1(x_1 + 2) \in J$, i.e. $\mathfrak{m}_O \subseteq \sqrt{J}$ and so $\mathfrak{m}_O = \sqrt{J}$, by the maximality of $\mathfrak{m}_O$ and the fact that $1 \notin \sqrt{J}$. Therefore J is not radical so $\frac{\mathbb{R}[x_1,x_2]}{J}$ is not a reduced ring, since x_2 is nilpotent in $\frac{\mathbb{R}[x_1,x_2]}{J}$. The geometric counterpart of the previous ideal analysis is that $Z_a(f)$ and $Z_a(g)$ are two real conics passing through O and sharing the same *tangent line* at O, which is the x_2-axis. The intersection at O between the two conics is said to be not *transverse*. The radical ideal replaces the system of equations of O cut-out by the to tangential conics with the two coordinate axes.

Solution 3.1

There exist V and W, $\mathbb{K}$-vector spaces both of dimension $n + 1$, such that $\mathbb{P}_1 = \mathbb{P}(V)$ and $\mathbb{P}_2 = \mathbb{P}(W)$. Set $P_i := [\underline{v}_i]$ and $Q_i := [\underline{w}_i]$. Since the points are in general position, then $\underline{v}_0, \ldots, \underline{v}_n$ is a basis for V and $\underline{w}_0, \ldots, \underline{w}_n$ is a basis for W. For any choice of $\lambda_0, \ldots, \lambda_n \in \mathbb{K} \setminus \{0\}$ there always exists an isomorphism $\psi : V \to W$, such that $\psi(\underline{v}_i) = \lambda_i \underline{w}_i$, $0 \leqslant i \leqslant n$. The problem is to determine suitable $\lambda_0, \ldots, \lambda_n \in \mathbb{K} \setminus \{0\}$ such that there also exists $\lambda_{n+1} \in \mathbb{K} \setminus \{0\}$ for which $\psi(\underline{v}_i) = \lambda_i \underline{w}_i$, $0 \leqslant i \leqslant n + 1$ and moreover $(\lambda_0, \ldots, \lambda_n, \lambda_{n+1}) \in (\mathbb{K} \setminus \{0\})^{n+2}$ is uniquely determined up to a constant multiplication factor. By the assumptions, $\underline{v}_{n+1}$ (respectively, $\underline{w}_{n+1}$) linearly depends on $\underline{v}_0, \ldots, \underline{v}_n$ (respectively, on $\underline{w}_0, \ldots, \underline{w}_n$). Therefore, there exist unique $a_0, \ldots, a_n \in \mathbb{K}$ and $b_0, \ldots, b_n \in \mathbb{K}$, such that

$$\underline{v}_{n+1} = \sum_{i=0}^{n} a_i \underline{v}_i \quad \text{and} \quad \underline{w}_{n+1} = \sum_{i=0}^{n} b_i \underline{w}_i.$$

Since the points are in general position, $a_i \neq 0$ (respectively, $b_i \neq 0$) for any $0 \leqslant i \leqslant n$. On the one hand, we have

$$\psi(\underline{v}_{n+1}) = \lambda_{n+1}\underline{w}_{n+1} = \lambda_{n+1}\left(\sum_{i=0}^{n} b_i\underline{w}_i\right),$$

on the other

$$\psi(\underline{v}_{n+1}) = \psi\left(\sum_{i=0}^{n} a_i\underline{v}_i\right) = \sum_{i=0}^{n} \lambda_i a_i\underline{w}_i.$$

Since $\underline{w}_0, \ldots, \underline{w}_n$ is a basis for W, the previous equality gives

$$\begin{cases} a_0\lambda_0 - b_0\lambda_{n+1} = 0 \\ a_1\lambda_1 - b_1\lambda_{n+1} = 0 \\ \qquad \cdots \qquad \qquad \cdots \\ \qquad \cdots \qquad \qquad \cdots \\ a_n\lambda_n - b_n\lambda_{n+1} = 0 \end{cases}$$

By standard Linear Algebra, the previous homogeneous linear system in the indeterminates $(\lambda_0, \ldots, \lambda_{n+1})$ is compatible with solutions which depend on a free parameter in $\mathbb{K}$.

Solution 3.2

One has $Z = Z_p(\mathcal{I})$, for some homogeneous ideal $\mathcal{I} \subset \mathcal{S}^{(2)}$. By the "Hilbert basis theorem", one has $\mathcal{I} = (F_1, \ldots, F_m)$, where F_i homogeneous polynomials, $1 \leqslant i \leqslant m$, since $\mathcal{I}$ is a homogeneous ideal (cf. Proposition 1.10.7). If $m = 1$, Z is a projective curve and we are done in this case. If $m = 2$ and F_1, F_2 have a non-constant greatest common divisor $G \in H(\mathcal{S}^{(2)})$, then $F_i = GA_i$, for some $A_i \in H(\mathcal{S}^{(2)})$, $1 \leqslant i \leqslant 2$. Then $Z_p(\mathcal{I}) = Z_p(G) \cup Z_p(A_1, A_2)$, where $Z_p(G)$ is a projective curve, as above, and where A_1 and A_2 have no non-constant common factor. Therefore, we can reduce to F_1 and F_2 with no non-constant common factor. In such a case, a necessary condition for a point $P = [p_0, p_1, p_2] \in \mathbb{P}^2$ to belong to $Z_p(F_1, F_2)$ is that $[p_0, p_1] \in Z_p(R_{X_2}(F_1, F_2))$, where $R_{X_2}(F_1, F_2)$ is the resultant polynomial of F_1 and F_2 as in Theorem 1.10.18. By the assumption on F_1 and F_2, this is a homogeneous polynomial of degree $\deg(F_1)\deg(F_2)$ in the indeterminates X_0 and X_1. By Proposition 1.10.19, one has only finitely many choices for $[p_0, p_1]$. Applying the same procedure also with respect to the other indeterminates, one deduces that $Z_p(F_1, F_2)$ consists of at most finitely many points. Recursively applying the previous arguments, one proves the general case for any possible m.

Solution 3.3

(i) For any α, $\beta \in \mathbb{K}$ and any f, $g \in \mathrm{Ann}(W)$ one has

$$(\alpha f + \beta g)(\underline{w}) = \alpha f(\underline{w}) + \beta g(\underline{w}) = 0,$$

so $\mathrm{Ann}(W)$ is a $\mathbb{K}$-subvector space of V^*. Let $\mathcal{B}_W = \{\underline{w}_1, \ldots, \underline{w}_{m+1}\}$ be a basis of W and let $\mathcal{B}_V = \{\underline{w}_1, \ldots, \underline{w}_{m+1}, \underline{v}_{m+2}, \ldots, \underline{v}_{n+1}\}$ a basis of V which extends $\mathcal{B}_W$. Let $\mathcal{B}_V^* = \{\underline{w}_1^*, \ldots, \underline{w}_{m+1}^*, \underline{v}_{m+2}^*, \ldots, \underline{v}_{n+1}^*\}$ be the dual basis of $\mathcal{B}_V$. Thus, $V^* = \mathrm{Span}(\mathcal{B}_V^*)$ and any $f \in V^*$ can be written as $f = \Sigma_{i=1}^{m+1} a_i \underline{w}_i^* + \Sigma_{j=m+2}^{n+1} a_j \underline{v}_j^*$, $a_i, a_j \in \mathbb{K}$. Note that $f \in \mathrm{Ann}(W)$ if and only if $f(\underline{w}_i) = 0$, for any $1 \leqslant i \leqslant m+1$, i.e. if and only if $a_1 = \cdots = a_{m+1} = 0$. Therefore, $\mathrm{Ann}(W) = \mathrm{Span}(\underline{v}_{m+2}^*, \ldots, \underline{v}_{n+1}^*)$ so $\dim(\mathrm{Ann}(W)) = n - m = \dim(V) - \dim(W)$.

(ii) $\Lambda^\perp = \{[L] \in \mathbb{P}(V)^* \mid \Lambda \subseteq Z_p(L)\}$, where $\Lambda = \mathbb{P}(W)$. Thus, $[L] \in \Lambda^\perp$ if and only if, for any $t \in \mathbb{K}$, $tL \in \mathrm{Ann}(W)$, namely $\Lambda^\perp = \mathbb{P}(\mathrm{Ann}(W))$.

Solution 3.4

(i) In the affine chart U_0 take affine coordinates $x_1 = \frac{X_1}{X_0}$ and $x_2 = \frac{X_2}{X_0}$, so that the trace $Z_0 := Z \cap U_0$ of Z in U_0 is simply $Z_a(x_1^2 + x_2^2 - x_1)$ which is a real ellipse since the intersection with the line at infinity of U_0 is empty. Let us denote by $z_0 = \frac{X_0}{X_1}$ and $z_2 = \frac{X_2}{X_1}$ the affine coordinates in the affine chart U_1; then the trace $Z_1 := Z \cap U_1$ in this chart is the affine conic given by $Z_a(1 + z_2^2 - z_0)$ which is a parabola, since Z is tangent to the line $X_1 = 0$. Finally, in the affine chart U_2, with affine coordinates $w_0 = \frac{X_0}{X_2}$ and $w_1 = \frac{X_1}{X_2}$, Z cuts out the affine conic $Z_2 = Z \cap U_2 = Z_a(w_1 + 1 - w_0 w_1)$ which is a hyperbola, being the line $X_2 = 0$ a secant line to the conic Z.

(ii) The equation of $C_a(Z)$ in $\mathbb{A}^3$, with affine coordinates X_0, X_1, X_2, is given by $Z_a(X_1^2 + X_2^2 - X_0 X_1)$ and it is a cone with vertex at the origin $O = (0,0,0) \in \mathbb{A}^3$ which projects, at its plane at infinity, the conic $Z \subset \mathbb{P}^2$. In particular $C_a(Z)_\infty = Z$. Similarly the equation of $C_p(Z)$ in $\mathbb{P}^3$, with homogeneous coordinates $[X_0, X_1, X_2, X_3]$, is once again given by $Z_p(X_1^2 + X_2^2 - X_0 X_1) \subset \mathbb{P}^3$. This is a cone with vertex at the fundamental point $P_3 = [0,0,0,1] \in \mathbb{P}^3$ which projects at the fundamental plane $H_3 = Z_p(X_3) \subset \mathbb{P}^3$ the conic $Z = Z_p(X_3, X_1^2 + X_2^2 - X_0 X_1)$.

Solution 3.5

Note that for any hypersurface $Z_p(F) \subset \mathbb{P}^n$, one has

$$Z_p(F) = Z_p(X_0 F, X_1 F, \ldots, X_n F).$$

If $I_p(X) = (F_1, \ldots, F_m)$, set $\overline{d} := \max_{1 \leqslant j \leqslant m} \{d_j\}$, where $d_j := \deg(F_j)$, $1 \leqslant j \leqslant m$. Thus,

$$X = Z_p(I_p(X)) = Z_p(X_0^{\overline{d}-d_1} F_1, \ldots, X_n^{\overline{d}-d_1} F_1, \ldots, X_0^{\overline{d}-d_m} F_m, \ldots, X_n^{\overline{d}-d_m} F_m)$$

and the latter are homogeneous polynomials all of the same degree $\overline{d}$.

Solution 4.1

Consider $f(x_1, x_2) := x_1^4 - 2x_1^2 + x_2^2 + 1 \in \mathbb{R}[x_1, x_2]$. Since $\mathbb{R}[x_1, x_2] \subset \mathbb{C}[x_1, x_2]$, note that in $\mathbb{C}[x_1, x_2]$ the polynomial f splits as $f = (x_1^2 - 1 + ix_2)(x_1^2 - 1 - ix_2)$, where each factor is irreducible in $\mathbb{C}[x_1, x_2]$. Indeed the corresponding AAS's, say $X_1 := Z_a(x_1^2 - 1 + ix_2)$ and $X_2 := Z_a(x_1^2 - 1 - ix_2)$, are both parabolas in $\mathbb{A}^2_\mathbb{C}$, for which $A(X_1) \cong \mathbb{C}[t] \cong A(X_2)$, where t is an indeterminate. Since $\mathbb{C}[x_1, x_2]$ is a UFD, factorization is uniquely determined up to invertible elements. Therefore, as an element of $\mathbb{R}[x_1, x_2]$, the polynomial f is irreducible. On the other hand, $Z_a(f) \subset \mathbb{A}^2_\mathbb{R}$ consists of (real) points given by $X_1 \cap X_2$, namely $Z_a(f) = \{P_1 = (1, 0), P_2 = (-1, 0)\}$ which is therefore reducible.

Solution 4.2

Note that $x_3(x_1 x_3 - x_2^2) - x_2(x_1^3 - x_2 x_3) = x_1(x_3^2 - x_1^2 x_2)$ which therefore belongs to J. Nonetheless $x_1 \notin J$ as well as $x_3^2 - x_1^2 x_2 \notin J$, therefore, J is not a prime ideal. Let us consider now the ideals

$$J_1 := (x_1 x_3 - x_2^2, x_1^3 - x_2 x_3, x_1) \quad \text{and} \quad J_2 := (x_1 x_3 - x_2^2, x_1^3 - x_2 x_3, x_3^2 - x_1^2 x_2).$$

Since $J \subseteq J_1$ and $J \subseteq J_2$, by reversing inclusion one has $Z_a(J_1) \subseteq Z_a(J)$ and $Z_a(J_2) \subseteq Z_a(J)$, so $Z_a(J_1) \cup Z_a(J_2) \subseteq Z_a(J)$. Note that $J_1 = (x_2^2, x_2 x_3, x_1)$ so $Z_a(J_1)$ is the line $\ell = Z_a(x_1, x_2) \subset \mathbb{A}^3$, which is irreducible being homeomorphic to $\mathbb{A}^1$. As for J_2, consider the map $\varphi : \mathbb{A}^1 \to \mathbb{A}^3$ given by $t \xrightarrow{\varphi} (t^3, t^4, t^5)$ and let $C := \operatorname{Im}(\varphi) \subset \mathbb{A}^3$. The map φ gives rise to the $\mathbb{K}$-algebra homomorphism

$$\varphi^* : \mathbb{K}[x_1, x_2, x_3] \to \mathbb{K}[t], \quad \varphi^*(x_1) = t^3, \quad \varphi^*(x_2) = t^4, \quad \varphi^*(x_3) = t^5.$$

Denote by $A(C) := \frac{\mathbb{K}[x_1, x_2, x_3]}{\mathrm{Ker}(\varphi)}$; since $A(C)$ is a subring of $\mathbb{K}[t]$ then $A(C)$ is an integral domain, i.e. the ideal $\mathrm{Ker}(\varphi)$ is a prime ideal. Moreover, by definition of φ^*, one has that $J_2 \subseteq \mathrm{Ker}(\varphi)$. To prove the other inclusion, since any $f \in \mathrm{Ker}(\varphi)$ is such that $f(t^3, t^4, t^5) = 0$, for any $t \in \mathbb{K}$, one easily observes that necessarily $f \in J_2$. Therefore, $J_2 = \mathrm{Ker}(\varphi)$ so $C = Z_a(J_2)$ is irreducible.

We are left to showing that $Z_a(J) = \ell \cup C$; if this is the case, C and ℓ will be the irreducible components of $Z_a(J)$. To prove this, from the generators of J note that, when $x_1 = 0$ then one finds the line ℓ. When otherwise $x_1 \neq 0$ then also $x_2 x_3 \neq 0$; therefore on $Z_a(J) \setminus Z_a(x_1, J)$ we can set $t := \frac{x_2}{x_1} = \frac{x_3}{x_2}$, therefore $t^2 = \frac{x_2}{x_1} \cdot \frac{x_3}{x_2} = \frac{x_3}{x_1}$, thus, $t^4 = \left(\frac{x_3}{x_1}\right)^2 = x_2$ since on $Z_a(J)$ one has $x_3^2 = x_1^2 x_2$; similarly $t^3 = \frac{t^4}{t} = \frac{x_2}{x_2/x_1} = x_1$ and moreover, from $x_2^2 = x_1 x_3$, we have $x_3 \frac{x_2^2}{x_1} = \frac{t^8}{t^3} = t^5$, which gives C as above.

Solution 4.3

Assume by contradiction that $Y = Y_1 \cup Y_2$ is reducible in $\mathbb{P}^n_{\mathbb{K}}$, where each Y_i is a closed proper subset of Y, $1 \leqslant i \leqslant 2$. By induced topology, there exist closed subsets $K_i \subseteq \mathbb{P}^n_{\mathbb{K}}$, such that $Y_i = Y \cap K_i$, $1 \leqslant i \leqslant 2$. Since $Y \subset X$, then $X_i := X \cap K_i \neq \emptyset$ is a closed subset of X such that $Y_i = X_i \cap Y$, $1 \leqslant i \leqslant 2$. Thus, Y would be reducible in X, a contradiction.

Furthermore, since X is locally closed in $\mathbb{P}^n_{\mathbb{K}}$, there exist an open set U of $\mathbb{P}^n_{\mathbb{K}}$ and a closed subset K of $\mathbb{P}^n_{\mathbb{K}}$ such that $X = U \cap K$. Similarly, Y is locally closed in X so there exist an open set U_X of X and a closed subset K_X of X such that $Y = U_X \cap K_X$. By induced topology, there exist an open set U' and a closed set K' of $\mathbb{P}^n_{\mathbb{K}}$ such that $U_X = U' \cap X$ and $K_X = K' \cap X$. Therefore,

$$Y = (U' \cap (U \cap K)) \cap (K' \cap (U \cap K)) = (U \cap U') \cap (K \cap K'),$$

where $U \cap U'$ is an open set of $\mathbb{P}^n_{\mathbb{K}}$ whereas $K \cap K'$ is a closed subset of $\mathbb{P}^n_{\mathbb{K}}$, so Y is locally closed in $\mathbb{P}^n_{\mathbb{K}}$.

Solution 4.4

Let $A := \frac{A^{(3)}}{I}$ denote the quotient ring. Any element of A is the residue, modulo I, of an element of the form $a(x_3) + b(x_3)x_1 + c(x_3)x_2 + d(x_3)x_1 x_2$ in $A^{(3)}$, where $a(x_3)$, $b(x_3)$, $c(x_3)$, $d(x_3) \in \mathbb{K}[x_3]$. Define the $\mathbb{K}$-algebra homomorphism

$$\alpha : A^{(3)} \to \mathbb{K}[t], \ \alpha(x_1) = t^9, \ \alpha(x_2) = t^6, \ \alpha(x_3) = t^4.$$

Its image gives rise to a subring $R := \mathrm{Im}(\alpha) \subset \mathbb{K}[t]$ and $\mathrm{Ker}(\alpha)$ is such that $I \subseteq \mathrm{Ker}(\alpha)$. Moreover, the morphism α induces a $\mathbb{K}$-algebra homomorphism $\varphi : A \to R$ for which $\varphi(f) = 0$ if and only if $f = 0$ in A. This implies that φ is injective; thus φ is an isomorphism, so $I = \mathrm{Ker}(\alpha)$. Since $R \subset \mathbb{K}[t]$, then R is an integral domain which implies that $I = \mathrm{Ker}(\varphi)$ is a prime ideal in $A^{(3)}$. Thus, $Z_a(I)$ is irreducible and I is radical, i.e. $I_a(Z_a(I)) = I$.

Solution 4.5

(i) Let $\mathrm{Sym}(3 \times 3; \mathbb{K})$ be the $\mathbb{K}$-vector space of symmetric matrices of order 3 with entries from $\mathbb{K}$. If $\underline{X}$ denotes the column vector of homogeneous coordinates of $\mathbb{P}^2$, then any $\Gamma \in \mathcal{C}$ has a cartesian equation of the form $\underline{X}^t A \underline{X} = 0$, for some $A \in \mathrm{Sym}(3 \times 3; \mathbb{K})$. Two different cartesian equations of a given conic $\Gamma \in \mathcal{C}$, say $\underline{X}^t A \underline{X} = 0$ and $\underline{X}^t A' \underline{X} = 0$, are such that $A' = \lambda A$, for some $\lambda \in \mathbb{K} \setminus \{0\}$. Then $\mathcal{C}$ can be identified with $\mathbb{P}(\mathrm{Sym}(3 \times 3; \mathbb{K}))$, where any $\Gamma \in \mathcal{C}$ is identified to the class of proportionality of matrices $[A]$ defining equivalent cartesian equations for Γ. Since $\dim_{\mathbb{K}}(\mathrm{Sym}(3 \times 3; \mathbb{K})) = 6$, then $\mathcal{C}$ is a projective space of dimension 5.

To fix natural homogeneous coordinates for $\mathbb{P}(\mathrm{Sym}(3 \times 3; \mathbb{K}))$, which therefore give isomorphism with the numerical projective space $\mathbb{P}_{\mathbb{K}}^5$, take $A := (a_{ij})$, $0 \leqslant i \leqslant j \leqslant 2$, a symmetric matrix whose entries are indeterminates over $\mathbb{K}$; homogeneous coordinates for $\mathbb{P}(\mathrm{Sym}(3 \times 3; \mathbb{K}))$ are given by the proportionality class $[\ldots, a_{ij}, \ldots]$ of the entries of A, set with lexicographic order on the indexes $0 \leqslant i \leqslant j \leqslant 2$.

(ii) Note that $\Gamma \in \mathcal{D}$ if and only if its (and so all) cartesian equation is determined by $[A] \in \mathbb{P}(\mathrm{Sym}(3 \times 3; \mathbb{K}))$ such that $\det(A) = 0$. Note that $\det(A) = \det(a_{ij})$ is a cubic polynomial in the entries a_{ij}'s of the symmetric matrix A; moreover if $A' = \lambda A$ then $\det(A') = \lambda^3 \det(A) = 0$. Therefore, $\mathcal{D}$ corresponds to the cubic hypersurface $\Sigma_3 := Z_p(\det(A)) \subset \mathbb{P}_{\mathbb{K}}^5$.

(iii) Let $S^{(5)} = \mathbb{K}[\ldots, a_{ij}, \ldots]$ be the homogeneous coordinate ring of $\mathbb{P}_{\mathbb{K}}^5$; since $F := \det(A) \in S_3^{(5)}$ then F is certainly reduced. We need to show that F is irreducible. Note that

$$F := \det(A) = a_{00}a_{11}a_{22} + 2a_{01}a_{02}a_{12} - a_{02}^2 a_{11} - a_{00}a_{12}^2 - a_{01}^2 a_{22}.$$

Take $\Lambda := Z_p(a_{11}, a_{02})$, which is a linear subspace of $\mathbb{P}_{\mathbb{K}}^5$ of dimension 3, i.e. $\Lambda \cong \mathbb{P}_{\mathbb{K}}^3$; note that $\Sigma_3 \cap \Lambda = Z_p(F, a_{11}, a_{02}) = Z_p(a_{00}a_{12}^2 + a_{01}^2 a_{22}, a_{11}, a_{02})$. Taking homogeneous coordinates $[a_{00}, a_{01}, a_{12}, a_{22}] = [Z_0, Z_1, Z_2, Z_3]$ on Λ, then $\Sigma_\Lambda := \Sigma_3 \cap \Lambda$ is a cubic hypersurface of Λ given by $\Sigma_\Lambda = Z_p(Z_0 Z_2^2 + Z_3 Z_1^2)$. Considering the affine open set $U_0 \subset \mathbb{P}_{\mathbb{K}}^3 \cong \Lambda$, which is the complement of the hyperplane $H_0 := Z_p(Z_0)$, then $\Sigma_\Lambda \cap U_0$ is the image in $\mathbb{A}_{\mathbb{K}}^3 \cong U_0$ of the continuous map

$$\mathbb{A}_{\mathbb{K}}^2 \to \Sigma_\Lambda \cap U_0 \subset U_0 \cong \mathbb{A}_{\mathbb{K}}^3, \quad (u, v) \to (u, uv, -v^2),$$

which implies that $\Sigma_\Lambda \cap U_0$ is irreducible, and so Σ_Λ is irreducible too. This implies that Σ_3 is irreducible: otherwise if it were $\Sigma_3 = H \cup Q$, where H a hyperplane and Q a (possibly reducible) hyperquadric in $\mathbb{P}_{\mathbb{K}}^5$, any linear section of Σ_3 would contain a linear subspace, contradicting the irreducibility of Σ_Λ.

Solution 5.1

The given map ϕ_1 is a homeomorphism, since one has

$$
\begin{array}{ccc}
\mathbb{A}^1 \supset W & \xrightarrow{\phi_1} & Y \quad \subset \mathbb{A}^2 \\
t & \to & (t, \frac{1}{t}) \\
x_1 & \leftarrow & (x_1, x_2).
\end{array}
$$

Moreover, since Y is an affine variety, then

$$\mathcal{O}_Y(Y) = A(Y) = \frac{\mathbb{K}[x_1, x_2]}{(x_1 x_2 - 1)} = \mathbb{K}[x_1, x_1^{-1}] \cong A_{x_1}^{(1)}.$$

Note that $A_{x_1}^{(1)}$ can be identified with a subring of $\mathcal{O}_W(W)$, since $\frac{1}{x_1} \in \mathcal{O}_W(W)$ and moreover $\mathbb{K}[x_1] = A(\mathbb{A}^1) \subset \mathcal{O}_W(W)$. Therefore, $\mathcal{O}_Y(Y) \subseteq \mathcal{O}_W(W)$ (in Example 6.2.5-(iv), we will more precisely show that $\mathcal{O}_Y(Y) \cong \mathcal{O}_W(W)$). As for function fields, one has $K(W) = K(\mathbb{A}^1)$, since W a principal open set of $\mathbb{A}^1$ (cf. Lemma 5.3.13). Thus, Y and W are not only homeomorphic as topological spaces, but moreover the function rings $\mathcal{O}_Y(Y) = A(Y)$ and $K(Y)$ are isomorphic to $\mathcal{O}_W(W)$ and $K(W)$, respectively (in Remark 6.1.4 we will show that ϕ_1 is indeed much more than a homeomorphism between topological spaces; it is actually an *isomorphism of algebraic varieties*). The map ϕ_1 is a *rational parametrization* of the hyperbola Y whereas ϕ_1^{-1} is nothing but the restriction to Y of the first projection of $\mathbb{A}^2$, i.e. the projection of the affine plane onto its x_1-axis.

The map ϕ_1 is the restriction of a natural projective homeomorphism. Indeed, identifying $\mathbb{A}^2$ with the affine chart U_0 of $\mathbb{P}^2$, the projective closure $\overline{Y}$ of Y is given by $Z_p(X_1 X_2 - X_0^2)$. The x_1-axis becomes the projective line $Z_p(X_2)$, which is homeomorphic to $\mathbb{P}^1$. In this way, one has the homeomorphism

$$\begin{array}{ccc}
\mathbb{P}^1 & \xrightarrow{\Phi_1} & \overline{Y} \quad \subset \mathbb{P}^2 \\
[\lambda, \mu] & \to & [\lambda\mu, \mu^2, \lambda^2] \\
[a_0, a_1] & \leftarrow & [a_0, a_1, a_2],
\end{array}$$

where $[a_0, a_1] = [\lambda\mu, \mu^2] = [\lambda, \mu]$ as it occurs on points of $Y \subset U_0$ (cf. Example 8.1.11 for more general motivations). The map Φ_1^{-1} naturally restricts to ϕ_1^{-1} and moreover it sends the two points at infinity of Y, i.e. $Q_1 = [0, 0, 1]$ and $Q_2 = [0, 1, 0]$, respectively to $[1, 0] \in \mathbb{A}^1 \setminus W$ and to $[0, 1]$, the point at infinity of $\mathbb{A}^1$. In the next chapter, we will show that Φ_1 is an *isomorphism of projective varieties*, which ensures as above that regular and rational functions on $\overline{Y}$ will be the same as those on $\mathbb{P}^1$. One can easily give a geometric interpretation of the map Φ_1^{-1}, observing that this is obtained by simply projecting $\overline{Y}$ to the line $Z_p(X_2) \subset \mathbb{P}^2$ from the point $Q_1 = [0, 0, 1] \in \overline{Y}$. Consider indeed the pencil of lines in $\mathbb{P}^2$ through the point Q_1; equation for this pencil is given by

$$Z_p(\lambda X_1 - \mu X_0) \subset \mathbb{P}^2,$$

for $[\lambda, \mu] \in \mathbb{P}^1$. For any point $P \in Y$, there is only one line of the pencil passing through P and conversely, since $\overline{Y}$ is a conic, points $P \in Y$ are in bijective correspondence with the subset of lines in the pencil consisting of all lines except for that passing through Q_2 and for the tangent line to $\overline{Y}$ at Q_1. The line of the pencil passing through Q_2 is given by $Z_p(X_0)$, whose trace over the X_1-axis in $\mathbb{P}^2$ is the point $[0, 1, 0]$, identified with $[0, 1] \in \mathbb{P}^1$ via Φ_1^{-1}, i.e. Q_2 corresponds via this projection to the point at infinity of the originary $\mathbb{A}^1$. Finally, the tangent line to $\overline{Y}$ at Q_1 is $Z_p(X_1)$, which intersects $\overline{Y}$ only at the point Q_1 with multiplicity 2. The trace of this tangent line over the X_1-axis is the point $[0, 0, 1]$, identified with $[1, 0] \in \mathbb{P}^1$ via Φ_1^{-1}, i.e. Q_1 corresponds via the projection to the origin $O \in \mathbb{A}^1 \setminus W$, where the originary parametrization ϕ_1 was not defined.

Solution 5.2

The map ϕ_2 is a homeomorphism, since one has

$$
\begin{aligned}
\mathbb{A}^1 &\xrightarrow{\phi_2} V \quad \subset \mathbb{A}^2 \\
t &\to (t, t^2) \\
x_1 &\leftarrow (x_1, x_2).
\end{aligned}
$$

Thus, V is homeomorphic to $\mathbb{A}^1$ (ϕ_2 will be more precisely an isomorphism of algebraic varieties, cf. Remark 6.1.4). The map ϕ_2 is a *polynomial parametrization* of V (recall (3.31)) and ϕ_2^{-1} is the restriction to V of projection of $\mathbb{A}^2$ onto the first axis x_1. As for the hyperbola, ϕ_2 is the restriction of the projective homeomorphism

$$
\begin{aligned}
\mathbb{P}^1 &\xrightarrow{\Phi_2} \overline{V} \quad \subset \mathbb{P}^2 \\
[\lambda, \mu] &\to [\lambda^2, \lambda\mu, \mu^2] \\
[a_0, a_1] &\leftarrow [a_0, a_1, a_2],
\end{aligned}
$$

where $\overline{V} \subset \mathbb{P}^2$ the projective closure of V and where $[a_0, a_1] = [\lambda^2, \lambda\mu] = [\lambda, \mu]$, as it occurs for points on V. The map Φ_2^{-1} naturally restricts to ϕ_2^{-1} and, similarly as for the hyperbola, it is nothing but the projection of $\overline{V}$ to $\mathbb{P}^1$ as one can show by using the pencil of lines through $[0, 0, 1]$ which is the point at infinity of V, i.e. where $\overline{V}$ is tangent to the line $Z_p(X_0)$. As for the hyperbola above, Φ_2 will be an isomorphism of projective varieties (cf. Chapter 6).

Solution 5.3

Since V is an affine rational curve with polynomial parametrization, then $V = Z_a(x_2 - x_1^2, x_3 - x_1^3, \ldots, x_d - x_1^d)$ as in (3.31). From Lemma 3.3.17, one has $I_a(V) = (x_2 - x_1^2, x_3 - x_1^3, \ldots, x_d - x_1^d)$, which is also prime, i.e. V is an affine variety in $\mathbb{A}^d$. Since V is affine, then $\mathcal{O}_V(V) = A(V) \cong \frac{A^{(d)}}{I_a(V)}$. From the generators of $I_a(V)$, one deduces that $A(V) \cong \mathbb{K}[t] = A^{(1)}$. Therefore, $K(V) = \mathcal{Q}(A(V)) \cong \mathbb{K}(t) = \mathcal{Q}^{(1)}$. Thus, $K(V)$ is a purely transcendental field extension of $\mathbb{K}$, with transcendence degree 1.

Solution 5.4

Note that

$$
Z_a(x_2) \cap V = Z_a(x_2, x_1 x_4) = Z_a(x_2, x_1) \cup Z_a(x_2, x_4) = \pi_1 \cup \pi_2,
$$

where π_1 and π_2 are planes contained in V. Therefore, if we set $U_1 := V \setminus (\pi_1 \cup \pi_2)$, then $U_1 \subset V$ is a dense open subset of V and certainly $f \in \mathcal{O}_V(U_1)$. On the other hand, from the equation defining V, one has $\frac{x_1}{x_2} = \frac{x_3}{x_4}$. As above,

$$Z_a(x_4) \cap V = Z_a(x_2 x_3, x_4) = Z_a(x_2, x_4) \cup Z_a(x_3, x_4) = \pi_2 \cup \pi_3,$$

where π_3 is another plane contained in V. The open set $U_2 := V \setminus (\pi_2 \cup \pi_3)$ is dense in V, $\frac{x_3}{x_4} \in \mathcal{O}_V(U_2)$, moreover $U_1 \cap U_2 \neq \emptyset$ and

$$\rho_{U_1 \cap U_2}^{U_1} \left(\frac{x_1}{x_2} \right) = \rho_{U_1 \cap U_2}^{U_2} \left(\frac{x_3}{x_4} \right).$$

Therefore,

$$U_1 \cup U_2 = V \setminus \pi_2, \quad \Phi = [U_1, f] \in \mathcal{O}_V(U_1 \cup U_2)$$

in particular $\mathrm{Dom}(f)$ contains $V \setminus \pi_2$. We want to show that this is an equality. Assume by contradiction that $V \setminus \pi_2 \subsetneq U' \subseteq \mathrm{Dom}(f)$ and let $\frac{g}{h} \in K(V)$ which extends on U' as a regular function the given function f. With no loss of generality, we may assume that g and h have no common non-constant irreducible factor. Thus, on U_1 one must have $\frac{g}{h} = \frac{x_1}{x_2}$, namely $x_2 g - x_1 h = 0$ on U_1 and so on V, as U_1 is dense in V. Since x_1 does not divide x_2 and x_2 does not divide x_1 in $\mathrm{A}(V)$, then x_2 divides h and x_1 divides g, therefore $\frac{g}{h} = \frac{x_1 g'}{x_2 h'}$ where x_2 does not divide g', otherwise g and h would have x_2 as a common irreducible factor, a contradiction. Thus, $\frac{g}{h}$ cannot be defined on an open set U' such that $V \setminus \pi_2 \subsetneq U'$. Therefore, one can conclude that $\mathrm{Dom}(f) = V \setminus \pi_2$.

Solution 5.5

Let $P_0 = [1, 0, 0, 0] \in \mathbb{P}^3$ be the fundamental point. Note that $P_0 \in Z$ and the projection of Z from P_0 gives the map

$$\pi_{P_0} : Z \setminus \{P_0\} \to Z_p(X_0) = H_0 \cong \mathbb{P}^2, \quad [X_0, X_1, X_2, X_3] \xrightarrow{\pi_{P_0}} [X_1, X_2, X_3].$$

This is a homeomorphism, whose inverse on $H_0 \setminus (H_2 \cap H_0)$ is given by

$$[a_1, a_2, a_3] \to \left[\frac{a_1^2}{a_2}, a_1, a_2, a_3 \right] = [a_1^2, a_1 a_2, a_2^2 a_2 a_3].$$

Thus, Z is a projective variety since it is irreducible, being homeomorphic to an open dense subset of a projective plane, and closed in $\mathbb{P}^3$. Since

$Z_0 := Z \cap U_0 = Z_a(x_1^2 - x_2) \subset \mathbb{A}^3$ is an open dense subset of Z, one has $K(Z) \cong K(Z_0)$. From the fact that Z_0 is an affine variety, one has $\mathcal{O}_{Z_0}(Z_0) = A(Z_0) = \frac{\mathbb{K}[x_1,x_2,x_3]}{(x_1^2-x_2)} \cong \mathbb{K}[x_1,x_3]$, therefore, $K(Z_0) = \mathcal{Q}(A(Z_0)) \cong \mathbb{K}(x_1,x_3)$, which is a purely transcendental extension of $\mathbb{K}$ with transcendence degree two.

Solution 6.1

The map φ is a morphism since it is a polynomial map with target an affine variety. Any point $P = (p_1, p_2) \in \mathbb{A}^2$ such that $p_1 \cdot p_2 \neq 0$ is such that $P \in \text{Im}(\varphi)$: namely $\varphi^{-1}(P) = \left(p_1, \frac{p_2}{p_1}\right)$. Note moreover that the line $Z_a(x_1) \subset \mathbb{A}^2$ is mapped via φ to the origin $O = (0,0)$ whereas any $P = (p_1, 0) \in Z_a(x_2)$ is such that $\varphi(P) = P$. In other words $\text{Im}(\varphi) = (\mathbb{A}^2 \setminus Z_a(x_1)) \cup \{O\}$, where $\mathbb{A}^2 \setminus Z_a(x_1)$ is a principal open set whereas O is Zariski closed. Thus, $\text{Im}(\varphi)$ is neither closed nor open in $\mathbb{A}^2$ but it is a constructible set (cf. Remark 6.1.10 and Corollary 11.1.3).

Solution 6.2

Note that $O = (0,0) \in C$. Consider the pencil of lines through O, namely $\{\ell_t := Z_a(x_2 - tx_1)\}_{t \in \mathbb{A}^1}$. The intersection with the curve C of any line of the pencil such that $t \neq \pm 1$ consists of the point O, with multiplicity $\mu(C, \ell_t; O) = 2$, and of an extra point $P_t := (t^2 - 1, t^3 - t)$. Whereas lines ℓ_1 and ℓ_{-1} intersect C only at O with $\mu(C, \ell_{\pm 1}; O) = 3$. This defines a map

$$\varphi : \mathbb{A}^1 \to C \subset \mathbb{A}^2, \quad t \xrightarrow{\varphi} (t^2 - 1, t^3 - t),$$

which is surjective and moreover it is a morphism, since it is a polynomial map with affine target. This proves that C is irreducible and so that C is an affine variety. Moreover, φ is the required polynomial parametrization. Observe that $A(C)$ is isomorphic to a proper subring of $A^{(1)}$, since on $A(C)$ one has $t^2 = 1$ and $t^3 = t$. Therefore, the natural map $\varphi^\# : A(C) \to \mathbb{K}[t] = A^{(1)}$ is injective but not an isomorphism so φ cannot be an isomorphism (indeed φ is surjective onto C but not injective).

Solution 6.3

For $[s,t] \in \mathbb{P}^1$, one can easily deduce homogeneous parametric equations of the conic $C \subset \mathbb{P}^2$ as $X_0 = s^2$, $X_1 = st$, $X_2 = t^2$. Composing with $\nu_{2,2}^{\text{st}}$ gives

$$\nu_{2,2}^{\text{st}}(C) = \{[s^4, s^3t, s^2t^2, s^2t^2, st^3, t^4] \mid [s,t] \in \mathbb{P}^1\} \subset \mathbb{P}^5.$$

Denoting by $[W_0, W_1, \ldots, W_5]$ homogeneous coordinates for the target space $\mathbb{P}^5$, note that $\nu_{2,2}^{st}(C)$ lies in the hyperplane section of the Veronese surface $V_{2,2}^{st}$ with the hyperplane $Z_p(W_2 - W_3) \cong \mathbb{P}^4$ and it has therein a homogeneous parametric representation of a standard rational normal quartic.

Solution 6.4

Similarly as in Exercise 6.3, the line ℓ has homogeneous parametric equations given by $\ell : X_0 = s$, $X_1 = t$, $X_2 = t$, $[s, t] \in \mathbb{P}^1$ so, composing with $\nu_{2,3}^{st}$, gives

$$\nu_{2,3}^{st}(\ell) = [s^3, s^2 t, s^2 t, s t^2, s t^2, s t^2, t^3, t^3, t^3, t^3], \quad [s, t] \in \mathbb{P}^1,$$

Denoting by $[W_0, W_1, \ldots, W_9]$ homogeneous coordinates for the target space $\mathbb{P}^9$, one notices that $\nu_{2,3}^{st}(\ell)$ is a standard projective twisted cubic in the linear subspace $Z_p(W_1 - W_2, W_3 - W_4, W_3 - W_5, W_6 - W_7, W_6 - W_8, W_6 - W_9) \cong \mathbb{P}^3$ in $\mathbb{P}^9$. Similarly,

$$\nu_{2,3}^{st}(C) = [s^6, s^5 t, s^4 t^2 s^4 t^2, s^3 t^3, s^3 t^3 s^2 t^4, s^3 t^3, s^2 t^4, s t^5, t^6], \quad [s, t] \in \mathbb{P}^1,$$

which shows that $\nu_{2,3}^{st}(C)$ is a standard projective rational normal sextic in the linear subspace $Z_p(W_2 - W_3, W_4 - W_6, W_5 - W_7) \cong \mathbb{P}^6$ in $\mathbb{P}^9$.

Solution 6.5

Since, up to projectivities, any $\mathfrak{g}_d^d$ on $\mathbb{P}^1$ is projectively equivalent to $\nu_{1,d}^{st}$, with no loss of generality we can reduce to this case with the choice of $P = [1, 0]$ a fundamental point of $\mathbb{P}^1$. Thus, $\mathrm{Im}(\nu_{1,d}^{st}) = V_{1,d}^{st}$ is the standard rational normal curve of degree d in $\mathbb{P}^d$ and the point P maps to $P_0 := [1, 0, \ldots, 0] \in V_{1,d}^{st}$. Divisors in the $\mathfrak{g}_d^d = \nu_{1,d}^{st}$ on $\mathbb{P}^1$ bijectively correspond to hyperplane sections of $V_{1,d}^{st} \subset \mathbb{P}^d$. Thus, the sublinear series $\mathfrak{g}_d^d(-P)$ consists of divisors which bijectively correspond to sections of $V_{1,d}^{st}$ with hyperplanes passing through P_0. Consider the projection

$$\pi_{P_0} : \mathbb{P}^d \setminus \{P_0\} \to \mathbb{P}^{d-1}, \quad [X_0, X_1, \ldots, X_d] \xrightarrow{\pi_{P_0}} [X_1, \ldots, X_d].$$

Its restriction π to $V_{1,d}^{st}$ is

$$\pi : V_{1,d}^{st} \setminus \{P_0\} \to \mathbb{P}^{d-1}, \quad [s^d, s^{d-1} t, \ldots, t^d] \xrightarrow{\pi} [s^{d-1} t, \ldots, t^d], \quad [s, t] \in \mathbb{P}^1.$$

Note that on $V_{1,d}^{st} \setminus \{P_0\}$ one has $t \neq 0$ therefore, dividing by t the expression of points on $\mathrm{Im}(\pi)$, one has $\mathrm{Im}(\pi) = \{[s^{d-1}, \ldots, t^{d-1}] \mid [s, t] \in \mathbb{P}^1\}$ which makes sense for any $[s, t] \in \mathbb{P}^1$. In other words $\overline{\mathrm{Im}(\mathfrak{g}_d^d(-P))} = \overline{\mathrm{Im}(\pi)} = V_{1,d-1}^{\mathrm{st}} \subset \mathbb{P}^{d-1}$, i.e. the standard rational normal curve of degree $d - 1$ is an *internal projection* of the standard rational normal curve $V_{1,d}^{\mathrm{st}} \subset \mathbb{P}^d$ from its point $P_0 = [1, 0, \ldots, 0]$.

Solution 7.1

The Segre variety $\Sigma_{1,1}$ is the *rank-four quadric surface* given by $Z_p(W_{00}W_{11} - W_{01}W_{10}) \subset \mathbb{P}^3$. Any fiber of the projection of $\mathbb{P}^1 \times \mathbb{P}^1$ onto its ith factor, $1 \leqslant i \leqslant 2$, is mapped via $\sigma_{1,1}$ to a line in $\Sigma_{1,1}$. Since $\sigma_{1,1}$ is an isomorphism onto its image, lines coming from fibers of the same projection do not meet, so they are skew in $\mathbb{P}^3$, whereas two lines coming from fibers of the two different projections are incident at just one point. More precisely, for any $P, Q \in \mathbb{P}^1$, $\Sigma_{1,1}$ contains lines

$$\ell_P := \sigma_{1,1}(\{P\} \times \mathbb{P}^1) \quad \text{and} \quad r_Q := \sigma_{1,1}(\mathbb{P}^1 \times \{Q\});$$

any two lines $\{\ell_P\}_{P \in \mathbb{P}^1}$ are skew, the same holds for any two lines $\{r_Q\}_{Q \in \mathbb{P}^1}$, whereas for any point $R \in \Sigma_{1,1}$, the lines ℓ_R and r_R meet at R. These two families of lines in $\Sigma_{1,1}$ are called *rulings* of the quadric surface, for this reason $\Sigma_{1,1}$ is said to be a *doubly ruled* surface. Note that $\Sigma_{1,1} \cap U_{00}^3 = Z_a(x_3 - x_1x_2)$ is the *hyperbolic paraboloid* in $\mathbb{A}^3$ (see figure below).

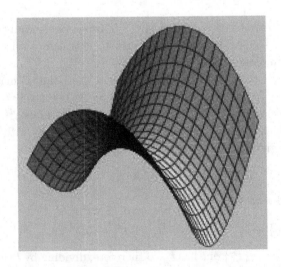

Solution 7.2

(i) Let $\gamma' : V' \xrightarrow{\cong} V$ and $\delta' : W' \xrightarrow{\cong} W$ be isomorphisms, which exist by the assumptions. With notation as in Theorem 7.4.2, take $X := V' \times W'$, $\rho_1 := \gamma' \circ \pi_{V'}$ and $\rho_2 := \delta' \circ \pi_{W'}$. Since X is a product, it satisfies assumptions as in Theorem 7.4.2 from which one deduces the isomorphism $V' \times W' \cong V \times W$.

(ii) Letting $X := W \times V$, with $\rho_1 = \pi_V$ and $\rho_2 = \pi_W$, since X is a product of algebraic varieties, the existence of an isomorphism $W \times V \cong V \times W$ is granted from Theorem 7.4.2.

(iii) This follows by simply recursively applying the same procedures as in (i) and (ii) above.

(iv) We focus on $\pi_V^{-1}(P)$, since the proof for $\pi_W^{-1}(Q)$ is similar. Note that, by definition of $\pi_V^{-1}(P)$, it is isomorphic to the algebraic variety $\{P\} \times W$. Moreover, since π_V is a morphism and $P \in V$ is closed in V, then $\pi_V^{-1}(P)$ is a closed subvariety of $V \times W$. The projection $\pi_W : \{P\} \times W \to W$ is an isomorphism which, composed with that between $\pi_V^{-1}(P)$ and $\{P\} \times W$, gives rise to an isomorphism $\pi_V^{-1}(P) \cong W$.

(v) If φ is a morphism, then $\pi_V \circ \varphi$ and $\pi_W \circ \varphi$ are obviously morphisms, since they are both compositions of morphisms. Conversely, assume that $\alpha := \pi_V \circ \varphi \in \text{Morph}(Z, V)$ and that $\beta := \pi_W \circ \varphi \in \text{Morph}(Z, W)$. Then $\alpha \times \beta$ is a morphism from Proposition 7.4.1. Moreover, since $\alpha \times \beta$ is the unique map which let the diagrams

$$Z \xrightarrow{\alpha \times \beta} V \times W \qquad \qquad Z \xrightarrow{\alpha \times \beta} V \times W$$
$$\searrow^{\alpha} \quad \swarrow^{\pi_V} \qquad \text{and} \qquad \searrow^{\beta} \quad \swarrow^{\pi_W}$$
$$V \qquad \qquad \qquad W$$

to be commutative, as φ does, one concludes that $\varphi = \alpha \times \beta$ and so that φ is a morphism.

Solution 7.3

The fact that $A \otimes_{\mathbb{K}} B$ is a $\mathbb{K}$-algebra of finite type directly follows from the definition of tensor product of $\mathbb{K}$-algebras as in Section 1.9.2. We need to show that it is an integral $\mathbb{K}$-algebra. Let $c, d \in A \otimes_{\mathbb{K}} B$ be such that $c \cdot d = 0$. We can write $c = \Sigma_i a_i \otimes b_i$ and $d = \Sigma_j \alpha_j \otimes \beta_j$, where the sets $\{b_i\}$ and $\{\beta_j\}$ of B can be taken to be linearly independent over $\mathbb{K}$. Let $h : A \to \mathbb{K}$

be any $\mathbb{K}$-algebra homomorphism. Then $h \otimes \mathrm{Id}_B : A \otimes_\mathbb{K} B \to B = \mathbb{K} \otimes_\mathbb{K} B$ is a $\mathbb{K}$-algebra homomorphism. Therefore, one has

$$0 = (h \otimes \mathrm{Id}_B)(c \cdot d) = (h \otimes \mathrm{Id}_B) \left(\sum_{i,j} a_i \alpha_j \otimes b_i \beta_j \right) = \left(\sum_{i,j} h(a_i \alpha_j) \otimes b_i \beta_j \right)$$

$$= (h \otimes \mathrm{Id}_B)(c) \cdot (h \otimes \mathrm{Id}_B)(d) = \left(\sum_i h(a_i) \otimes b_i \right) \cdot \left(\sum_j h(\alpha_j) \otimes \beta_j \right) \in B.$$

Since B is integral, one of the two factors has to be zero e.g. $(\Sigma_i h(a_i) \otimes b_i) = 0$. Since $\{b_i\}$ is linearly independent over $\mathbb{K}$, this means that $h(a_i) = 0$, for any i. Therefore, the same holds for $h(a_i \cdot \alpha_j) = h(a_i) \cdot h(\alpha_j)$, for any i and j. Since this holds for any $\mathbb{K}$-algebra homomorphism $h : A \to \mathbb{K}$ and since $A = \frac{\mathbb{K}[x]}{I}$ for some ideal I in a suitable polynomial ring $\mathbb{K}[x]$, this implies that $a_i \cdot \alpha_j = 0$ in A: indeed, by the Hilbert "Nullstellensatz"-weak form, to give any $\mathbb{K}$-algebra homomorphism $h : A \to \mathbb{K}$ is equivalent to giving a maximal ideal $\mathfrak{m}_P \in \mathrm{Specm}(A)$ corresponding to a point $P \in Z_a(I)$, such that

$$A = \frac{\mathbb{K}[x]}{I} \xrightarrow{h} \mathbb{K} \cong \frac{A}{\mathfrak{m}_P}, \quad h(a) = \mathrm{ev}_P(a),$$

where a a residue class of a polynomial in $\mathbb{K}[x]$ modulo I, thus a polynomial a which vanishes at each point of $Z_a(I)$ is the zero-polynomial of A by the Hilbert "Nullstellensatz"-strong form. Since A is an integral $\mathbb{K}$-algebra, $a_i \cdot \alpha_j = 0$, for any i and j, implies that either $a_i = 0$ for any i or $\alpha_j = 0$ for any j.

Finally, if $V \subseteq \mathbb{A}^n$ and $W \subseteq \mathbb{A}^m$ are affine varieties defined over an algebraically closed field, we already know that $V \times W \subseteq \mathbb{A}^n \times \mathbb{A}^m \cong \mathbb{A}^{n+m}$ is a closed immersion. Moreover, we proved that $A(V \times W) \cong A(V) \otimes_\mathbb{K} A(W)$. Since $A(V)$ and $A(W)$ are both integral $\mathbb{K}$-algebras of finite type and since $\mathbb{K}$ is algebraically closed, from the previous part $A(V) \otimes_\mathbb{K} A(W)$ is an integral $\mathbb{K}$-algebra of finite type, so $V \times W$ is irreducible therefore it is an affine variety such that, as a set, $V \times W = \mathrm{Specm}(A(V) \otimes_\mathbb{K} A(W))$.

Solution 7.4

Let $\mathbb{K} = \mathbb{R}$ and let $\mathbb{C} = \frac{\mathbb{R}[x]}{(x^2+1)}$. Thus, $\mathbb{C}$ is an integral $\mathbb{R}$-algebra of finite type. Nonetheless $\mathbb{C} \otimes_\mathbb{R} \mathbb{C}$ is a $\mathbb{R}$-algebra of finite type which is not integral, indeed $(1 \otimes i - i \otimes 1) \cdot (1 \otimes 1 - i \otimes i) = 0$.

Solution 7.5

Let $V = \mathbb{A}^2$ and $Z = \mathbb{A}^1$, with $\varphi = \pi_x$ the projection onto the first coordinate, whereas let $W = \{0\}$ and $\psi = \iota_W$ be the closed immersion of the point $\{0\}$ in $\mathbb{A}^1$. In this case

$$\mathbb{A}^2 \times_{\mathbb{A}^1} \{0\} = \{(P, \{0\}) \in \mathbb{A}^2 \times \{0\} \mid \pi_x(P) = 0\} \cong \pi_x^{-1}(0),$$

which is isomorphic to $\mathbb{A}^1$ as $\pi_x^{-1}(0) = \{(0, p_2) \mid p_2 \in \mathbb{A}^1\}$. In this case the fiber-product turns out to be the fiber over $0 \in \mathbb{A}^1$ of the projection of $\mathbb{A}^2$ onto the first coordinate. On the contrary, if we take $V = Z_a(x_1^2 + x_2^2 - 1) \subset \mathbb{A}^2$, $W = \{0\}$, $Z = \mathbb{A}^1$, $\varphi = \pi_x \circ \iota_V$ whereas ψ as above, then $V \times_Z W = \varphi^{-1}(0) = \{(0, 1)\} \cup \{(0, -1)\} \subset V$, which is a reducible algebraic subset of V.

Solution 8.1

Let $U := \mathbb{A}^{n-1} \setminus Z_a(x_1 x_2 \cdots x_{n-1})$. On the open set U we can consider the map

$$\phi : U \to \mathbb{A}^n, (x_1, x_2, \ldots, x_{n-1}) \xrightarrow{\phi} \left(x_1, x_2, \ldots, x_{n-1}, \frac{1}{x_1 x_2 \cdots x_{n-1}} \right),$$

which is a morphism, by Proposition 6.2.2, whose image is X. Thus, X is irreducible therefore it is an affine variety. Moreover ϕ defines a rational map $\Phi : \mathbb{A}^{n-1} \dashrightarrow X$. Consider now the multi-index $I = (1, \ldots, n-1)$ and the corresponding projection $\pi_I : \mathbb{A}^n \to \mathbb{A}^{n-1}$, then $\psi := \pi_I \circ \iota_X \in \text{Morph}(X, U)$, where ι_X denotes the closed immersion of X in $\mathbb{A}^n$. The morphism ψ defines the rational inverse of the rational map Φ.

Solution 8.2

For $n = 1$, we already know that, e.g. the hyperbola $Y_1 = Z_a(x_1 x_2 - 1) \subset \mathbb{A}^2$ is birationally equivalent but not isomorphic to $\mathbb{A}^1$. Therefore, for any $n \geqslant 2$, we can consider $Y_n := (\mathbb{A}^1 \setminus \{0\}) \times \mathbb{A}^{n-1}$. Recall that $\mathbb{A}^1 \setminus \{0\}$ is an affine variety, being isomorphic to the hyperbola Y_1 above, therefore, Y_n is an affine variety which is birational to $\mathbb{A}^1 \times \mathbb{A}^{n-1} \cong \mathbb{A}^n$, for any $n \geqslant 2$. On the other hand Y_n is not isomorphic to $\mathbb{A}^n$, since $A(\mathbb{A}^n) = A^{(n)}$ whereas $\frac{1}{x_1} \in \mathcal{O}_{Y_n}(Y_n) \setminus A^{(n)}$.

Solution 8.3

Note that $O \in C$. The pencil of lines $\{Z_a(x_2 - mx_1)\}_{m \in \mathbb{A}^1}$ through O gives a parametrization $\mathbb{A}^1 \to C$, $m \to (m^3 - 1, m^4 - m)$ which is injective off the closed subset $Z_a(m^3 - 1) \subset \mathbb{A}^1$. Its rational inverse $C \setminus \{O\} \to \mathbb{A}^1$ is given by $(x_1, x_2) \to \frac{x_2}{x_1}$. On the other hand the previous maps are not isomorphisms, since $\frac{x_2}{x_1}$ is not regular at O.

Solution 8.4

From the previous exercise, we already know that C is birational but not isomorphic to $\mathbb{A}^1$. Intersecting C with the lines of the pencil gives

$$x_2 - mx_1 = 0 = x_1^3(x_1 + 1 - m^3).$$

In $\mathbb{A}^3$, with coordinates (x_1, x_2, x_3), the proper transform of C after blowing-up $\mathbb{A}^2$ at the origin O is given by the curve $C' \subset \mathbb{A}^3$ of parametric equations $x_1 = m^3 - 1$, $x_2 = m^4 - m$, $x_3 = m$. This gives a polynomial parametrization $\mathbb{A}^1 \to C'$, namely $m \to (m^3 - 1, m^4 - m, m)$, which is therefore a morphism and whose inverse morphism is given by the projection onto the x_3-axis. Thus, C' is isomorphic to $\mathbb{A}^1$ and birational (but not isomorphic) to C.

Solution 8.5

From the expression of q, one notices that q is an *involution* on the open set $U := U_0 \cap U_1 \cap U_2 \subset \mathbb{P}^2$, namely $q|_U$ is an isomorphism such that $q|_U^{-1} = q|_U$, in particular q is a birational transformation of $\mathbb{P}^2$ into itself. Note that q can also be written as

$$q([X_0, X_1, X_2]) = [X_1 X_2, X_0 X_2, X_0 X_1],$$

which shows that q coincides with the rational map ν_L as in (6.16), given by the linear system associated to $L = \mathrm{Span}\{X_1 X_2, X_0 X_2, X_0 X_1\} \subset \mathcal{S}_2^{(2)}$. Thus, the indeterminacy locus of q coincides with the base locus $B := \{P_0 = [1, 0, 0], P_1 = [0, 1, 0], P_2 = [0, 0, 1]\}$ of the linear system of conics $\mathbb{P}(L)$. The points in B, which are the fundamental points of $\mathbb{P}^2$, are called *fundamental points of q*. Denoting by $H_i := Z_p(X_i)$, $0 \leqslant i \leqslant 2$, the fundamental lines of $\mathbb{P}^2$, note that q contracts $H_i \setminus (H_i \cap B)$ to the fundamental point P_i, for any $i \in \{0, 1, 2\}$. For this reason, the lines H_i are called *exceptional lines* of q, $0 \leqslant i \leqslant 2$. Since q is an involution as a birational map, one says that q *blows-up* the fundamental points to the exceptional lines. Any line $\ell = Z_p(\Sigma_{i=0}^2 a_i X_i)$ not intersecting B maps via q to the conic $Z_p(a_0 X_1 X_2 + a_1 X_0 X_2 + a_2 X_0 X_1) \subset \mathbb{P}^2$. Whereas any line through one of the fundamental points is instead mapped to another line; more precisely, take, e.g. $\ell = Z_p(\lambda X_1 + \mu X_2)$ any line through the point P_0, with $[\lambda, \mu] \in \mathbb{P}^1$. Consider $\ell \setminus \{P_0\}$, whose points are given by $[\mu, \mu, -\lambda]$ with $\mu \neq 0$; then $q([\mu, \mu, -\lambda]) = [-\mu\lambda, -\mu\lambda, \mu^2] = [\lambda, \lambda, -\mu]$, i.e. the projective closure of $q(\ell \setminus \{P_0\})$ is the line $r_\ell = Z_p(\mu X_1 + \lambda X_2)$, which cuts out on H_0 the point $[0, \lambda, \mu]$. Thus the points on the exceptional line H_0 are in one-to-one correspondence with the directions of lines through the fundamental point P_0; the same occurs for the other fundamental points (exceptional lines, respectively).

Solution 9.1

(i) Since $\mathbb{A}^1 \times \mathbb{A}^1 \cong \mathbb{A}^2$, then Z is the hyperbola in the affine plane which is therefore a closed affine subvariety of $\mathbb{A}^2$.

(ii) One has $\pi_2(Z) = \mathbb{A}^1 \setminus \{0\}$ which is not closed in $W = \mathbb{A}^1$. From Definition 9.1.1, it follows that $\mathbb{A}^1$ is not a complete variety.

(iii) Take $V = \mathbb{P}^1$ in such a way that $\mathbb{A}^1$ is identified with the affine open chart U_0 so that $t = \frac{X_1}{X_0}$, where $[X_0, X_1]$ homogeneous coordinate on $\mathbb{P}^1$. Then the closure $\widehat{Z}$ of Z in $\mathbb{P}^1 \times \mathbb{A}^1$ is given by $Z(X_1 s - X_0)$ and the projection π_2 is given by the restriction to Z of $\pi_{\mathbb{A}^1} : \mathbb{P}^1 \times \mathbb{A}^1 \to \mathbb{A}^1$, which reads $([X_0, X_1], s) \to s$. Note π_2 extends to $\widehat{Z}$ since $\pi_{\mathbb{A}^1}^{-1}(0) \cap \widehat{Z} = \{([X_0, X_1], 0) \mid X_1 \cdot 0 - X_0 = 0\} = \{([0, 1], 0)\}$ so $\widehat{Z}$ surjectively maps onto $\mathbb{A}^1$, i.e. $\pi_{\mathbb{A}^1}(\widehat{Z}) = \mathbb{A}^1$ is closed.

Solution 9.2

Considering the affine chart $U_0 \cong \mathbb{A}^3$, the trace of $\bar{\ell}$ in U_0 is the affine line ℓ passing through $P = (1, 0, 1) \in \mathbb{A}^3$ with direction $\underline{v} = (-1, 1, 0)$, namely $\ell : x_1 = 1 - t$, $x_2 = t$, $x_3 = 1$, $t \in \mathbb{A}^1$. Take now $f = x_1 x_2 - x_3 \in A^{(3)}$. Then $\ell \cap Z_a(f) = \emptyset$, since $t^2 - t + 1 \in \mathbb{R}[t]$ has no roots in $\mathbb{R}$. Moreover, $\overline{Z_a(f)}^p = Z_p(X_0 X_3 - X_1 X_2) \subset \mathbb{P}^3$ is such that $\bar{\ell} \cap \overline{Z_a(f)}^p = Z_p(X_1 + X_2 - X_0, X_3 - X_0, X_0 X_3 - X_1 X_2)$. By the previous discussion about the affine part of this intersection, we only need to check for possible intersections when $X_0 = 0$; however in this case, we have $Z_p(X_1 + X_2, X_3, X_1 X_2, X_0) = \emptyset$.

Solution 9.3

One can identify $\mathbb{P}(\mathcal{S}_d^{(n)})$ with the numerical projective space $\mathbb{P}^N$, where $N = \binom{n+d}{d} - 1$. Let $\mathcal{R}_k \subset \mathbb{P}^N$ be the locus of $[F] \in \mathbb{P}^N$ s.t. there exist an integer k, $1 \leqslant k \leqslant d - 1$, and homogeneous polynomials $F_k' \in \mathcal{S}_k^{(n)}$ and $F_{d-k}'' \in \mathcal{S}_{d-k}^{(n)}$ s.t. $[F] = [F_k' \cdot F_{d-k}'']$. One has $\mathcal{R} = \cup_{k=1}^{d-1} \mathcal{R}_k$. For any $1 \leqslant j \leqslant d - 1$, set $N_j := \binom{n+j}{j} - 1$ and $\mathbb{P}^{N_j} := \mathbb{P}(\mathcal{S}_j^{(n)})$. One has a natural multiplication map

$$\mathbb{P}^{N_k} \times \mathbb{P}^{N_{d-k}} \xrightarrow{m_k} \mathbb{P}^N, \quad ([F_k'], [F_{d-k}'']) \xrightarrow{m_k} [F_k' \cdot F_{d-k}''],$$

where $\text{Im}(m_k) = \mathcal{R}_k$. Since $\mathbb{P}^{N_k} \times \mathbb{P}^{N_{d-k}} \cong \Sigma_{N_k, N_{d-k}}$ is the Segre variety of indexes N_k and N_{d-k}, which is a projective variety, and since m_k is a morphism, by completeness of projective varieties, $\mathcal{R}_k$ is a projective variety in $\mathbb{P}^N$. Then $\mathcal{R}$ is a closed algebraic set in $\mathbb{P}^N$ and any $\mathcal{R}_k$ is one of its irreducible component, $1 \leqslant k \leqslant d - 1$. Since in $\mathcal{S}_d^{(n)}$ exist infinitely many irreducible polynomials, then $\mathcal{R}$ is a proper closed subset of $\mathbb{P}^N$.

Solution 9.4

(i) Up to replacing V and W, respectively, with the affine open sets $\varphi^{-1}(U_P)$ and U_P of the definition, with no loss of generality we may assume that V and W are affine varieties and therefore that φ is surjective. Up to isomorphism, we may assume that $V \subseteq \mathbb{A}^n$, for some non-negative integer n so $A(V)$ is a quotient, modulo some prime ideal, of $\mathbb{K}[x_1, \ldots, x_n]$. Let $A(V) = \mathbb{K}[\overline{x}_1, \ldots, \overline{x}_n]$ where we denote by $\overline{x}_1, \ldots, \overline{x}_n$ the images in $A(V)$ of the indeterminates in $\mathbb{K}[x_1, \ldots, x_n]$. By the assumption on φ, any $\overline{x}_j \in A(V)$ satisfies an identity of the form

$$(\overline{x}_j)^{d_j} + a_{j,d_j-1}(\overline{x}_j)^{d_j-1} + \cdots + a_{j,0} = 0, \quad a_{j,i} \in A(W).$$

Let $Q \in W$; points in $\varphi^{-1}(Q)$ have coordinates $\overline{x}_1(Q), \ldots, \overline{x}_n(Q) \in V \subseteq \mathbb{A}^n$, such that

$$(\overline{x}_j(Q))^{d_j} + a_{j,d_j-1}(Q)(\overline{x}_j(Q))^{d_j-1} + \cdots + a_{j,0}(Q) = 0, \quad a_{j,i}(Q) \in \mathbb{K}.$$

Since any polynomial $T^{d_j} + a_{j,d_j-1}(Q)(T)^{d_j-1} + \cdots + a_{j,0}(Q) \in \mathbb{K}[T]$ has finitely many roots in $\mathbb{K}$, then $\varphi^{-1}(Q)$ is a finite set.

(ii) Consider the plane section of Z with the plane $Z_a(x_1 - x_2)$; it is the conic $C := Z_a(x_1 - x_2, x_3^2 + x_1^2 - 1)$ for which $A(C) \cong \frac{\mathbb{K}[x_1, x_3]}{(x_1^2 + x_3^2 - 1)}$ is an integral $\mathbb{K}$-algebra thus C is irreducible; since Z is a quadric surface and since C is an irreducible conic plane section of Z, one deduces that Z is irreducible. Set now $f := x_3^2 + x_3(x_1 - x_2) + x_1^2 - 1$, which is the polynomial defining Z. We observe that φ is surjective onto $\mathbb{A}^2$, indeed for any $P = (p_1, p_2) \in \mathbb{A}^2$, $\varphi^{-1}(P)$ contains points $Q = (p_1, p_2, p_3) \in Z$, where $p_3 \in \mathbb{K}$ is determined by the finitely many solutions of the polynomial equation $F_P(T) = 0$, where $F_P(T) := T^2 + (p_1 - p_2)T + (p_1^2 - 1) \in \mathbb{K}[T]$. In particular, $\varphi(Z) = \mathbb{A}^2$ which is closed in $\mathbb{A}^2$. The surjectivity of φ implies that we have an injective $\mathbb{K}$-algebra homomorphism $\mathbb{K}[x_1, x_2] \overset{\varphi^{\#}}{\hookrightarrow} A(Z)$. Let $\bar{x}_3$ be the image of the indeterminate x_3 in $A(Z)$. Then for $\bar{x}_3$ one has that $\bar{x}_3^2 + \bar{x}_3(x_1 - x_2) + x_1^2 - 1 = 0$, i.e. $\bar{x}_3$ is integral over $\mathbb{K}[x_1, x_2]$. Therefore, the ring inclusion is an integral extension of $\mathbb{K}$-algebras so φ is a finite morphism.

Solution 9.5

Up to a projective transformation, we may assume that $P = P_n = [0, 0, \ldots, 0, 1]$, so that the projection from P is simply $[X_0, X_1, \ldots, X_{n-1}, X_n] \dashrightarrow [X_0, X_1, \ldots, X_{n-1}]$. From the assumption on V, we may write $V = Z_p(X_n^d + F_1 X_n^{d-1} + \cdots + F_d)$, where $F_j \in \mathbb{K}[X_0, \ldots, X_{n-1}]_j$, $1 \leqslant j \leqslant d$. Consider $H_i = Z_p(X_i) \subset \mathbb{P}^{n-1}$ and $U_i = \mathbb{P}^{n-1} \setminus H_i$ the affine chart of $\mathbb{P}^{n-1}$, $0 \leqslant i \leqslant n-1$. Now, $\varphi^{-1}(U_i)$ is a hypersurface in an affine space $\mathbb{A}_i^n$, with affine coordinates $(y_0, y_1, \ldots, y_{i-1}, y_{i+1}, \ldots, y_{n-1}, y_n)$, given by $\varphi^{-1}(U_i) = Z_a(y_n^d + \delta_i(F_1)y_n^{d-1} + \cdots + \delta_i(F_d))$, where δ_i the dehomogeneization w.r.t. the indeterminate X_i, $0 \leqslant i \leqslant n-1$. The hypersurfaces $\varphi^{-1}(U_i)$ determine an affine open covering of V and the restriction of φ to $\varphi^{-1}(U_i)$ is simply the restriction to $\varphi^{-1}(U_i)$ of the projection from $\mathbb{A}_i^n$ to $\mathbb{A}_i^{n-1}$ given by

$$(y_0, y_1, \ldots, y_{i-1}, y_{i+1}, \ldots, y_{n-1}, y_n) \overset{\pi_i}{\longrightarrow} (y_0, y_1, \ldots, y_{i-1}, y_{i+1}, \ldots, y_{n-1}),$$

i.e. the projection onto the first $(n-1)$ coordinates of $\mathbb{A}_i^n$. Therefore, we are reduced to the case of an affine hypersurface $Z_a(f) \subset \mathbb{A}^n$, where

$$f = x_n^d + f_1(x_1, \ldots, x_{n-1})x_n^{d-1} + \cdots + f_d(x_1, \ldots, x_{n-1}) \in A^{(n)},$$

and where $\varphi : V \to \mathbb{A}^{n-1}$ is the restriction to V of the projection $\pi_I : \mathbb{A}^n \to \mathbb{A}^{n-1}$, with $I = (1, 2, \ldots, n-1)$. From the expression of f, the morphism φ is surjective, therefore it induces an injective $\mathbb{K}$-algebra homomorphism $\varphi^{\#} : A^{(n-1)} \hookrightarrow \frac{A^{(n)}}{(f)}$. Note that $\bar{x}_n \in \frac{A^{(n)}}{(f)}$ is integral over $A^{(n-1)}$ as it is clear from the expression of f. Therefore, the ring inclusion is an integral $\mathbb{K}$-algebra extension which implies that φ is a finite morphism (cf. Ciliberto, 2021, § 10.2, pp. 125–128).

Solution 10.1

Up to replacing V and W with the affine open sets $\varphi^{-1}(U_P)$ and U_P, respectively, we may assume V and W to be affine variety. Since $A(W) \subseteq A(V)$ is an integral extension, then $K(W) \subseteq K(V)$ is an algebraic extension. Therefore, $\mathrm{trdeg}_{\mathbb{K}}(K(V)) = \mathrm{trdeg}_{\mathbb{K}}(K(W))$ and we conclude.

Solution 10.2

If $\Lambda \subset \mathbb{P}^n$ is a linear subspace of dimension s, then it is cut-out by $n - s$ independent hyperplanes $W_1, \ldots, W_{n-s}$ in $\mathbb{P}^n$. Since $\Lambda \cap V = \emptyset$, then $\dim(V) \leqslant n - s - 1$; indeed if by contradiction $\dim(V) \geqslant n - s$, from Theorem 10.3.1-(ii), we would have $V \cap \Lambda \neq \emptyset$.

Solution 10.3

The hypersurface V is isomorphic to $\mathbb{A}^2$: indeed the map $\psi : \mathbb{A}^2 \to \mathbb{A}^3$, defined as $(u, v) \xrightarrow{\psi} (u, uv, v)$, maps $\mathbb{A}^2$ onto V and it is a morphism, since it is a polynomial map, moreover the restriction to V of the projection $\pi_J : \mathbb{A}^3 \to \mathbb{A}^2$, where $J = (1, 3)$, defines ψ^{-1}; thus V is irreducible. Now $\varphi : V \to \mathbb{A}^2$ simply reads $(u, uv, v) \to (u, uv)$; in particular φ is not surjective onto $\mathbb{A}^2$ as points $(0, k)$ with $k \neq 0$ are not in the image of φ. On points $(p_1, p_2) \in \mathbb{A}^2$ such that $p_1 \cdot p_2 \neq 0$, one has $\varphi^{-1}(p_1, p_2) = (p_1, p_2, \frac{p_2}{p_1}) \in V$ so $\dim(\varphi^{-1}(p_1, p_2)) = 0$. If $P = (h, 0)$ then $\varphi^{-1}(h, 0) = (h, 0, 0) \in V$ so $\dim(\varphi^{-1}(h, 0)) = 0$. However, $(0, 0) \in \mathrm{Im}(\varphi)$ and $\varphi^{-1}(0) = \{(0, 0, k) \in \mathbb{A}^3 \mid \forall \, k \in \mathbb{K}\}$ is a line. Therefore, the open set where the fibers have constant dimension equal to 0 is $(\mathbb{A}^2 \setminus Z_a(x_1))$. From Exercise 9.4-(i), we deduce that φ cannot be a finite morphism.

Solution 10.4

From the equation of ℓ, it is clear that the line is contained in $\mathfrak{Q}$. Let $\mathcal{S}(\mathfrak{Q})$ be the homogeneous coordinate ring of $\mathfrak{Q}$, which is a graded, integral

$\mathcal{S}^{(3)}$-module and a $\mathbb{K}$-algebra of finite type. Note that the homogeneous ideal $I_{p,\mathfrak{Q}}(\ell) \subset \mathcal{S}(\mathfrak{Q})$ of ℓ in $\mathfrak{Q}$ is generated by the images of the two linear generators of the homogeneous ideal $I_p(\ell) = (X_0, X_1)$ of the line ℓ in $\mathbb{P}^3$ under the quotient epimorphism $\pi : \mathcal{S}^{(3)} \twoheadrightarrow \mathcal{S}(\mathfrak{Q})$. Therefore, ℓ is not complete intersection in $\mathfrak{Q}$. The line ℓ is not even set-theoretically complete intersection in $\mathfrak{Q}$; indeed $\mathrm{PGL}(4, \mathbb{K})$ contains $\mathrm{Aut}(\mathfrak{Q})$ which is an algebraic group of dimension 6, thus chosen any two distinct lines ℓ and ℓ' is $\mathfrak{Q}$ there always exists a projectivity in $\mathrm{Aut}(\mathfrak{Q})$ sending ℓ in ℓ' and viceversa. Therefore, if ℓ were set-theoretically complete intersection in $\mathfrak{Q}$ then also ℓ' would be set-theoretically complete intersection in $\mathfrak{Q}$. On the other hand, if ℓ and ℓ' are lines of the same ruling of $\mathfrak{Q}$ we know that $\ell \cap \ell' = \emptyset$ (cf. Exercise 7.1), contradicting Proposition 10.4.2-(ii).

Solution 10.5

Replacing $\mathfrak{Q}$ with $\mathfrak{Q}' = Z_p(X_0 X_2 - X_1^2)$, the same reasoning as in Exercise 10.4 shows that the line $\ell = Z_p(X_0, X_1)$ is contained in $\mathfrak{Q}'$ and that it is not complete intersection in the quadric cone. On the other hand, taking the image of $X_0 \in \mathcal{S}^{(3)}$ via the quotient epimorphism $\pi : \mathcal{S}^{(3)} \twoheadrightarrow \mathcal{S}(\mathfrak{Q}')$, we see that $\ell = Z_{\mathfrak{Q}'}(\pi(X_0))$ so ℓ is set-theoretically complete intersection in $\mathfrak{Q}'$.

Solution 11.1

Let $V = Z_p(X_0 X_2 - X_1^2) \subset \mathbb{P}^2$ which is an irreducible conic and let $W = \mathbb{P}^1$. Let φ be the restriction to V of the projection $\pi_{P_2} : \mathbb{P}^2 \dashrightarrow \mathbb{P}^1$, where $P_2 = [0, 0, 1] \in V$, $\pi_{P_2} : [X_0, X_1, X_2] \overset{\pi_{P_2}}{\dashrightarrow} [X_0, X_1]$. For any $Q \in \mathbb{P}^1 \setminus \{[0, 1]\}$, the fiber $\varphi^{-1}(Q)$ consists of two distinct points, one of which is P_2, whereas $\varphi^{-1}([0, 1]) = \{P_2\}$. In particular this also shows that φ is surjective.

Solution 11.2

Let $V = Z(2X_0 X_1 + t X_2^2) \subset \mathbb{P}^2 \times \mathbb{A}^1$, which is a closed subset (cf. Proposition 7.2.7-(ii)). First we show that V is irreducible: taking $U_0^2 \subset \mathbb{P}^2$ the affine chart with affine coordinates $x_1 = \frac{X_1}{X_0}$ and $x_2 = \frac{X_2}{X_0}$, one has $V_0 := V \cap (U_0^2 \times \mathbb{A}^1) = Z_a(2x_1 + t x_2^2) \subset U_0^2 \times \mathbb{A}^1 \cong \mathbb{A}^3$ and $A(V_0) \cong \mathbb{K}[t, x_2]$ which implies that V_0 and so $V = \overline{V_0}$ are irreducible. Now the map

$$\varphi : V \to \mathbb{A}^1, \ ([X_0, X_1, X_2], \ t) \overset{\varphi}{\longrightarrow} t$$

is surjective and, for any $t \in \mathbb{A}^1$, $\varphi^{-1}(t) = Z_p(2X_0X_1 + tX_2^2) =: C_t \subset \mathbb{P}^2$ is a conic. Therefore, all the fibers of φ are of pure dimension 1. For $t \neq 0$, the symmetric matrix of the conic C_t is of maximal rank, therefore, C_t is irreducible for $t \neq 0$. On the other hand, C_0 consists of two incident lines $Z_p(X_0) \cup Z_p(X_1) \subset \mathbb{P}^2$.

Solution 11.3

Note first that neither W can be a point, because $\dim(W) > 0$ by assumption, nor φ can be a constant morphism, since otherwise $\varphi^{-1}(P) = V$ contradicting the assumptions on $\varphi^{-1}(P)$ and on $\dim(V) > 0$. Since V is irreducible and φ is a morphism, then $\mathrm{Im}(\varphi)$ is irreducible and $\dim(\mathrm{Im}(\varphi)) = \dim(\overline{\mathrm{Im}(\varphi)})$. Therefore, we need to show that $\dim(\overline{\mathrm{Im}(\varphi)}) = \dim(V)$. Since W is projective, $\overline{\mathrm{Im}(\varphi)}$ is a projective variety and therefore $\varphi \in \mathrm{Morph}(V, \overline{\mathrm{Im}(\varphi)})$ is a surjective morphism between projective varieties (cf. Corollary 9.1.2). Thus, from Theorem 11.2.2-(ii), there exists a nonempty open set $U \subseteq \overline{\mathrm{Im}(\varphi)}$ where all the fibers have constant dimension equal to 0, since $\varphi^{-1}(P)$ consists of finitely many points. Thus, one has $\dim(V) - \dim(U) = 0$. One can conclude from the fact that $\dim(U) = \dim(\overline{\mathrm{Im}(\varphi)})$.

Solution 11.4

The arguments are similar as in the proof of Proposition 10.2.3. Indeed, from Corollary 6.4.3, there is a bijective correspondence between closed subvarieties $Z \subset V$ containing W and prime ideals of $\mathcal{O}_{V,W}$. The length of maximal chains of such subvarieties is $\dim(V) - \dim(W) = \mathrm{codim}_V(W)$ which equals therefore the Krull-dimension of $\mathcal{O}_{V,W}$.

Solution 11.5

Since any algebraic variety has an affine open covering, we can reduce to the affine case. Therefore, up to isomorphism, let $V \subseteq \mathbb{A}^n$ be an affine variety and let $A(V)$ be its affine coordinate ring. Any irreducible element $f \in A(V)$ gives rise to an irreducible hypersurface $Z_V(f) \subset V$. Thus, if V is an affine curve, P is the requested irreducible hypersurface not containing Q. Assume therefore $\dim(V) \geqslant 2$. Let $\mathfrak{m}_{V,P} \neq \mathfrak{m}_{V,Q}$ be the maximal ideals in $A(V)$ corresponding to the points $P \neq Q$. Since they are distinct maximal ideals, there exists $f \in \mathfrak{m}_{V,P} \setminus \mathfrak{m}_{V,Q}$; for any such a f, by reversing inclusion,

$Z_V(\mathfrak{m}_{V,P}) \subset Z_V(f)$, i.e. $P \in Z_V(f)$ whereas $Q \notin Z_V(f)$. If f is irreducible, then take $W = Z_V(f)$ and we are done. If otherwise $f = f_1 \cdots f_r$, where f_i's the irreducible factors of f, then $Q \notin Z_V(f_i)$ for any $1 \leqslant i \leqslant r$ whereas there exists $i_0 \in \{1, \ldots, r\}$ such that $P \in Z_V(f_{i_0})$ and one sets $W = Z_V(f_{i_0})$.

Solution 12.1

Note that C is the union of the three coordinate axes in $\mathbb{A}^3$; in particular $O = (0,0,0) \in C$. Thus, $\mathrm{T}_{C/\mathbb{A}^3,O} \cong \mathbb{K}^3$ whereas, for any affine plane curve $D \subset \mathbb{A}^2$ and for any point $Q \in D$, one has $\dim(\mathrm{T}_{D/\mathbb{A}^2,Q}) \leqslant \dim(\mathrm{T}_{\mathbb{A}^2,Q}) = 2$.

Solution 12.2

Note that $\frac{\partial F}{\partial X_0} = X_1$, $\frac{\partial F}{\partial X_1} = X_0$ and $\frac{\partial F}{\partial X_2} = -2X_2$. Let $[U_0, U_1, U_2]$ be homogeneous coordinates in $\mathbb{P}^2$, the target of φ; so we have

$$U_0 = X_1, \quad U_1 = X_0, \quad U_2 = -2X_2.$$

Using the equation of C, one finds $\varphi(C) = Z_p(4U_0U_1 - U_2^2) \subset \mathbb{P}^2$, i.e. the dual curve of C is once again an irreducible conic which is therefore smooth.

Solution 12.3

$Z = Z_p(X_0F_{d-1} + F_d)$, where $F_j \in \mathbb{K}[X_1, \ldots, X_n]_j$, $j = d-1, d$. Any line passing through P_0 has homogeneous parametric equations

$$L : X_0 = \lambda, \ X_1 = \mu v_1, \ldots, X_n = \mu v_n, \ [\lambda, \mu] \in \mathbb{P}^1, \ (v_1, \ldots, v_n) \neq (0, \ldots, 0).$$

Thus, $Z \cap L$ is determined by the polynomial equation

$$\mu^{d-1} \left(\lambda F_{d-1}(v_1, \ldots, v_n) + \mu F_d(v_1, \ldots, v_n) \right) = 0.$$

The solution $\mu = 0$, i.e. $[1,0] \in \mathbb{P}^1$, is of multiplicity $d-1$ and corresponds to the point P_0 on Z.

Solution 12.4

By Euler's formula, $v(Z)$ is defined by the vanishing of the partial derivatives of order $(d-1)$ of the homogeneous polynomial $F \in \mathcal{S}_d^{(n)}$, such that, $Z = Z_p(F)$. These partial derivatives are linear forms.

Solution 12.5

For any $P \in v(Z)$, let $Q \in Z$ s.t. $Q \neq P$. Let $L = P \vee Q \subset \mathbb{P}^n$ be the line through the two distinct points of Z. Then $\mu(Z, L; P) \geqslant d$ and $\mu(Z, L; Q) \geqslant 1$. From (12.7), one must have $L \subset Z$. This holds for any $P \in v(Z)$ and any $Q \in Z \setminus \{P\}$.

Bibliography

Atiyah M. F. and Macdonald I. G., *Introduction to Commutative Algebra*, Addison-Wesley Publishing Co., Reading, Mass.-London-Don Mills, Ont. 1969.

Barth W. P., Hulek K., Peters C. A. M. and Van de Ven A., *Compact Complex Surfaces*, Second edition, Ergebnisse der Mathematik und ihrer Grenzgebiete. **3** Folge. A Series of Modern Surveys in Mathematics, **4**. Springer-Verlag, Berlin, 2004.

Beauville A., *Complex Algebraic Surfaces*, Second edition, London Mathematical Society Student Texts **34**. Cambridge University Press, Cambridge, 1996.

Bourbaki N., *Elements of Mathematics. General Topology*, Hermann, Paris; Addison-Wesley Publishing Co., Reading, Mass.-London-Don Mills, Ont., 1966.

Caporaso L., *Introduction to Moduli of Curves*, School Pragmatic 2004 (course notes) Università di Catania, August/September 2004, available at http://www.mat.uniroma3.it/users/caporaso/modari.pdf.

Catanese F., *Moduli of Algebraic Surfaces*, Pubbl. Dipartimento di Matematica dell'Univ. di Pisa, Pisa, 1986.

Ciliberto C., *An Undergraduate Primer in Algebraic Geometry*, Unitext, **129**. La Matematica per il 3 + 2. Springer, Cham, 2021.

Clemens C. H. and Griffiths P. A., The intermediate Jacobian of the cubic threefold, *Ann. Math.*, **95**(2), 281–356, 1972.

Cohn P. M., *Algebraic Numbers and Algebraic Functions*, Chapman and Hall Mathematics Series. Chapman & Hall, London, 1991.

Cohn D., Little J. and O'Shea D., *Ideals, Varieties and Algorithms. An Introduction to Computational Algebraic Geometry and Commutative Algebra*. Fourth edition. Undergraduate Texts in Mathematics. Springer, Cham, 2015.

Dolgachev I., *Introduction to Algebraic Geometry*, Lecture Notes 2013, University of Michigan at Ann Arbor, available at http://www.math.lsa.umich.edu/~idolga/631.pdf.

Eisenbud D., *Commutative Algebra. With a View Toward Algebraic Geometry*, Graduate Texts in Mathematics, **150**. Springer-Verlag, New York, 1995.

Fulton W., *Algebraic Curves. An Introduction to Algebraic Geometry*, 2008 (slightly modified version of the text *Algebraic Curves. An Introduction to Algebraic Geometry*, Advanced Book Classics. Addison-Wesley Publishing Company, Advanced Book Program, Redwood City, CA, 1989), available at http://www.math.lsa.umich.edu/~wfulton/CurveBook.pdf.

Harris J., *Algebraic Geometry. A First Course*, Graduate Texts in Mathematics, **133**. Springer–Verlag, New York, 1995.

Hartshorne J., *Algebraic Geometry*, Graduate Texts in Mathematics, **52**. Springer–Verlag, New York-Heidelberg, 1977.

Hassett B., *Introduction to Algebraic Geometry*, Cambridge University Press, Cambridge, 2007.

Hulek K., *Elementary Algebraic Geometry*, Student Mathematical Library, **20**. American Mathematical Society, Providence, RI, 2003.

Lang S., *Algebra*, Revised third edition. Graduate Texts in Mathematics, **211**. Springer-Verlag, New York, 2002.

Matsumura H., *Commutative Algebra*, Second edition. Mathematics Lecture Note Series, **56**. Benjamin/Cummings Publishing Co., Inc., Reading, Mass., 1980.

Milne J. S., *(Basic First Course in) Algebraic Geometry*, v. 6.02, 2017, available at http://www.jmilne.org/math/CourseNotes/ag.html.

Miranda R., *Algebraic Curves and Riemann Surfaces*, Graduate Studies in Mathematics, **5**. American Mathematical Society, Providence, RI, 1995.

Mumford D., *The Red Book of Varieties and Schemes*, Lecture Notes in Mathematics, **1358**. Springer-Verlag, Berlin, 1988.

Mumford D., *Algebraic Geometry I. Complex Projective Varieties*, Classics in Mathematics. Springer-Verlag, Berlin, 1995.

Reid M., *Undergraduate Algebraic Geometry*, London Mathematical Society Student Texts, **12**. Cambridge University Press, Cambridge, 1988.

Sernesi E., *Appunti del corso di Geometria Algebrica, a.a. 1991/92*, Dip. di Matematica *Guido Castelnuovo*, Univ. di Roma La Sapienza, 1992.

Sernesi E., *Deformations of Algebraic Schemes*, A Series of Comprehensive Studies in Mathematics, **334**. Springer–Verlag, 2006.

Shafarevich I. R., *Basic Algebraic Geometry 1. Varieties in Projective Space*, Second edition, Springer–Verlag, Berlin, 1994.

Ueno K., *An Introduction to Algebraic Geometry*, Translations of Mathematical Monographs, **166**. American Mathematical Society, Providence, RI, 1997.

Verra A., *Lectures on Cremona Transformations*, School in Algebraic Geometry (course notes), Università di Catania, Politecnico di Torino, September 19–26, 2005, available at http://www.mat.uniroma3.it/users/verra/other.html.

Index

Printed in the United States
by Baker & Taylor Publisher Services